Howard Thurman
10705 Forest Circle

FLAVOR TECHNOLOGY:
PROFILES, PRODUCTS, APPLICATIONS

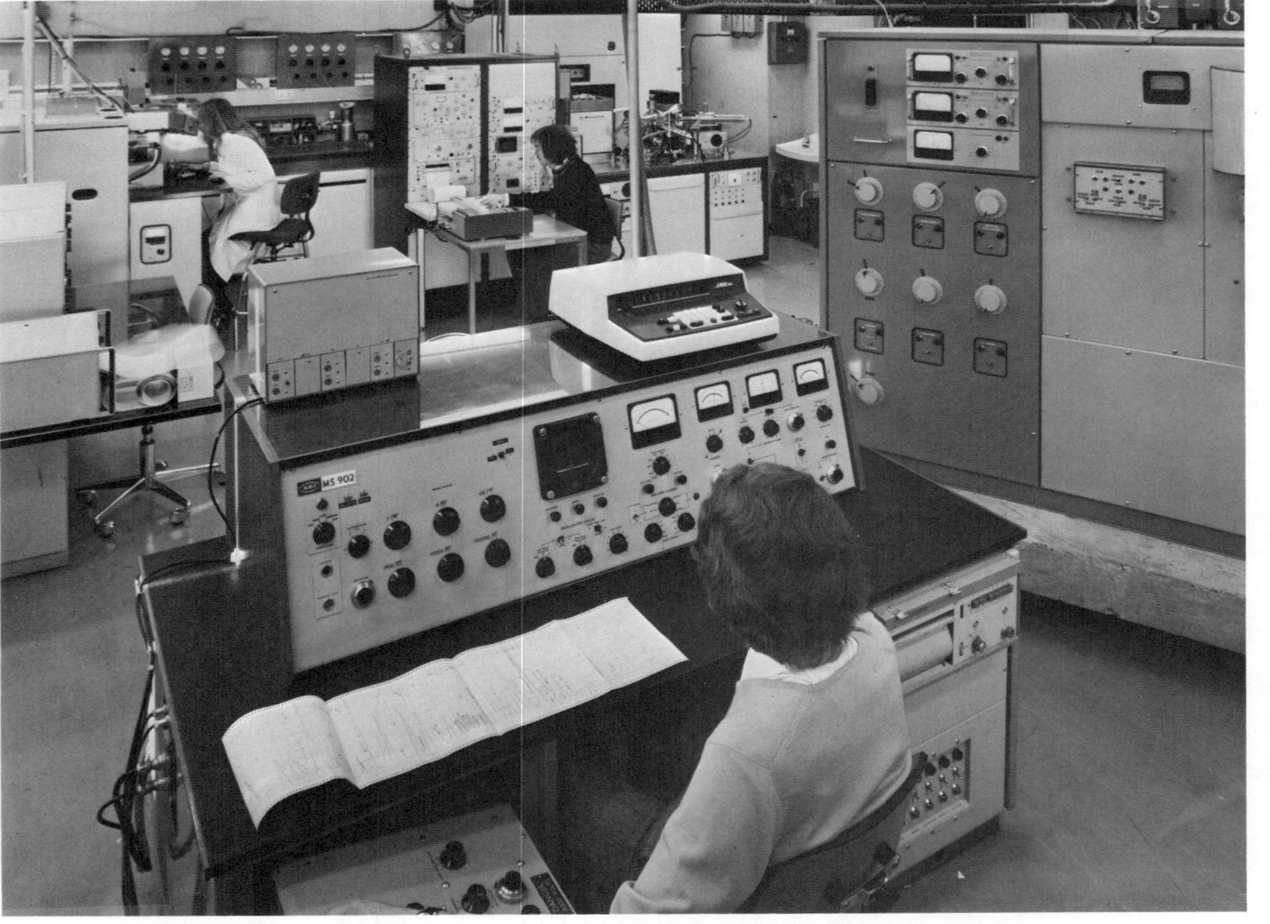

GLC-MASS SPECTROMETER-COMPUTER USED TO EXAMINE COMPLEX AROMATIC MIXTURES

FLAVOR TECHNOLOGY:
PROFILES, PRODUCTS, APPLICATIONS

Henry B. Heath, M.B.E., B. Pharm.
(London)

Bush Boake Allen, Ltd.,
A Division of Albright & Wilson,
Ltd., London, England

AVI PUBLISHING COMPANY, INC.,
Westport, Connecticut

Library of Congress Cataloging in Publication Data

Heath, Henry B.
Flavor technology.

Includes index.
1. Flavoring essences. I. Title.
TP418.H4 664'.5 77-28595
ISBN 0-87055-258-9

Printed in the United States of America

Preface

The food, beverage and other industries producing goods for consumption all have an essential involvement with the flavor of their end-products. There are throughout these industries, as well as in the specialist flavor manufacturers and dealers who service them, a considerable number of people who are required to have a working knowledge of flavoring materials in relation to the aroma and flavor of the products they manufacture or handle. These individuals may be highly skilled technical staff whose training has already given them an insight into flavor chemistry, the ramifications of flavor creation or development, flavor assessment, etc., or they may be nontechnical staff involved in the purchasing, promotion, etc., of raw materials and/or finished goods. In many research departments, both academic and industrial, there are those whose work is directly or indirectly concerned with flavor and flavoring materials; for such, an understanding of aromatic chemicals in relation to odors and flavors may be helpful. All too often in research papers the sensory aspects of the work are neglected, maybe because of an inadequate appreciation of the language of sensory assessment and the way in which flavor profiles are evolved. Then there are many students and trainees in development and control laboratories throughout industry who require a basic knowledge of food technology in relation to the many applications of flavorings as well as the problems likely to be encountered in their appraisal and effective use. It is for all of these that this book is written.

The technology described is selective and intended to give sufficient information on processing techniques and conditions to enable the flavorist, the food technologist and those involved in technical service to discuss, develop and apply flavors to achieve the optimum effect in the end-product. For those requiring a fuller exposition of processing methods, plants, etc., reference should be made to the numerous informative books and articles quoted in the appropriate bibliographies.

The text has been divided into three sections. The first sets out in an ordered sequence the aromatic profiles of the main natural flavoring materials widely used in foods, beverages, confectionery, etc. These profiles were evolved by a team of expert assessors over a long period and represent average findings after much discussion. It would be an impossible task to define the profiles of every source of herbs, spices and other aromatic flavoring materials and only those of major importance have been treated comparatively. For others, the profiles are intended as a guide to assessment and use, leaving the individual to evolve a personal appraisal of any particular product.

The second section deals with the nature and preparation of flavoring materials and products for use in food processing. It sets out their characteristics and the considerations necessary for the correct selection of flavorings for specific product groups. This section embodies some 27 years of experience by the author in this field.

The final section is a review of the technology and use of flavorings in the main branches of the industries concerned. There are doubtless innumerable variations of detail when it comes to processing conditions. Indeed, it is not intended that the operating conditions cited shall be anything more than a guide and an indication of the parameters dictating flavor application. The specialist literature and trade journals covering each particular industry normally carry articles written by experts in the field and the reader is referred to these in the bibliographies which accompany each subsection.

The food and related consumer industries are restless and work in an environment of constant development and change involving new technology, new processing plants and techniques, new product concepts and marketing. The book will have achieved its purpose if the reader attains a better basic understanding of the nature of flavor and flavorings in what we eat and drink and the problems of providing attractive, safe and flavorful products for discriminating consumers.

The author wishes to acknowledge the advice and assistance given by his colleagues in Bush Boake Allen Ltd., during many years in flavor development and technical service.

Special thanks are due to associates, too numerous to mention individually, who participated in, what must have appeared at the time as interminable, odor and taste panel sessions, resulting in the profiles included in section one. Also to the specialists in the flavors applications laboratories of the company and to the members of the

Food Study Group, for their technical contributions over the years.

The author is indebted to Dr. Martin Peterson, Consultant, Natick, Massachusetts, for his initial interest and encouragement and to Mrs. Lucy Long, Senior Editor of AVI Publishing Company, for invaluable advice on the presentation of the material.

Finally, the author thanks the Managing Director and Board of Albright and Wilson, Ltd., for permission to publish this work which embodies so much experience gained over 27 years with Bush Boake Allen, Ltd., in the U.K. and overseas.

Henry B. Heath

LaSalle, Quebec
November 1, 1977

For Eileen

Contents

PREFACE

SECTION I: Flavor Profiles

1 Introduction 3
2 The Culinary Herbs 26
3 The Spices 73
4 Aromatic Vegetables and Mushrooms 167
5 The Citrus Fruits 182
6 Vanilla 198

SECTION II: Flavor Products

7 Flavoring Materials
PART 1 Introduction 211
PART 2 Materials Used to Make Beverages 212
PART 3 Herbs and Spices 223
PART 4 Fruit Juices, Concentrates and Pastes 252
PART 5 Essential Oils 261
PART 6 Imitation Flavorings 295
8 Creation and Development of Flavorings 368

SECTION III: Flavor Applications

9 Flavors in Food Processing 387
10 Meat, Poultry and Fish Products 392
11 Baked Goods and Bakery Products 414
12 Snack Foods 427
13 Sugar and Chocolate Confectionery 434
14 Pickles and Sauces 452
15 Soups 469

16 Ice Cream and Frozen Goods 479
17 Soft Drinks and Beverages 493
18 Quality Assurance of Highly-Flavored Products 507
Appendix
English, French and German Terms Used in Sensory Descriptive Analysis 523
Bibliographical Index 527
Subject Index 529

SECTION I
FLAVOR PROFILES

1

Introduction

The realm of flavors in food and anything else we eat or drink spreads across several disciplines and for flavor application to be effective involves specialists working together in a team towards one end—the production of really attractive, nutritionally valuable, and safe products to satisfy an ever-increasing consumer demand. The team may include scientists and nonscientists with very different levels of knowledge and technical skills but all dedicated to contributing their particular expertise: There is the creative flavorist with intimate knowledge of the aromatic profiles of hundreds of natural and synthetic flavoring components, who has to be more of an artist than a technologist. There is the raw materials specialist who must have a detailed understanding of the availability and costs of such widely differing products as black pepper from Sumatra or Malabar to synthetic organic chemicals such as 4,4 dibutyl-butyrolactone. There is the flavor technologist who, with a wide knowledge of food technology and processing conditions, ensures that any flavoring is suited to its end use and is stable and quite safe for consumption. And finally, there is the food technologist who, as part of yet another team including nutritionalists, chefs, home economists, marketeers, package designers, etc., formulates, manufactures and ultimately puts on the market the food, beverage or confectionery products in which the flavor appears. The whole development process has to be carried out in the clear understanding that the end-product will be perfectly safe and wholesome whenever it is consumed and be sufficiently attractive to appeal to a wide spectrum of consumers to ensure its commercial success.

Inevitably, each aspect of the subject is capable of review in depth but here only the broad principles involved have been dealt with and guidance given on further specialist reading in the bibliographies. Those directly concerned with flavor creation and application in its many forms will readily appreciate that the subject is currently attracting much research and that the results of this are likely to accelerate the pace of change within the flavor industry itself and also within the food and related industries which it serves. However, present legislative views are tending to slow down innovation in the interests of safety so that many changes, particularly those involving the use of chemical additives and flavorings, will be reflected very slowly in end-products.

CLASSIFICATION OF FLAVORS

The flavor of food is all-important. Much of our diet by itself can be unattractive, unappetizing and often so lacking in intrinsic flavor that it is insipid and quite unpalatable. If the aroma and flavor of food is enticing and satisfying to the eater then it is digested better and results in an overall feeling of satisfaction. It does not matter how attractive food is to look at or how balanced it may be nutritionally, if it does not smell and taste good then these essential reflexes are not activated and eating remains a mere chore rather than the positive pleasure which it should be.

The flavor of any food product, confectionery or beverage will, of course, depend upon all the ingredients used in its preparation and the nature of the processing involved. Therefore a knowledge of the components of any product is essential. These may be classified as:

(A) Prime ingredients which determine the character of the product.

(B) Flavorings.

(C) Coloring.

(D) Additives necessary to achieve an acceptable product form or stability.

In this book only those items directly affecting the flavor of the end-product will be considered.

Natural Flavors

Some of our staple foods are almost devoid of flavor in their raw state whereas others are intensely flavorful. Natural food materials may generally be classified into one of three quite distinct groups:

Low Flavor Impact.—Highly nutritious but intrinsically not very flavorful staple food items; e.g., raw meat, fish, milk, cereals and most root vegetables.

Medium Flavor Impact.—Of moderate nutritional value and adequately flavorful to make eating pleasurable; e.g., fruits, nuts and some aromatic vegetables.

High Flavor Impact.—Of low or no nutritional value but having a powerful flavor content; e.g., herbs, spices, vanilla and cacao.

Those in the first group form the basis of a staple diet providing a full range of proteins, fats, carbohydrates and vitamins necessary to maintain the body in good health. However, it is necessary to improve the flavor of this group of foodstuffs and this can most readily be achieved simply by cooking or, more convincingly, by the addition of some of the more aromatic materials found in the other two groups or by the use of added compounded flavorings.

Those in the second group occupy a half-way place in that they are usually eaten in their natural state either as part of a meal to produce an entirely different flavor sensation in the mouth, or at other times as a snack. Because of their attractiveness and wide appeal the flavor of fruits is often incorporated into products which would otherwise have little or no flavor; e.g., sugar confectionery, soft drinks, etc. In such cases the added flavoring may be derived directly from the fruits or may be an imitation including or entirely composed of synthetic aromatic chemicals.

Natural products such as the herbs and spices having a very high aromatic content are of considerable importance to the food processor and it is the character of these natural flavorings which form the subject of this section. It is the blending of these various flavor factors which is the cornerstone of *haute cuisine* and, on the commercial scale, food product development, for it is flavor attributes which to a large extent determine the acceptability or rejection of food and drink.

Changes in Natural Flavors

Because they are characteristically flavorful, many fruits and vegetables are delicious when eaten in their raw state and there is little to compare with the fine flavor of fresh strawberries, a succulent pear, or a crisp stick of celery; but cook these and the flavor character is entirely changed. For many persons this resultant flavor is less acceptable than that of the fresh fruit. Canned strawberries or pears have a totally different flavor profile from that of the fresh fruit. If strawberries are cooked at a higher temperature to make strawberry jam the resultant profile is again further changed. It is a knowledge of these different profiles which is so importnt in the creation of food products. Again, the fruits demonstrate a spectrum of flavor profile which depends upon their state of ripeness at the time of gathering. Preference for one or the other is a matter of individual choice but to the food processor an appreciation of the different flavor qualities and strengths which result on this score is essential when handling fruit as a flavoring component in a product.

To some extent the same considerations apply to meat although in this case the raw flesh has little appeal per se. To be palatable meat must be cooked. The flavor of raw meat is due to blood, mineral salts and certain organic substances dissolved in the tissue juices. All meats contain broadly similar components but as each meat has a characteristic flavor there must be significant differences either in the proportions of the substances present or in the way in which they react when heated. The differences in flavor due to cooking are a function of temperature and the degree of water present which determine the course of the Maillard reaction between the amino acids, peptides and reducing sugars which form the main flavor precursors in meat. Whereas humans have a preference for cooked meat, animals (on the other hand) show the reverse preference and pet food manufacturers have for many years been searching for a flavor additive which could be incorporated into canned or cooked meat products to cover the changed flavor and give back to the contents the same fresh meat attractiveness which animals prefer.

The above considerations enable flavors to be further classified as follows:

(A) Flavor not present per se in the natural product:
   (1) Flavor developed by heat (cooking): meat, cacao, vanilla.
   (2) Flavor developed by enzymic action in the cold: onion, mustard.

(B) Flavor present in the natural product and recoverable by physical means (i.e., extraction, distillation, etc.): herbs and spices, fruits.

Years of patient study have resulted in an insight into how plant materials can be cultivated to give high yields of high quality food products. At the same time, chemists have acquired an ability to analyze and ultimately to imitate nature synthetically, particularly with regard to those components responsible for aroma and flavor. More and more, technology is taking the guesswork out of what we eat and drink and making us progressively independent of the many causes of shortages in the supply of natural raw materials. However, in spite of the continuous advances in technology there are those who clamor for a return to more leisurely ways of preparing foods without recourse to synthetic additives. There are many who regard the terms "chemicals" and "additives" with grave suspicion and consider a return to "natural" foods, cooked and prepared in traditional ways, as necessary to the continuing health of mankind. Food needs flavor and in a modern society there is a real need for both natural and imitation flavorings. The modern flavorist and technologist are fully aware of their responsibilities for safety-in-use in the long-term interests of the consumer.

Definition of Flavor

Flavor is a complex appreciation of the total sensations perceived whenever food or drink is consumed. Over the years many attempts have been made to define flavor in precise terms. In 1969, the U.S. Society of Flavor Chemists proposed the following definitions:

> *Flavor* is the sensation caused by those properties of any substance taken into the mouth which stimulates one or both of the senses of taste and smell and/or also the general pain, tactile and temperature receptors in the mouth.
>
> *A flavor* is a substance which may be a single chemical entity or a blend of chemicals of natural or synthetic origin whose primary purpose is to provide all or part of the particular flavor or effect to any food or other product taken into the mouth.

The use of the term "flavoring" is becoming more widely accepted to distinguish between the intrinsic flavor of a product and any added flavorful ingredients used to modify or impose a new character on the flavor profile of the end-product.

The definition of flavor is further complicated by the popular misuse of related words such as aroma, odor, smell, fragrance, taste and flavor all of which are applied indiscriminately in conversation and writing with little or no reference to the more restrictive but accurate definition given above.

Definition of Taste

Taste is defined by Amerine, Pangborn and Roessler (1965) as one of the senses, the receptors for which are located in the mouth and are activated by a large variety of different compounds and solutions. Most investigators usually limit gustatory qualities to four: saline, sweet, sour and bitter. Taste is an essential element of flavor for if one removes the aromatic fraction of foods or beverages or blocks the odor receptors in the nose then one is left with a more or less characterless mass having a sweet-salty-bitter backnote and possibly a tart lingering after-taste.

Other sensory responses may also influence the overall flavor profile and such effects as astringency, pungency, cooling, warming and "metallic" sensations can be isolated as specific component attributes.

NATURE OF FLAVOR COMPONENTS

Plant materials used in foods depend for their characteristic aroma and flavor on a complex blend of organic chemicals produced in the plant tissues during its normal growth.

Considerable research has been directed towards separating and identifying the chemicals responsible. Progress in this field has been rapid and is accelerating as fuller use is made of modern instrumental techniques coupled with computerised data assessment and recovery. Hundreds of chemicals present in natural foods and flavorings have already been positively identified but some still defy categorization. This increased knowledge of plant biochemistry has, in turn, given rise to other problems. The most pressing of these is the need to establish meaningful correlations between objective and subjective assessments of the relative contribution of each component to the overall aromatic profile. This is further complicated by the absence of a comprehensive and universally understood flavor language to enable meaningful descriptions of the aromatic attributes of the chemical compounds isolated and characterized. Parallel with these predominantly subjective problems is the urgent need to establish relative toxicity and safety in use for, as is well known, it is wrong to assume that everything found in nature is automatically safe for consumption.

Nature is complex and does not readily reveal many of her secrets. The techniques necessary to separate aromatic compounds from inert plant tissues are often involved and tedious. One cannot always be sure that essential flavoring components have not been lost or modified in the process of extraction and concentration; or that entirely different artifacts have not been created. Many of the chemicals which have the most significance on the odor and flavor profile are known to be present only in trace quantity and often demonstrate a very limited stability when isolated and purified. All of these factors make research into the chemistry of flavor components one of extreme complexity but, in spite of difficulties, considerable progress has been made. Modern research in this field has materially helped us understand some of the biochemical processes which exist in the plant and the extent to which these are responsible for the flavor of our food. This knowledge not only enables flavors to be recreated synthetically but also enables plant geneticists to create the right growing environment to increase yields and flavor naturally.

There is still an underlying mistrust of anything "chemical" and many consumers dislike the idea of added flavorings which have been compounded from synthetic organics even when it can be demonstrated that these are of the same chemical constitution and purity as those found in nature. Synthetic vanillin, for example, is the same chemical entity (3-methoxy-4-hydroxybenzaldehyde) as vanillin separated and isolated from vanilla beans. The character of the natural flavor is, of course, much more complex owing to the presence of other aromatics. To most palates this natural flavor is preferable to one based solely on vanillin; that fact is not disputed; but, each is equally safe when used as a food flavoring. A quite satisfactory imitation flavor based entirely on synthetic organics can readily be compounded to match very closely the natural vanilla extract flavor profile and to give the same overall effect in a food product or ice cream. Most of the chemicals used will be "nature identical"; but in spite of this, there is the feeling of mistrust in the minds of many when they read the description "imitation flavoring," "artificial flavor" or "synthetic flavor" in the ingredient list on the product wrapper. This feeling is totally unjustified.

The nature of the chemicals found in food flavors and used in the compounding of flavorings will be further discussed in a later chapter.

NATURAL VERSUS SYNTHETIC FLAVORS

Natural or Imitation Flavorings—Which to Use?

This is a question to which there is no one answer. Price, availability of raw materials, permissibility under current legislation and the type of end-product in which the flavoring is to be used are all factors which determine whether it is better to use an entirely natural flavor or one compounded using synthetic organic chemicals or a mixture of the two. A survey of the technical literature over the past decade bears ample evidence of the considerable and rightful importance of flavor in what we eat and drink. Our understanding of the chemistry and toxicology of our food and flavor adjuncts is constantly improving but in spite of this there is still a lack of appreciation in the lay mind of the true nature of flavor whether it be intrinsic to the food or added during processing. There is a tendency to assume a higher degree of confidence in the wholesomeness and safety of natural foods and natural flavors than those based on chemicals. There is a marked prejudice against imitation flavors compared with those of natural origin even where it can be demonstrated that chemically they are identical. To the layman the word "chemical" is highly emotive and is equated with "second rate," "poor quality," "substitute" or the very meaningful German word "*ersatz*." It is difficult to understand the reason for this. Such opinions are frequently without any foundaton and are usually stirred up by articles and opinions expressed in the media.

To those directly concerned with the drafting of legislation governing what may be used with safety in foodstuffs and beverages, such ill-founded opinions form a poor basis for decision. In spite of vociferous clamor there is an increasing pressure to provide more food for an ever-increasing world population and it is necessary to widen one's views and base one's opinions of real or imaginary dangers not on mere feelings but on the results of well-designed biological tests conducted under carefully controlled conditions, supported by a wealth of human experience.

Throughout his evolution man has sought those products in his environment which he found were able to convert his fairly monotonous staple diet into food which was a pleasure to eat. The use of such flavoring ingredients was almost certainly discovered by the age-old process of "suck it and see," during the course of which it is equally certain that many highly attractive berries, roots, fungi, etc., would have proved fatal when eaten in spite of their being "natural." In consequence, the experience of what was toxic and what was safe was quickly learned and passed down through the centuries. It is unreasonable to ignore such well-founded human experience. The original Generally Regarded as Safe (GRAS) lists were an expression of this. Tasteless foods cannot be tolerated for any length of time by most people. Historically, man has added salt, sugar, herbs and spices and other aromatic vegetables to give relish to his diet. This being so, we are likely to continue to use added flavorings of natural origin without any qualms on the part of the consumer. It is when we come to imitation flavors that the troubles start.

Most people recognize that all matter is made up of chemicals of varying degrees of complexity but prefer to ignore the fact that the substances which are responsible for odor and flavor in nature are also chemicals and, indeed, embrace virtually all classes and functional groups of both inorganic and organic compounds. Such chemicals have the same properties whether they be made by natural biosynthesis in the plant or by synthesis in a chemical laboratory or manufacturing plant. Plants and animals are, after all, really extremely efficient chemical factories capable of carrying out very complex chemical reactions with a minimum of effort and in a far simpler way than can be achieved in the laboratory. Many of the most prized natural flavors arise out of the natural metabolic functions of the plant or animal and may indeed be mere waste products of that process. In some cases the flavor most sought after is produced by subsequent treatment (e.g., vanilla flavor which is only produced when the beans are cured). In most cases the flavoring chemicals represent only a minor part of the natural plant material so that concentration is necessary before they can be of any value as an added flavoring.

Problems of Using Natural Flavors

The flavorist has available to him a wide spectrum of natural flavoring materials which by blending can produce an almost unlimited range of effects; so much so, that one may wonder why it is necessary to use synthetic versions at all. The reasons are clear:

(A) Many natural flavors have a low intensity and in consequence have to be used at a high dosage rate in order to produce the desired effect in the end-product. This often results in an unsatisfactory texture and poor stability.

(B) Concentration of natural flavors is, in most cases, possible by extraction or evaporation but such processes are usually accompanied by significant changes in the flavor profile.

(C) Natural flavors exhibit variations in strength and quality the difference being accentuated by the source of the material, its ripeness at harvest and its subsequent handling.

(D) The supply of natural materials is becoming progressively more uncertain and the quantity available now falls far short of demand. This situation is likely to be accentuated as native populations turn to more economically attractive products.

(E) Most natural flavors are unstable and undergo changes during postharvest handling, processing or storage.

(F) Many natural products contain enzyme systems which may result in the formation of off-notes or taints and a fall-off of flavoring power.

(G) The toxicity of many natural products has yet to be established with the same degree of detail as that applied to synthetic chemicals used in flavorings; when they are, the results are likely to be surprising.

Entirely natural flavors, therefore, pose many problems to the food manufacturer and, in order to overcome these difficulties, it is becoming increasingly necessary to use desirable natural flavoring effects but recreated from synthetic components of known aromatic character, purity and safety. In this respect a knowledge of the chemical composition of natural flavors is of inestimable value to the flavorist although direct reproduction of a list of chemicals shown to be present in a given flavor rarely results in an acceptable imitation.

Disadvantages of Using Imitation Flavors

The need for imitation versions of naturally occuring flavors is undoubtedly present; but before considering their virtues it is also well to establish their disadvantages. Imitation flavorings based on synthetic organics may present the following problems:

(A) The original natural flavor is often, but not always, of a more subtle and acceptable character whereas the imitation version may be described as "chemical."

(B) The imitation flavor may pose difficulties in labeling and must always be formulated so as to comply with any legislation in the country in which the end-product is to be consumed.

(C) Many natural flavors have a built-in reservoir of flavor precursors which, under certain conditions, can result in the generation of additional flavor—a feature which is absent in imitation flavors.

(D) Imitation flavors generally require the use of either a solvent or a carrier. These may be restricted by legislation or may pose problems with texture in the end-product.

Advantages of Using Imitation Flavors

Imitation flavorings have several marked advantages, namely:

(A) In flavoring power they are usually much cheaper than the equivalent natural product necessary to produce the same flavoring effect (even where this is possible) and are usually much less sensitive to changing costs.

(B) They are stable and have a very long shelf-life.

(C) They can be designed to withstand severe processing conditions.

(D) They are highly concentrated and can be produced in a variety of forms (e.g., alcohol-based, oil-based, or other permitted solvent-based solutions or encapsulated powders) suitable for specific applications.

(E) Within modern manufacturing constraints they are generally readily available being independent of natural cropping, seasonal or other supply considerations.

(F) They can be tailor-made to give the optimum desired flavor effect. This flexibility leads to an ability to create product distinctiveness. Any fruit—or indeed fruit at a particular stage of ripeness—can be imitated; and at the same time, any undesirable characters in the natural flavor profile can be modified or omitted.

(G) They have a consistency of quality and flavoring effect.

This is a significant list of attributes in favor of imitation flavorings and has long been appreciated by food processors. Imitation flavors are becoming more balanced and much less "chemical" than was once the case and as the public palate becomes more sophisticated many of the older generation of artificial flavors are falling into disuse. The crude fruit flavors composed of a few harsh esters such as amyl acetate are things of the past and have almost entirely been replaced by compounded flavors having an aromatic profile closely similar to that of the fruit they are imitating.

Today one hears much about flavor. It is used as the "centerpiece" of high-pressure advertising; its apparent lack in many processed foodstuffs is regretted by those who, rightly or wrongly, claim they can remember when everything tasted far better than it does today. Flavor is the subject of innumerable articles in the daily press and glossy magazines. Large sums of money are spent each year on research into the chemistry of flavors. This interest in flavor is likely to increase as a larger proportion of our food is factory-produced in order to cope with the demands of an increasing population.

Undoubtedly there is a real need for added flavorings in almost everything we eat and drink. It is clearly recognized that, however good natural flavors may be, when it comes to food processing they are far from adequate and are technically unsatisfactory. If the food industry is to provide a sufficiency of flavorful products then these will, of necessity, require the addition of flavorings which will be composed mostly, if not entirely, of synthetic organic chemicals the safety of which has been established beyond reasonable doubt. Attitudes are changing and the term "imitation flavor" is now much more widely accepted. The consuming public is still apprehensive but no longer automatically rejects the idea of added "artificial flavor." The arguments for and against natural or imitation flavors will continue and may even intensify but unless one is prepared to forego many of the excellent food products currently available through the use of imitation flavors then these will continue to be used. There is a need to educate the public, to allay fears, suspicion and alarm and to apply sound judgement on what is aesthetically and technologically necessary for any particular product.

DESCRIPTION OF AROMATIC PROFILES

Much has been published on the history, nature and processing of herbs, spices and other aromatic plant materials used as food flavorings but one subject which has received little attention is that of their sensory characteristics. What do they actually smell and taste like? What quantitative and qualitative contribution can one expect them to give to the total flavor complex of any product in which they are used? How can one describe the observable differences in aroma and flavor? The absence of any really informative articles covering this important aspect of flavoring is not surprising when one realizes how extremely difficult it is to achieve meaningful descriptions of well-known but purely sensory effects (e.g., the flavor of a banana).

Courtesy of A. D. Little, Inc.

FIG. 1.1. ODOR- AND NOISE-FREE PANEL ROOM SET UP FOR FLAVOR PROFILE EVALUATION

Courtesy of A. D. Little, Inc.

FIG. 1.2. BRANDY SNIFTERS PROVIDE AN EXCELLENT MEANS FOR ODOR CONFINEMENT

Certain descriptive terms have become well established by use and understood by the majority of those likely to be called upon to evaluate aromatic materials. Generally, however, the mere reading of a descriptive profile gives little idea of the effect obtained. It is fair to say that no odor or flavor can yet be described verbally in any language in such a way that the uninitiated layman can immediately recognize and visualize the material and be able to identify it when presented with a sample. Even the individual words used do not necessarily convey the correct impression. For instance, one may have a reasonably clear understanding of the word "aromatic" as something which has both a hedonic and nonhedonic connotation; i.e., it is both pleasing and sweet. On the other hand, the term "green," which is also very frequently used in describing aromas and flavors, is far less precise, ranging from the effect one associates with freshly cut grass ("grassy") to that of damp leaves ("leafy") or even of freshly cut garden herbs ("herbaceous"); all of which are quite different. This problem is accentuated when one has to translate terms into other languages. Frequently, a single word replacement is not understood or may, in fact, give a totally wrong impression. In most cases one has to carry out an evaluation at first hand in order to appreciate fully the differences in aromatic character or create a picture of the total profile.

The meaning of words in this context has been and is likely to remain the cause of constant discussion. The problems of using comprehensible terminology are difficult enough for experts but when even experts disagree how much more difficult the problem becomes when one is trying to convey a sensory impression to someone with perhaps little or no first hand experience and asking them to understand words having special meanings when applied to the product under discussion. This situation often arises between the research chemist and the flavorist or between the technical product developer and the lay marketeer. Two specialists rarely come up with the same descriptive profile. However, as with so many aspects of sensory assessment, training and practice can do much to resolve these differences. A meaningful vocabulary, using selected aromatic materials for reference if possible, can be established by discussion amongst members of a panel of assessors. This, of necessity, must be done before practical evaluation of the material under consideration and, consequently, any published profiles should be regarded as a basis for discussion and agreement between panel members to ensure that all are conversant with the terms used.

The description of odors has been investigated by Harper and his colleagues (1968) and, in trying to establish an effective list of descriptive terms and phrases, four well defined precepts were stated:

(A) Abstract terms are generally not acceptable as these are frequently meaningless.

(B) Descriptions are best defined in terms of specific substances or effects which are frequently well known or demonstrable.

(C) There is a bias towards first describing undesirable or off-odors or off-flavors in preference to those of more pleasing aspects.

(D) Certain commonly used words have a spectrum of meaning and may need clarification by reference to the particular context of their use.

In the widely accepted descriptive flavor profile method of sensory assessment it is first necessary to evaluate the odor and flavor of the material systematically and as a result to evolve an appropriate vocabulary of terms which may adequately describe the attributes present by discussion and resolution among the panel members concerned. It is this test method which has been found most satisfactory as the basis for the examination of a whole range of natural aromatic materials and products derived from them. The terms most widely used are listed in the Appendix.

The flavor of certain essential oils is better evaluated by making a 1% ethanolic solution and diluting this 1 to 250 with a 10% sucrose solution. This is equal to a dosage of 40 ppm. For concentrated and terpeneless oils dilute to 4 ppm and for the alliaceous oils to 1 ppm.

It cannot be overstressed that the profiles presented in the following chapters should only be regarded as a guide and used as the basis for discussion and agreement between those involved in any such sensory evaluation. This is a particularly subjective area of judgement and a clear understanding of the meaning of each term is best established by personal appraisal and, if possible, aided by reference to some well-defined substance which displays the odor or flavor attribute to be described.

In the profiles given in the following chapters the impressions have been tabulated in order of their appearance and the preferred descriptive terms are shown in capital letters. Supplementary or synonymous words are given but in lower-case letters. No attempt has been made to quantify the effects described as the relative intensity of each attribute demonstrates wide variability depending on the source of the material, the age of the sample and other contributing factors all of which make the establishment of acceptable quantitative anchor points extremely difficult.

Generally, odors and flavors are assessed as a whole so that one gets an overall impression. However, in creating profiles for individual natural flavors it is necessary to be more analytical, assessing each aromatic "note" as it is perceived. Even so, as Harper (1968) points out, trained assessors rarely provide a fully effective description involving all the attributes to be found in an odor or flavor, hence the need to form a composite opinion based on several individual assessments within a panel of adequate size. Most natural flavoring materials display wide variations not only in their total aromatic content but also in the relative proportions of specific aromatic components depending to a large extent upon agronomic influences. These variations may result in quite different profiles although, generally, any particular material is still recognizable as such, e.g., sage from Dalmatia and sage from either Greece or Italy are all quite different in detailed profile but are still readily recognized as "sage."

Every aromatic plant has its own characteristic odor by which it may be recognized. It is this factor which is of great importance in quality assessment as it allows one to differentiate one sample of the product from another based on a recognizable norm. The case of recognition is largely conditioned by experience and one's odor and flavor memory. It is all too easy to use the word "characteristic" in describing an odor or flavor. This is largely without real meaning, for to say that basil herb smells like "basil" is begging the question.

Not all aromatic effects are of equal strength as was well demonstrated by Sjostrum *et al* as far back as 1957. Some have the most effect on the nose giving to the food a light pleasing aroma but little flavor when ingested; others are dull and heavy contributing little to the smell but very considerably to the fullness of flavor. These latter often have a marked contribution to bitterness and pungency.

The flavor intensity of herbs and spices in particular differs very widely. Some spices, such as capsicum, are intensely pungent; others, such as fenugreek, are mild and require a large dosage before any flavor contribution becomes significant. The placing of some relative value upon these effects has been attempted by Rietz (1961) in his gastronomic chart, but the effective blending of herbs and spices is still one of individual appreciation and preference.

Again, the flavor of an herb or spice may significantly alter during postharvest handling and subsequent processing prior to use. For instance, the flavor effect of raw green herbs is much higher than that of the dried rubbed version owing to volatile losses and changes which occur during dehydration. The essential oil distilled from the freshly harvested herb has quite a different profile from that obtained by distilling the dried herb. The flavor profile of the extracted oleoresin will depend not only upon the method of processing but upon the solvent used and the degree of heat to which it is subsequently subjected in order to remove the solvent. In other words, a description of the profile of freshly gathered rosemary herb does not give a good indication of the profile of the dried commercial product nor of oil of rosemary or oleoresin rosemary.

The picture is further complicated by interaction of aromatic effects during ultimate processing of food products, for conditions of pH and temperature can result in significant changes in the odor and flavor profiles of the components present.

Flavoring Components and Profile

Considerable research has been directed into the nature of the constituents responsible for the aromatic character of plant materials, most prominent of which are the essential oils. All herbs and spices, with the exception of the capsicums, contain some proportion of volatile oil which can be recovered from the prepared material by distillation. However, it is well recognized that most spices as well as other natural flavoring materials contain nonvolatile components which may also make a significant contribution to the overall flavor profile.

Guenther's standard text on the essential oils is still a valuable source of information on their chemical constitution but modern techniques reviewed by Jennings (1972) are increasingly revealing their true complexity and the importance of trace components, many of which are still to be positively identified. But chemical classification per se has little relevance from the point of view of odor or flavor assessment other than in the broadest terms as a means of grouping materials having predominantly similar constituents which may be classified on the basis of their chemical constitution as follows:

(A) Hydrocarbons of the general formula $(C_5H_8)_n$; where n = 2 these are called terpenes, where n = 3 they are called sesquiterpenes and where n = 4 diterpenes.

(B) Oxygenated components derived from these hydrocarbons including alcohols, aldehydes, esters, ethers, ketones, phenols, oxides, etc.

(C) Other specific compounds containing either nitrogen or sulfur.

It is necessary to recognize that each component has a characteristic profile and that every flavor is a combination of the profiles of its constituents; some acting in unison, others competing to the point of elimination. In certain instances, one main constituent may predominate to the extent of 85% or more (e.g., clove bud oil contains 85–95% eugenol); in others, there is a balance of components so that no one characteristically predominates (e.g., alcohols in oil of sweet

Courtesy of Bush Boake Allen, Ltd.

FIG. 1.3. A COUPLED GLC-MASS SPECTROMETER-COMPUTER LAYOUT FOR THE EXAMINATION AND CHARACTERIZING OF COMPLEX AROMATIC MIXTURES

marjoram). The presence of trace components, even those which are as yet unidentified, often modify the odor and flavor to a significant extent. Although unusual components may be present in essential oils derived from different geographical sources or even from different parts of the same plant, it is more often differences in the relative proportions of normal components that give rise to differences in the aromatic profile.

An examination of the odor profile of an essential oil may be compared with those of its known constituents and in Table 1.1 this has been done for Spanish oil of rosemary. A further odor pattern may be determined by the sensory evaluation of the effluent from a GLC preparative column (Fig. 1.4). Normally, one's first and generally diagnostic impression of the odor of any essential oil is given by the lighter terpenoid fractions, the pattern changing gradually as these evaporate and are replaced, in turn, by the higher boiling components present. The techniques used in establishing the profiles which follow were:

TABLE 1.1
PROFILES OF OIL OF ROSEMARY AND ITS MAIN CONSTITUENTS

Oil of Rosemary	α-Pinene	Cineole 17–30%	Borneol 16–20%	Camphor about10%	Bornyl Acetate 2–7%
Odor					
Fresh	Fresh	Fresh		Ethereal	Fresh
Cooling		Cooling		Cooling	
Slightly warming	Warm			Warm	
Penetrating		Penetrating	Pungent	Penetrating	
Strong		Strong			Strong
Eucalyptus-like (cineolic),		Eucalyptus-like		Slightly minty	
Aromatic					Sweet, balsamic
Camphoraceous, piney	Resinous	Camphoraceous	Camphoraceous		Piney
	Piney				
	Turpentine-like				
Dry-out					
Top notes quickly lost	Very poor tenacity	Poor tenacity		Poor tenacity	
Residuals:					
Herbaceous					Herbaceous
Dry			Dry		
Woody			Woody		
Peppery			Peppery		

Odor: Smelling strip examined immediately; then after 1 hour; 2 hours; 6 hours; and overnight.

Flavor: 5% dispersion of the essential oil on salt added to a premade neutral soup and evaluated at 55°C.

Each sample was assessed for odor, differentiating between "top note," "body," "persistence" and "dry-out." Then for flavor and after-

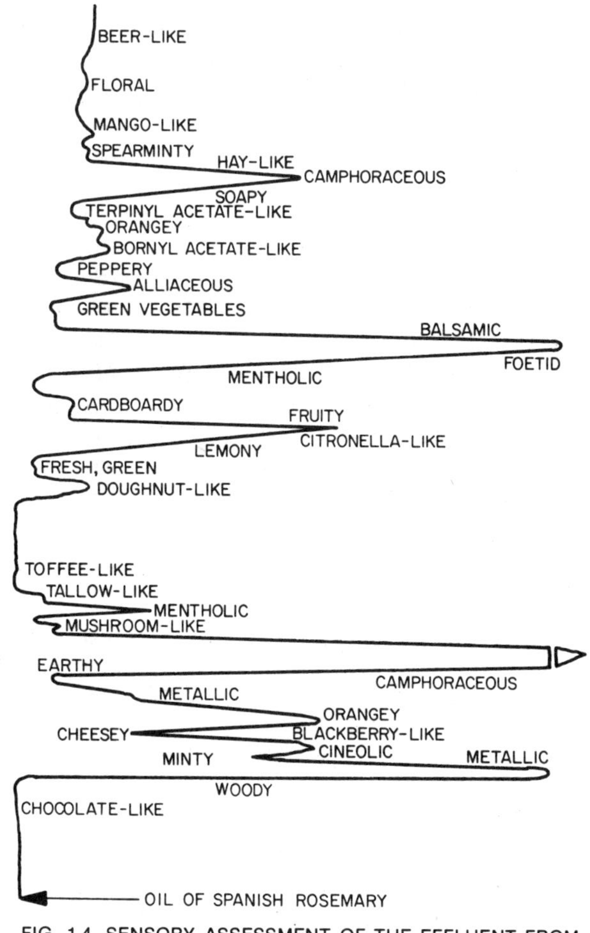

FIG. 1.4. SENSORY ASSESSMENT OF THE EFFLUENT FROM GLC OF SPANISH OIL OF ROSEMARY

taste, care being taken to ensure that adequate time was allowed between tests so as to avoid any carryover. It is appreciated that there is a considerable difference between the profiles of the ground herbs and spices and their essential oils but it is more generally by their essential oil content that the spices are judged. The differences in aromatic components between oils distilled directly from the herbs and spices and those obtained from the equivalent oleoresin have been the subject of a paper by Eiserle and Rogers (1973) and their findings are of importance in the quality control of foods containing spice extractives rather than traditional ground spices.

The need to establish realistic profiles for natural flavoring materials has been expressed by the International Standards Organization (ISO) which is actively investigating the problems associated with odor and flavor descriptions for incorporation into international specifications. As a first step, lists of acceptable terms for use in sensory analysis have already been published by several member nations. Of necessity, this work is very subjective and the profiles offered in this book reflect the consolidated views of the panels concerned. One still cannot avoid using the word "characteristic" in the recognition of any aromatic plant material but the attributes described should assist greatly in the more effective blending of these products to achieve desirable flavor profiles in any processed foods in which they are used. They will also act as a guide for the quality control of these materials. Refinements in techniques and the evolution of more precise descriptive terms are possible and such topics are likely to occupy the thoughts of flavor chemists for many years to come.

BIBLIOGRAPHY

Classification of Odors and Flavors

ANON. 1969. Definition of flavor. Food Technol. *23*, No. 11, 28.

ANON. 1970. The correct type of flavor to use and its correct usage. A glossary of types of flavoring agents. Flavour Ind. April 1, 223–224.

ARCTANDER, S. 1960. Perfumery and Flavor Materials of Natural Origin. Stephen Arctander, Elizabeth, N.J.

HARDER, U. 1973. Verbal classification of odors. Parfum. Kosmet. *54*, No. 4, 106–112. (German)

HODGE, W. H. 1975. Survey of flavor producing plants. Int. Flavours Food Additives *6*, No. 4, 244–245.

JENNINGS, W.G. 1972. The changing field of flavour chemistry. Food Technol. *26*, No. 11, 25,27,30,32,34.

KASTNER, D. 1973. Description and classification of odors. Parfum. Kosmet. *54*, No. 4, 97–105. (German)

MOSKOWITZ, H. I. 1977. Psychological qualities of odor. I. Profiling systems. Perfumer Flavorist *2*, No. 4, 13–16,18.

NOVAK, G. 1973. Odor classification. Riechst., Aromen, Köerperpflegem. *23*, No. 9, 279–280,282,284–286. (German)

OHLOFF, G. 1972. Classification and genesis of food flavours. Flavour Ind. *3*, 501-508.
PANGBORN, R. M. 1968. Interrelationship of odor, taste and flavor. Food Prod. Dev. *2*, No. 4, 74-76,81.
PEYRON, L. 1973. Flavors for human nutrition—The isolation of flavors Riv. Ital. Essenze, Profumi, Piante Off. *55*, No. 9, 617-630 (277 refs). (French)
RIETZ, C. A. 1961. A Guide to the Selection, Combination and Cooking of Foods, Vol. 1 and 2. AVI Publishing Co., Westport, Conn.

Flavoring Components

AYRES, J. and CLARK, W. R. E. 1975. Instrumental techniques in flavour research. Food Manuf. *50*, No. 1, 19-20,22.
BAJAI, K. L. and BHATIA, I. S. 1975. Flavor constituents of foods. Riechst., Aromen, Köerperpflegem. *25*, No. 2, 46-47; *25*, No. 4, 100-103. (German)
CHANG, S. S. 1973. Overcoming problems in flavor component identification. Food Technol. *27*, No. 4, 27-28,30,32,34,36,39.
CREVELING, R. K. and JENNINGS, W. G. 1970. Volatile components of Bartlett pear. J. Agric. Food Chem. *18*, No. 1, 19-24.
FENAROLI, G. 1969. Influence of small amounts of some constituents, on sensory properties of aromas. Fr. Ses Parfums. *12*, No. 66, 401-406. (French)
FENAROLI, G. 1970. Studies of some active principles derived from aromatic plants used for flavoring foods and beverages in the Common Market and in the United States. Fr. Ses Parfums *13*, No. 69, 267-270. (French)
MARONI, G. and FENAROLI, G. 1973. Fruits, vegetables, spices, natural flavors—chemical composition. Riv. Ital. Essenze, Profumi, Piante Off. *55*, No. 3, 168-198, (274 refs). (Italian)
MERRITT, C., Jr. *et al.* 1974. A combined GLC-mass spectrometer-computer system for the analysis of the volatile components of foods. J. Agric. Food Chem. *22*, No. 5, 750-755.
MORRIS, W. W. 1973. High resolution infrared spectra of fragrance and flavor compounds. J. Assoc. Off. Anal. Chem. *56*, No. 5, 1037-1064.
NURSTEN, H. 1975A. Four areas of aroma research explored. Int. Flavours Food Additives *6*, No. 4, 213-214,216.
NURSTEN, H. 1975B. Chemistry of flavours—past, present and future. Int. Flavours Food Additives *6*, No. 2, 75-82.
PALMER, J. K. 1973. Symposium on biogenesis of flavor components. J. Agric. Food Chem. *21*, No. 4, 559.
PALMER, J. K. 1973. Separation of components of aroma concentrates on the basis of functional groups and aroma quality. J. Agric. Food Chem. *21*, No. 5, 923-925.
PICKETT, J. A., COATES, J., and SHARPE, F. R. 1975. Distortion of essential oil composition during isolation by steam distillation. Chem. Ind. No. 13, July 5, 671-672.
ROHAN, T. A. 1970. Food flavor volatiles and their precursors. Food Technol. *24*, No. 11, 29-32,37.
SOLMS, J. and NEUKOM, H. (Editors). 1967. Aroma- and Flavor-Producing Substances in Foods. A.G. Forster-Verlag, Zurich, Switzerland. (German)
SLOAN, J. L., BILLS, D. S., and LIBBEY, L. M. 1969. Heat-induced compounds in strawberries. J. Agric. Food Chem. *17*, No. 6, 1370-1372.
TERANISHI, R. 1970. High resolution gas chromatography in aroma research. Flavour Ind. *1*, 35-40.
TERANISHI, R., HORNSTEIN, I., ISSENBERG, P., and WICK, E.L. 1971. Flavour Research. Principles and Techniques. Marcel Dekker, New York.

Aromatic Profiles

AMERINE, A. M., PANGBORN, R. M., and ROESSLER, E. B. 1965. Principles of Sensory Evaluation of Food. Academic Press, New York and London.

ANON. 1964. Sensory testing guide for panel evaluation of foods and beverages. Food Technol. *18*, No. 8, 1135–1141.

ANON. 1971. Sensory evaluation—the path to better flavors. Red Seal Letter *21*, June, Warner-Jenkinson Company, St. Louis.

ANON. 1973. Odor glossary *T-453*, Barnebey-Cheney, Columbus, Ohio.

BROWER, K. R. and SCHAFER, R. 1975. The recognition of chemical types by odor: The effect of steric hindrance at the functional group. J. Chem. Educ. *52*, No. 8, 538–540.

CAUL, J. F. 1970. Sensory testing in new product development. Cereal Sci. Today *16*, No. 1, 13–16.

CAUL, J. F. *et al.* 1958. The flavor profile in review. *In* Flavor Research and Food Acceptance, Arthur D. Little, Inc., Reinhold Publishing Corporation, New York.

DIXON, M. P. 1970. Sensory descriptive analysis; the flavour profile. Flavour Ind. *1*, January, 45–46.

EISERLE, R. J., and ROGERS, J. A. 1973. The composition of volatile oils derived from oleoresins. J. Am. Oil Chem. Soc. *49*, 573–577.

FENAROLI, G. 1969. The influence of small amounts of some constituents on sensory properties of aromas. Fr. Ses Parfums *12*, No. 60, 423–425. (French)

GREGSON, R. A. M. and MITCHELL, M. J. 1974. Odour quality similarity scaling and odour word profile matching. Chem. Senses Flavour. *1*, No. 1, 95–101.

HARPER, R. 1975. Some chemicals representing particular odour qualities. Chem. Senses Flavour. *1*, No. 3, 353–357.

HARPER, R. 1975. Terminology in the sensory analysis of food. Int. Flavours Food Additives *6*, No. 4, 215–216.

HARPER, R., BATE-SMITH, E. C., and LAND, D. G. 1968. Odour. *In* Description and Odour Classification. J. A. Churchill, London.

HARPER, R., BATE-SMITH, E. C., LAND, D. G., and GRIFFITHS, N. M. 1968. A glossary of odour stimuli and their qualities. Perfum. Essent. Oil Rec. *59*, No. 1, 22–27.

HARPER, R., LAND, D. G., GRIFFITHS, N. M., and BATE-SMITH, E. C. 1968. Odour qualities: A glossary of usage. J. Psychol. *59*, No. 3, 231–252.

HIRSH, N. L. 1970. Attempts at quantitating flavor differences. Food Prod. Dev. *4*, No. 2, 22–23, 26.

LAWLESS, H. T. and CAIN W. S. 1975. Recognition memory for odours. Chem. Senses Flavour *1*, No. 3, 339–352.

MOSKOWITZ, H. R. 1974. Combination rules for judgements of odor quality. J. Agric. Food Chem. *22*, No. 5, 740–743.

MOSKOWITZ, H. R., DRAUNIERS, A., CAIN, W. S., and TURK, A. 1974. Standardized procedure for expressing odour intensity. Chem. Senses Flavour *1*, No. 2, 235–237.

MOSKOWITZ, H. R. and VON SNYDOW, E. 1975. Computer derived perceptive maps of flavors. J. Food Sci. *40*, No. 3, 788–792.

MÜLLER, A. 1973. Odor of plants. II. Citrus scents. Riechst. Aromen, Koerperpflegem. *23*, No. 10, 316. (German)

MÜLLER, A. 1975. Odor of plants. VI. Spicy odors. Riechst. Aromen, Koerperpflegem. *25*, No. 8, 218–220. (German)

NOVAK, G. 1974. Sensory description of some plants. Riechst. Aromen, Koeperpflegem. *24*, No. 1, 14–17 and No. 2, 37–40.

O'MAHONY, M. and THOMPSON, B. 1977. Taste quality descriptions: Can the subject's response be affected by mentioning taste words in the instructions? Chem. Senses Flavour *2*, 283–298.

PALMER, D. H. 1974. Multivariate analysis of flavor terms used by experts and non-experts for describing teas. J. Sci. Food Agric. *25*, No. 2, 153–164.

PARLIAMENT, T. H. and SCARPELLINO, R. 1977. Organoleptic techniques in chromatographic food flavor analysis. J. Agric. Food Chem. *25*, 97–99.

PFAFFMAN, C. 1974. The sensory coding of taste quality. Chem. Senses Flavour *1*, No. 1, 339–352.

RANDEBROCK, R. E. M. 1973. Objectification and evaluation of olfactory impressions using computer techniques. Parfums, Cosmet. Savons Fr. *3*, No. 7, 382–392. (French)

SAWYER, F. M. 1971. Interaction of sensory panel and instrumental measurement. Food Technol. *25*, No. 3, 51–52.

SCHWOB, R. 1971. Standardization of terminology in the field of organoleptic properties. Flavour Ind. *2*, November, 627–629.

SFIRAS, J. 1969. Olfactometry and chromatography of odorous emissions. Fr. Ses Parfums *12*, No. 66, 393–398. (French)

SJÖSTRUM, L. B., CAINCROSS, S. E. and CAUL, J. F. 1957. Methodology of the flavor profile. Food Technol. *11*, No. 9, 20–25.

SPENCER, H. W. 1971. Techniques in the sensory analysis of flavours. Flavour Ind. *2*, May, 293–302.

WILLIAMS, A. A. 1975. The development of a vocabulary and profile assessment method for evaluating the flavor contribution of cider and perry aroma constituents. J. Sci. Food Agric. *26*, No. 5, 567–582.

ZAUSCH, G. Th. 1969. Sensory teminology. Am. Perfum. Cosmet. *84*, 57–58.

ZAUSCH, G. Th. 1970A. Representation of sensory impressions of single pure fragrance materials as standard references. Riechst. Aromen, Koerperpflegem. *20*, No. 12, 479, 480, 482, and 484. (German)

ZAUSCH, G. Th. 1970B. Standardization of terminology in sensory analysis. Milchwissenschaft *25*, 34–36. (German)

Flavors

DOWNEY, W. J. and EISERLE, R. J. 1970. Substitutes for natural flavors. J. Agric. Food Chem. *18*, No. 6, 983–987.

HALL, R. L. 1975. GRAS—concept and application. Food Technol. *29*, No. 1, 48, 50, 52–53.

HEATH, H. B. 1971. Why natural—why not synthetic? Flavour Ind. *2*, November, 630–632.

WIGGERS DE VRIES, V. 1972. Modern trends in flavor development. Am. Cosmet. Perfum. *87*, No. 11, 37–40.

2

The Culinary Herbs

DEFINITION OF HERBS

Botanically herbs are soft-stemmed plants whose main stem dies down to the root and re-grows each year. Culinary herbs popularly include annual and biennial plants as well as the leaves of some bushes or even trees. Herbs may or may not be strongly aromatic in character but those of value in food flavoring have a quite distinctive character. They contain only low levels of essential oil which is responsible for their odor and flavor profile. For culinary use the whole herbaceous tops are gathered and may be used either fresh or after drying. In the latter case it is usual to remove the leaves, flower heads, seeds, etc., from the heavier, harder and less aromatic stems by screening. Such herbs are called "rubbed" or "broken" and it is in this form that they are sold for domestic use. The flavor intensity of the freshly-cut green herb is appreciably higher and of a different character from that of its dried equivalent. A considerable proportion of the lighter essential oil fractions are either lost or modified during the dehydration process. The fine, clean notes associated with the fresh herb are, in the dried version, generally overlaid by a dull hay-like aroma.

CLASSIFICATION OF HERBS

In studying a range of aromatic materials, such as the herbs, it is desirable to classify them into groups having some character in common. A botanical classification is of interest but the system which gives the most meaningful comparative grouping is one based on sensory attributes related to the prime constituents of the essential oil. Although the essential oil is important and is responsible for the characteristic odor and flavor of the plant, it does not represent the total flavor complex; nonvolatile constituents may play a significant part in rounding off the profile and contributing taste elements, particularly sharpness and bitterness.

Arctander (1960) proposed 88 groups of related natural raw materials used in food flavorings and fragances of which 22 groups contained herbs or spices. A simple system of only 5 groups has been adopted in this chapter, the herbs being classified as follows:

Part 1—Those containing cineole:
bay laurel, rosemary, spanish sage.

Part 2—Those containing thymol and/
or carvacrol: thyme, origanum, wild
marjoram, sweet savory, Mexican sage,
oregano.

Part 3—Those containing sweet alcohols:
sweet basil, sweet marjoram, tarragon.

Part 4—Those containing thujone:
Dalmation sage, Greek sage, English
sage.

Part 5—Those containing menthol:
peppermint, corn mint, spearmint,
garden mint.

AROMATIC PROFILES

Each of the herbs included in the above classifications will be reviewed with particular emphasis on the odor and flavor profile of its essential oil. The impressions tabulated are in order of their appearance and each profile was built up by discussion after several panel evaluations. The terms used were limited to those given in Appendix A.

Appell (1968) examined the volatility patterns of a range of essential oils, including those of several herbs, and quantified their volatility index and evaporation characteristics. In the following profiles the initial impact has been rated on a four-point scale: 1—slight; 2—moderate; 3—strong; 4—powerful.

Part 1: Herbs Containing Cineole

The distinctive note of these herbs is that of eucalyptus although each has its own quite characteristic background notes. They are used in seasonings chiefly to give a "lift" to the overall flavor profile but when used in excess they can be too powerful and quite easily dominate the seasoning blend and swamp any other natural flavor present.

BAY LAUREL, SWEET BAY (*Laurus nobilis*, L.)

Sweet Bay is a tree which is cultivated throughout Turkey, Israel, Russia, Italy and France although, increasingly, commercial crops of leaves are being exported from Greece and Mexico. A similar tree is found in China but the variety is different and the leaves contain an essential oil having a marked phenolic odor not unlike that of West Indian Bay (*Pimenta racemosa*, Miller). The leaves yield about 2% essential oil when steam distilled.

FIG. 2.1. SWEET BAY LAUREL (*Laurus nobilis*, L.)

Profile of Oil of Sweet Bay Laurel

Origin Distilled from Turkish Bay leaves.

Odor Impact: 2

	Initial:	Strongly EUCALYPTUS-like.
		FRESH, initially WARMING but later COOLING.
		PENETRATING; irritating, slightly pungent.
		CAMPHORACEOUS.
		SWEET, slightly GREEN; leafy.
		Pleasantly SOOTHING; soft, slightly oily.
		MEDICINAL.
	Persistence:	Good, very slow evaporation rate.
	Dry-out:	Increasingly GREEN, HERBACEOUS.
		Markedly PEPPERY, DRY.
Flavor		TINGLING.
		Initially pleasantly SWEET, SPICY.
		Later BITTER which is unpleasantly lingering.
	After-taste:	BITTER at the back of the mouth.
Note	When warmed the oil acquires a marked FRUITY, slightly LEMONY character and the GREEN notes are much intensified.	

Components of the Essential Oil

α-pinene	cineole (45–50%)
β-phellandrene	1-linalool
	1-α-terpineol
	geraniol
	eugenol
	methyl eugenol
	geranyl and eugenyl esters

ROSEMARY (*Rosmarinus officinalis*, **L.**)

Rosemary is a small evergreen shrub which grows wild throughout Spain, France, Algeria, Tunisia, Italy, Jugoslavia and Morocco. The main Spanish crop is cut during March and July yielding 0.3 to 2% essential oil depending on the dryness of the material at the time of distillation. The essential oil is pale yellow to almost colorless, the odor varying considerably with origin.

FIG. 2.2. ROSEMARY PLANT (*Rosmarinus officinalis*, L.)

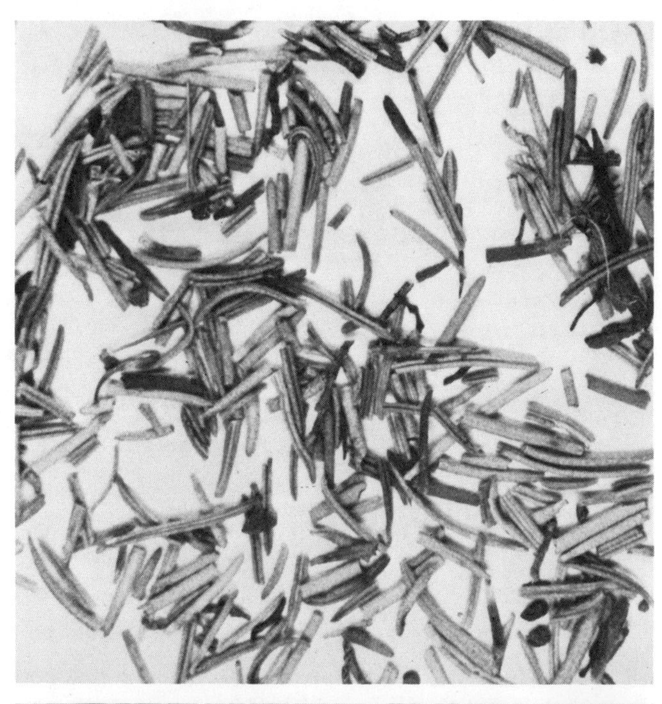

FIG. 2.3. ROSEMARY NEEDLES (MAGNIFIED X 2)

Profile of Oil of Rosemary

Origin	Distilled from Spanish Rosemary.	
Odor	Impact:	3
	Initial:	FRESH, initially COOLING but later WARMING; light.
		PENETRATING; irritating, pungent.
		EUCALYPTUS-like; aromatic.
		CAMPHORACEOUS; turpentine-like, piny.
	Persistence:	The pleasing top-notes are quickly lost and are initially replaced by a heavier camphoraceous, herbaceous quality which is lingering and tenacious.
	Dry-out:	PEPPERY, slightly WOODY.
Flavor		TINGLING followed by slight NUMBING.
		ASTRINGENT.
		BITTER.
		WARMING (the cooling effect is absent in the flavor).
	After-taste:	BITTER.

Components of the Essential Oil

α-pinene	cineole (17–30%)
camphene	borneol (16–20%)
	camphor (about 10%)
	bornyl acetate (2–7%)
	terpineol
	verbenone

SPANISH SAGE (*Salvia lavandulaefolia*, **L.**)

Sage grows wild throughout the arid mountain slopes of southern Spain and often in close proximity to the cultivation of Spike Lavender, particularly in the provinces of Granada, Jaen and Murcia. Little of the herb reaches the market as it is usually harvested and distilled locally to give about 0.7% of a yellowish essential oil. Because of the marked similarity of odor, this oil is often used as an extender of the much more expensive oil of Spike Lavender.

Profile of Oil of Spanish Sage

Origin	Distilled in Spain.	
Odor	Impact:	2
	Initial:	FRESH, light. SHARP, PENETRATING; irritating. EUCALYPTUS-like; piny. SPIKE LAVENDER-like. CAMPHORACEOUS. Pleasantly GREEN, HERBACEOUS.
	Persistence:	The cineolic top-notes are quickly lost and replaced by the heavier camphoraceous, herbaceous notes.
	Dry-out:	Strongly GREEN, HERBACEOUS. Slightly IRRITATING. PINY.
Flavor		Very BITTER. Slightly TINGLING. GREEN, strongly HERBACEOUS. SHARP but not pungent.
	After-taste:	BITTER.
Note:	When warmed the oil has a marked FRUITY note but not specifically a named fruit.	

Components of the Essential Oil

α-pinene	cineole (25–35%)
	d-camphor (about 20%)
	1-linalool (about 20%)
	linalyl acetate
	sabinyl acetate
	α-terpinyl acetate

Part 2: Herbs Containing Thymol and/or Carvacrol

The establishment of aromatic profiles for members of this group is complicated by the considerable confusion over nomenclature. The botany of the plants in the family *Labiatae*, which are variously known as "thyme" or "origanum," is very involved; much of the misunderstanding is due to the fact that Spanish thyme is often origanum and vice versa and the term "marjoram" or the French word "marjolaine" is applied to several aromatic plants of different species. In commerce, the position is not improved as the essential oils from these various plants are often blended and offered under both names. However, there is now general agreement for the following nomenclature.

THYME is *Thymus vulgaris*, L. or *Thymus zygis*, L. The essential oil from these plants has a total phenol content of 40–60% of which not less than 90% is crystallizable thymol.

WILD THYME is usually regarded as *Thymus serpyllum*, L. and is not widely available except in Russia.

ORIGANUM is *Thymus capitatus* (Hoff & Link) and some other species of *Thymus* or *Origanum* the essential oil of which contains 60–75% of total phenols consisting mainly of noncrystallizable carvacrol.

WILD MARJORAM is either *Origanum vulgarae*, L. or *Thymus masticina*, L. and is so described in *Nomenclature of Spices* by the International Standards Organization.

SWEET MARJORAM is *Marjorana hortensis*, Mnch., the essential oil of which does not contain any phenols. In consequence, this herb is classified in the following group.

WHITE THYME OIL is not a prime essential oil from a different plant but is a redistillation of red thyme oil of commerce.

THYME (*Thymus vulgaris*, **L.** or *T. zygis*, **L.** var. *gracilis* or *floribunda*)

The rubbed thyme herb of commerce is derived from the flowering tops of two species of thyme which grow wild on the vast heathlands of southeast Spain, along the coast of the Mediterranean and on the southern slopes of the Atlas mountains in Morocco. The entire herb is harvested when the flowers are fully open during July and August, the cut herb being distilled in field stills to give 0.6 to 1.0% of a yellowish-red essential oil. The color of the oil may darken considerably due to iron contamination from the stills.

Profile of Red Thyme Oil

Origin	Commercial Spanish red thyme oil.	
Odor	Impact:	4
	Initial:	RICH, SWEET, pleasingly AROMATIC. Initially WARMING then COOLING with a slight anaesthetic effect.

FIG. 2.4. THYME PLANT (*Thymus vulgaris*, L.)

FIG. 2.5. THYME HERB CLOSE-UP

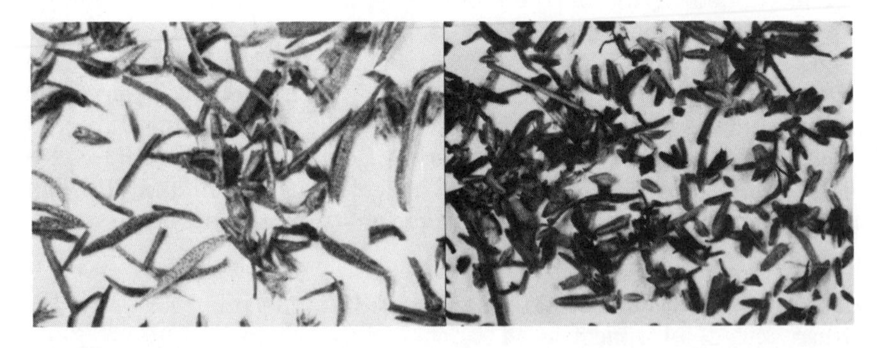

FIG. 2.6. SPANISH THYME HERB BROKEN FOR DOMESTIC USE (MAGNIFIED X 2)

FIG. 2.7. FRENCH THYME HERB BROKEN FOR DOMESTIC USE (MAGNIFIED X 2)

		PUNGENT; irritating, astringent.
		GREEN, HERBACEOUS.
		SPICY, slightly WOODY.
	Persistence:	Lingering with very little change in character.
	Dry-out:	SWEET, PHENOLIC; medicinal.
Flavor		SHARP; biting.
		WARM, SPICY, RICH, full-bodied.
		HERBACEOUS.
	After-taste:	Lingering SHARPNESS.
Note	Poor quality and old oils may display a strong TAR-like profile with slightly BURNT and SMOKY notes. Commercial oils may be admixed with oil of rosemary but EUCALYPTUS-like top-notes which are atypical would be detected.	

Profile of White Thyme Oil

Origin	Commercial sample.	
Odor	Impact:	3
	Initial:	SWEET.
		PENETRATING; irritating.
	Compared with Red Thyme Oil it is:	
		Less TERPENEY, SHARP or HERBACEOUS.
		More SMOOTH and pleasingly AROMATIC.

Composition of the Essential Oil

α-pinene	thymol (about 50%)
camphene	carvacrol
terpinene	geraniol
β-caryophyllene	terpen-4-ol
	borneol
	1-linalool

ORIGANUM (*Thymus capitatus*, **Hoff & Link and other species**)

There are several species of *Thymus* which yield essential oils characterized by a high total phenol content almost all of which is noncrystallizable carvacrol. The principal species of commercial value are:

Thymus capitatus from Spain, Israel, Turkey and Greece.
Thymus virens from Morocco.
Thymus vulgare from Russia, Bulgaria and Italy.

Profile of Spanish Oil of Origanum

Origin	Commercial oil distilled in Spain.	
Odor	Impact:	4
	Initial:	Very HARSH. FRESH, GREEN, HERBACEOUS. CAMPHORACEOUS. TAR-like; cade-like, burnt. DRY, WOODY; full bodied. PHENOLIC; medicinal, clinical.
	Persistence:	The initially unpleasant character airs-off quickly, the herbaceous notes intensify and are lingering.
	Dry-out:	DRY. TAR-like.
Flavor		Strongly unpleasant. BURNING but not pungent, WARMING. PHENOLIC; medicinal. RICH, SPICY; full-bodied. HERBACEOUS; green.
	After-taste:	Lingering PHENOLIC, the HERBACEOUS; back-note is slow to develop.

Components of the Essential Oil

α-pinene	carvacrol (60–80%)
dipentene	thymol (about 6%)
p-cymene	bornyl acetate
camphene	borneol
γ-terpinene	linalool
	geraniol

Comparison of the Odors of Thyme and Origanum

Thyme	Origanum
SWEET	HEAVY
COOLING	PHENOLIC
FRESH	HARSH
LIGHT	BURNT

WILD MARJORAM (*Origanum vulgare*, **L.**)

So-called "wild marjoram" is really *Origanum vulgare* which grows freely in temperate regions and is also cultivated all over the world. Most commercial crops and essential oils distilled from them originate in those countries bordering the Meditarranean and in Mexico. What is known in commerce as "Oil of Wild Marjoram" is usually obtained from *Thymus masticina*, L., an herb which grows in the foothills of Seville and Almeira in Spain. This plant is usually harvested when in full bloom during June/August and distilled in field stills to give about 0.2% of a brownish essential oil. Two types of oil of wild marjoram are recognized—the thymol-type and the carvacrol-type, depending on the constituent phenols present.

FIG. 2.8. ORIGANUM (WILD MARJORAM) BROKEN FOR DOMESTIC USE (MAGNIFIED X 2)

Profile of Thymol-type Wild Marjoram Oil

Origin Commercial oil distilled in Spain.
Odor Impact: 4

	Initial:	STRONG but LIGHT, FRESH; clean.
		Initially WARMING but later COOLING.
		PENETRATING; irritating.
		EUCALYPTUS-like, camphoraceous.
		SWEET, SPICY, slightly FLORAL; aromatic.
		Slightly anise-like.
	Persistence:	The odor remains strong and characteristically PHENOLIC.
	Dry-out:	DRY, SPICY.
Flavor		WARMING; burning, astringent.
		Very BITTER.
		Pleasantly AROMATIC; spicy.
	After-taste:	BITTER changing to a fresh SWEETNESS.

Components of the Essential Oil

α-pinene (7-8%)	thymol (about 50%)
ρ-cymene	cineole
dipentene	linalool
β-caryophyllene	linalyl acetate

Profile of Carvacrol-type Wild Marjoram Oil

Origin	Commercial oil distilled in Spain.	
Odor	Impact:	4
	Initial:	RICH; full bodied.
		Strongly PENETRATING; irritating, pungent.
		PHENOLIC, TAR-like, BURNT.
		SPICY, slightly WOODY.
	Persistence:	Lingering BURNT notes.
	Dry-out:	HARSH, HEAVY, slightly TAR-like.
		BURNT.
Flavor		SHARP; harsh, burning.
		SPICY.
		HERBACEOUS.
		BITTER.

After-taste:	BITTER with lingering GREEN notes.

Components of the Essential Oil

terpenes (30–40%)	total phenols about 50% mainly carvacrol

SWEET SAVORY (*Satureia hortensis*, **L.** or *Satureia montana*, **L.**)

There are two distinct species of the herb savory which yield essential oils having a marked origanum-like quality:

Summer Savory (*Satureia hortensis*).—An annual plant cultivated throughout France and the Mediterranean region; harvested when in flower in June/July and distilled to give about 0.2% of a yellowish-brown essential oil.

Winter Savory (*Satureia montana*).—A small hardy shrub growing wild in southern France; harvested when in bloom, field-dried and distilled to give about 0.25% of a deep orange-brown essential oil.

The "Oil of Savory" of commerce is mostly derived from *Satureia hortensis.*

Profile of Summer Savory Oil

Origin	Commercial oil distilled in France.	
Odor	Impact:	3
	Initial:	FRESH. PINY; medicinal, disinfectant-like. PUNGENT; sharp, irritating. SWEETLY SPICY; smooth, slightly fruity. HERBACEOUS; green. PHENOLIC. Initially very slightly WARMING but later slightly COOLING.
	Persistence:	Lingering with an increase in in the HERBACEOUS notes.
	Dry-out:	MEDICINAL with marked GREEN note. SOAPY, markedly perfumed.

FIG. 2.9. SUMMER SAVORY PLANT (*Satureia hortensis*, L.)

FIG. 2.10. WINTER SAVORY PLANT (*Satureia montana*, L.)

Flavor	SHARP, BURNING; biting. BITTER. HERBACEOUS more pronounced at low usage. PHENOLIC.

After-taste:	BITTER with a slightly lingering PUNGENCY.

Composition of the Essential Oils

α-pinene (7–8%)	carvacrol (30–45%)
ρ-cymene (about 30%)	thymol
dipentene	1-linalool
	terpineol
	borneol
	1-carvone

Comparison of the Odors of Summer and Winter Savory Oils

Summer Savory	Winter Savory
FRESH	CINEOLIC
CLEAN	Markedly EARTHY, HEAVY.
DISINFECTANT-like	SPICY

OREGANO

A number of botanicals are classified under the title "Oregano" largely because of the lack of precision in translating this word into different languages where common usage is not identical. On the continent of Europe the word "Oregano" is common to most of the languages and is usually taken as descriptive of wild marjoram or origanum but its Federal Drug Administration definition in the United States GRAS list is:

> **Oregano** (Mexican Oregano; Mexican Sage) usually *Lippia graveolens* although species of *Origanum, Coleus, Lantana* and *Hyptio* may be similarly described.

Such a wide interpretation is indicative of the confusion which exists. The designation "oregano" may be taken to refer not to any one particular plant species but rather to a type of herb flavor, the interpretation of which will vary in different parts of the world. For the purpose of establishing a characteristic profile for oregano a commercial sample of the herb was obtained from Mexico. This was probably *Lippia graveolens* but was not specifically characterized.

Profile of Oil of Mexican Sage (Oregano)

Origin	Commercial Mexican Sage herb distilled in the United Kingdom.	
Odor	Impact:	2
	Initial:	SWEET, slightly FLORAL. COOLING.

FIG. 2.11. MEXICAN SAGE (OREGANO) BROKEN FOR DOMESTIC USE (MAGNIFIED X 2)

		Slightly PENETRATING; slightly pungent.
		FRUITY; guava-like.
	Persistence:	Rapid change in character, the light top-notes being lost and replaced by the following profile which is lingering.
		HARSH, WOODY; spicy.
		HEAVY, WARM; tar-like.
	Dry-out:	Unpleasantly SWEATY, UNFRESH.
		PLASTICINE-like.
		WOODY.
Flavor		SHARP.
		WARM, HERBACEOUS; full-bodied.
		FRUITY.

	After-taste:	Lingering WOODY note which is unpleasantly UNFRESH.
Note	The flavor is much more acceptable when fully diluted when the warm, herbaceous and slightly fruity character is more evident.	

Part 3: Herbs Containing Sweet Alcohols

The "sweet" herbs—basil, marjoram and tarragon—are very widely used because of their attractive fragrance and superb blending properties when included in a seasoning. Each has a quite distinctive profile and is more pungently fragrant when freshly cut than after drying. The dried, rubbed herb does, however, retain its flavoring quality very well and in consequence the hay-like overtones are much less obvious.

SWEET BASIL (*Ocimum basilicum*, **L.**)

Due to profuse cross-pollination and consequent hybridization there are several strains and varieties of the herb which, when distilled, yield essential oils of quite different aromatic profiles. Two are of commercial importance. They tend to be offered indiscriminately as "Oil of Basil" but are very different in their flavoring strength and quality. The two varieties concerned are:

(a) Sweet Basil (Mediterranean-type), which is cultivated in southern France and to a lesser extent in Italy, Hungary, Spain and the United States.

(b) Exotic basil (Réunion-type) which is cultivated in the Comoro Islands, the Seychelles and the Malagasy Republic.

The aromatic character of these two essential oils makes them readily distinguishable: Sweet Basil (Mediterranean-type) has a sweet floral linalool-like odor whereas the Exotic variety has a pronounced camphoraceous-anise-like note.

Two other distinct forms of basil are recognized but are much less well known commercially:

(a) Bulgarian basil (Methyl cinnamate-type), possibly from *Ocimum canum*, Sims (Guenther 1949), which grows in West Africa, the East Indies and Indonesia.

(b) Java basil (Eugenol-type), from *Ocimum gratissimum*, L., obtained from Java and the south Pacific islands.

Profile of Sweet Basil Oil (Mediterranean-type)

Origin Commercial oil distilled in southern France.

FIG. 2.12. SWEET BASIL PLANT (*Ocimum basilicum*, L)

FIG. 2.13. BASIL HERB BROKEN FOR DOMESTIC USE (MAGNIFIED X 2

Odor	Impact:	3
	Initial:	PENETRATING; slightly irritating. SWEET; aromatic, floral. Pleasantly COOLING, FRESH. Slightly GREEN, HERBACEOUS. Slightly MINTY. Faintly BALSAMIC. Slightly SMOKY; spicy, woody. Delicately ANISE-like.
	Persistence:	The character of the oil changes very slowly, the sweet linalool-like notes becoming more pronounced initially and then the anise-like character predominates.
	Dry-out:	SWEETLY HERBACEOUS with lingering ANISE-like back-note.
Flavor		WARM, SPICY. ANISE-like. HERBACEOUS; green. Pleasantly FRESH.
	After-taste:	HERBACEOUS but slightly BITTER.

Note

1. The freshly-gathered herb has a much more pronounced MINTY note with a hint of CLOVE. It is more pleasingly SWEET than the essential oil.
2. The principal constituent of sweet basil oil is linalool and it is not unusual for commercial oils to be sophisticated with this synthetic compound in order to cheapen the price. However, trace components in the natural oil make this a very powerful and well-balanced flavoring agent and, hence, more economical in use than a sophisticated oil which may have an apparently stronger aroma.

Composition of the Essential Oil

ρ-cymene	d-linalool (about 40%)
myrcene	methyl chavicol (about 25%)
sesquiterpene hydrocarbons	eugenol
	cineole

geraniol
α-terpineol
methyl heptenone

Profile of Exotic Basil Oil (Réunion-type)

Origin	Commercial oil distilled in the Malagasy Republic.	
Odor	Impact:	3
	Initial:	HARSH, strongly PENETRATING.
		CAMPHORACEOUS.
		Strongly ANISE-like; tarragon-like.
		PHENOLIC; anaesthetic, clove-like.
		WARM, HEAVY; spicy.
	Persistence:	The harsh camphoraceous notes of this oil are quickly diminished, the profile becoming more SPICY and ANISE-like.
	Dry-out:	HEAVY, AMBER-like, lingering ANISE note.
Flavor		Strongly ANISE-like.
		BITTER.
		GREEN, HERBACEOUS.
	After-taste:	Lingering BITTER.

Components of the Essential Oil (Lawrence *et al.* 1971)

α-pinene	methyl chavicol (about 85%)
Δ-3-carene	1-linalool (about 0.7%)
limonene	camphor
caryophyllene	borneol
	eugenol
	cineol
	bornyl acetate
	methyl eugenol

Comparison of the Odors of Oil of Sweet Basil (French) and Exotic Basil (Malagasy Republic)

Sweet Basil	Exotic Basil
SWEETLY ANISE-like	HARSHLY CAMPHOR-ACEOUS
COOLING	HEAVY, SPICY
FLORAL	PHENOLIC

Oil of Basil is also available commercially from North African sources. This is mostly derived from *Ocimum gratissimum*, L., which differs from the eugenol-type in that the oil contains up to 40% of phenols which are mostly thymol and some carvacrol. The odor is readily distinguished from French basil oil:

North African *O. gratissimum*	French *O. basilicum*
WARMLY SPICY	SWEETLY ANISE-like
HERBACEOUS	COOLING
PHENOLIC, medicinal	FLORAL

SWEET MARJORAM (*Majorana hortensis*, **Mnch.**)

The considerable confusion which exists over herbs designated as "marjoram" or in the French as "marjolaine" has already been commented upon under "origanum." Sweet marjoram may refer to several culinary herbs belonging to different species of which the following are the most widely known:

(a) *Origanum majorana*, L.—A perennial herbaceous shrub growing throughout the Mediterranean region.

(b) *Majorana hortensis*, Mnch.—An annual herb cultivated in Europe, North Africa and to a limited extent in the United States.

(c) *Origanum vulgare*, L.—The commonly-called "wild marjoram" which also grows profusely in the countries bordering the Mediterranean.

In this section it is oil of sweet marjoram derived from *Majorana hortensis* cultivated in the south of France which is the variety under consideration. Sweet marjoram is a plant which is extremely sensitive to cold and is generally grown as an annual crop. In southern France the flowering tops are cut at two main periods—June and August. The cut plant is field dried prior to distillation to give about 0.8% essential oil.

Profile of Oil of Sweet Marjoram

Origin Oil distilled from the fresh herb in the south of France.

Odor Impact: 2

Initial: SWEET, FRESH; aromatic, floral.
PENETRATING but in no way irritating; smooth.

FIG. 2.14. SWEET MARJORAM PLANT (*Majorana hortensis*, M.)

Courtesy of Energis-Verlag, Heidelberg

FIG. 2.15. (LEFT) FRENCH MARJORAM HERB (*Origanum majorana*, L.)

		COOLING, slightly anaesthetic.
		GREEN, HERBACEOUS.
		Slightly WOODY; musty.
		Slightly FRUITY; lemony.
	Persistence:	The oil airs-off with little change in character, it is lingering and tenacious without being overpowering.
	Dry-out:	HERBACEOUS and finally SWEET, WOODY.
Flavor		WARM, SPICY.
		Slightly BITTER.
		HERBACEOUS but not predominantly GREEN.
		Smoothly AROMATIC; floral.
	After-taste:	Slightly BITTER.

Components of the Essential Oil

terpenes (about 40%) mainly terpinene	d-α-terpineol (about 25%)
	l-linalool (about 6%)
	geraniol (about 19%)
	eugenol (about 8%)
	terpinen-4-ol (about 10%)

Comparison of the Odors of Several Commercial Varieties of Oil of Marjoram.

French	Spanish	Cyprus
SWEET	LIGHT	HEAVY
COOLING	EUCALYPTUS-like	THYMOL-like
HERBACEOUS	PENETRATING	HERBACEOUS
		Slightly BURNT

TARRAGON (Estragon) (*Artemisia dracunculus*, **L.**)

Tarragon is a hardy perennial which grows wild throughout central Europe and Asia but is also widely cultivated owing to its popularity in tarragon vinegar-based sauces and dressings. Two distinct varieties are recognized:

(a) French Tarragon—Cultivated particularly in the south of France, the Netherlands and in parts of the United States.

(b) Russian Tarragon—Cultivated in Germany and in several eastern European countries.

The plants are harvested twice a year, most of the crop being

FIG. 2.16. (ABOVE LEFT) FRENCH TARRAGON PLANT (*Artemisia dracunculus*, L.)

FIG. 2.17. (ABOVE RIGHT) TARRAGON PLANT CLOSE-UP

FIG. 2.18. TARRAGON HERB BROKEN FOR DOMESTIC USE (MAGNIFIED X 2)

field-dried, and distilled to give about 0.8 to 1% of a light yellow to pale greenish-yellow essential oil.

Profile of Oil of Tarragon

Origin	Commercial oil distilled in Grasse (S. France).	
Odor	Impact:	3
	Initial:	SWEET (to some SICKLY SWEET). FRESH, COOLING. ANISE-like, BASIL-like. PENETRATING, becoming SHARP; harsh. BALSAMIC, slightly WOODY. HERBACEOUS. AROMATIC; spicy.
	Persistence:	The initial sweet odor rapidly airs-off and the fresh herbaceous notes predominate and linger. After several hours the profile is markedly TOBACCO-like.
	Dry-out:	Very little residual odor.
Flavor		FRESH, slightly SWEET, AROMATIC. HERBACEOUS. ANISE-like.
	After-taste:	Almost none, very slightly ANISE-like and MEDICINAL.

Components of the Essential Oil

ocimene	methyl chavicol (about 65%)
β-phellandrene	ρ-methoxy cinnamaldehyde

Note As methyl chavicol (estragole) is the principal component of the essential oil, the synthetic compound may be used to extend the natural oil. Such compounded oils generally are THIN and air-off very quickly leaving virtually no dry-out notes after only a very few hours.

Part 4: Herbs Containing Thujone—the Sages

Many varieties of *Salvia officinalis*, L., and other *Salvia* species are known simply as "sage" being differentiated by the place name from which they were derived. The dried, highly aromatic leaves form an essential ingredient of many seasonings and the essential oil, to which the plant owes so much for its flavor and distinctive character, is also produced in commercial quantities by steam distillation of the freshly-harvested herb. By far the largest source of oil of sage is the Dalmatian coast of Yugoslavia. The yield of oil is about 2.5%. Climatic conditions as well as soil and methods of harvesting result in oils having widely different aromatic profiles.

The following are the most important to the flavor technologist:

(a) Dalmatian sage (*Salvia officinalis*, L.)
(b) Spanish sage (*Salvia lavandulae folia*, L.)
(c) Greek sage (*Salvia triloba*, L.)
(d) English sage (*Salvia officinalis*, L.)

DALMATIAN SAGE (*Salvia officinalis*, L.)

Dalmatian Sage is a small bush which grows wild on the rocky sunny hillsides of the Dalmation islands and the adjacent mainland coast. The main distillation of sage takes place on the islands of Cherso and Krk although the Dubrovnik district produces the major commercial quantity of the finest quality. The leaves, stripped from the harder woody stems, yield 1.5 to 2.5% essential oil.

A sage of Dalmatian character is also grown in the eastern United States and along the Pacific coast. The quality of this herb and essential oil is good and the profile very similar to that of the genuine oil from Yugoslavia.

Profile of Dalmatian Oil of Sage

Origin	Commercial oil obtained from Yugoslavia.	
Odor	Impact:	4
	Initial:	Strongly AROMATIC, pleasantly SWEET (to some SICKLY SWEET). Initially COOLING then WARMING, SPICY. PENETRATING; irritating, slightly anaesthetic. GREEN; herbaceous. Slightly CAMPHORACEOUS. Slightly CINEOLIC. MUSTY; heavy, dull slightly unfresh.

FIG. 2.19. DALMATIAN SAGE PLANT (*Salvia officinalis*, L.)

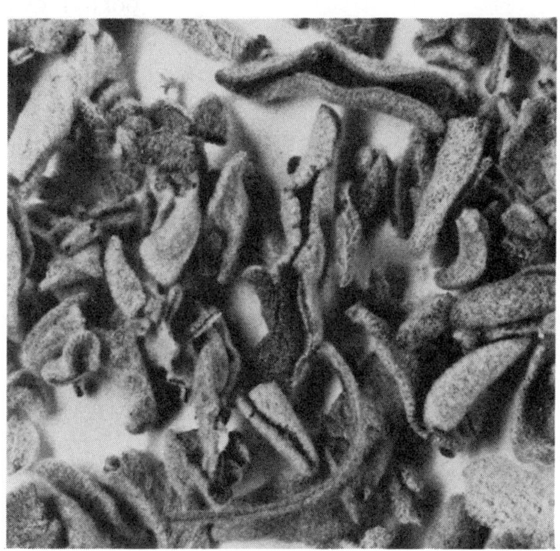

FIG. 2.20. DALMATIAN SAGE LEAVES BROKEN FOR DOMESTIC USE (MAGNIFIED X 2)

	Persistence:	The characteristic profile is very persistent and after several days on a smelling-strip the overall odor is little changed.
	Dry-out:	HERBACEOUS but much SWEETER and less fresh.
Flavor		BURNING. Overpoweringly HERBACEOUS. SWEET to the point of being SICKLY. EUCALYPTUS-like.
	After-taste:	Strongly HERBACEOUS with lingering BITTER-SWEET note.

Components of the Essential Oil

salvene (thujane)	cineole (about 15%)
α-pinene	α-thujone, β-thujone } (40–60%)
	borneol (7–16%)
	bornyl esters (1–4%)
	d-camphor

GREEK SAGE (*Salvia triloba*, **L.**)

Profile of Greek Oil of Sage

Origin	Laboratory distilled from Greek sage herb.	
Odor	Impact:	3
	Initial:	HARSH, PENETRATING; irritating. ROSEMARY-like; EUCALYPTUS-like. HERBACEOUS; green. CAMPHORACEOUS.
	Persistence:	Very tenacious after the top-notes have evaporated.
	Dry-out:	The residual character is like that of Dalmatian sage.
Flavor		Pleasantly AROMATIC, FRESH. HERBACEOUS; green. SWEET.

		Slightly WARMING.
	After-taste:	HERBACEOUS, very slightly BITTER.

ENGLISH SAGE (*Salvia officinalis*, **L.**)

Much of the sage grown in the United Kingdom is cultivated in the Vale of Evesham and in parts of the eastern counties. Normally, the plants are grown from seed. The mature plants are harvested before flowering by cutting off the top 4 in. of young leaves. The leaves are then either dried and rubbed free of stalk for domestic use or distilled fresh to give an almost colorless essential oil.

Courtesy of Bush Boake Allen, Ltd.

FIG. 2.21. A FIELD OF ENGLISH SAGE HERB READY FOR HARVESTING

Profile of English Oil of Sage

Origin	Distilled in Suffolk from locally-grown and freshly-harvested herb.	
Odor	Impact:	3
	Initial:	FRESH, light. Pleasingly SWEET.

		PENETRATING but not irritating.
		GREEN, HERBACEOUS.
		EUCALYPTUS-like; piny.
		Slightly CAMPHORACEOUS.
	Persistence:	The light cineolic top-notes are quickly lost but the underlying herbaceous notes persist.
	Dry-out:	SWEETLY CAMPHORACEOUS with a marked HERBACEOUS character.
Flavor		Pleasingly AROMATIC, SWEET.
		HERBACEOUS; green.
		WARMING at first but then ASTRINGENT.
	After-taste:	Very slightly BITTER.

Comparison of the Odors of Sage Oils from Different Sources

Dalmatian	Greek	English	Spanish
SWEET HEAVY DULL WARMING	HARSH PENETRATING HERBACEOUS EUCALYPTUS-like	FRESH, LIGHT SWEETLY GREEN CAMPHORACEOUS	FRESH, LIGHT PENETRATING EUCALYPTUS-like

Part 5: Herbs Containing Menthol—the Mints

Guenther (1949) and Fenaroli (1975), in describing the several varieties of mint, reveal considerable uncertainty and disagreement among botanists on the origins and classification of these members of the family *Labiatae*. The following are recognized as of major commercial importance as flavoring materials either in the form of the fresh chopped leaves, dried and rubbed flakes or as the essential oil:

1. *Peppermint (Mentha piperita,* L.).—This species is considered to be a natural hybrid:

M. piperita, L. = *M. aquatica*, L.
= *M. viridis*, L. = *M. silvestris*, L.
M. rotundifolia, L.

Two main varieties are cultivated on a commercial scale:

Black mint or "Mitcham mint" var. *vulgaris.*

White mint var. *officinalis.*

2. Corn Mint (*Mentha arvensis*, L.).—The corn mints are characterized by containing an essential oil very rich in menthol. Again the classification of the species is in doubt but the two main varieties of commercial value are:

var. *piperascens*, Holmes (Japan and Brazil)

var. *glabrata*, Holmes (China)

3. Spearmint (*Mentha spicata*, L. and *M. cardiaca*, Ger.)—Spearmint oils are characterized by a high carvone content which is responsible for their distinctive odor and flavor.

4. Culinary Mint.—Domestically grown mints represent a very wide spectrum of varieties and strains, the following being the main species involved:

M. viridis, L. (Garden mint)

M. rotundifolia, L. (Apple mint)

M. gentilis, var. *variegata* (American Apple mint)

M. spicata, var. *crispata* (Curly mint)

A knowledge of the aromatic qualities of these various mints is most important as their flavoring characteristics show considerable variation depending on source, growth conditions, harvest and post-harvest handling, drying and/or distillation techniques employed, etc. Peppermint and spearmint oils are of considerable commercial value and as they are used mainly to give a specific flavor character to a product their quality is of prime concern. All mint oils are raw when freshly distilled but the GREEN, WEEDY notes air-off on standing. For most uses, the prime oil does not have a sufficiently smooth flavor profile and most commercial samples have been rectified to remove undesirable attributes. The subject of the redistillation of peppermint oils is dealt with in a later chapter.

PEPPERMINT (*Mentha piperita*, L.)

Of all the flavors competing for the public taste, that of peppermint is one of the most popular and the consumption of oil of peppermint in the making of sugar confectionary is on the increase. Of all the peppermint oils produced, the so-called "Mitcham" oil, still grown to a very limited extent in parts of southern England, is considered to have the finest flavor. Oil distilled from "white mint" has a very fine flavor profile but the plants of this variety are not hardy and are subject to disease; in consequence, this variety has largely been replaced by the much hardier "black mint." Mint is propagated by cuttings and the original Mitcham variety was introduced successfully into many growing areas of the world. Large crops are now cultivated in the United States, Russia, Italy, France, Bulgaria, Poland, Argentina, Hungary, Morocco, the Netherlands, Spain, Yugoslavia, Germany, India and Australia. Not all of these areas produce high quality oils and the largest commercial source is from northwestern United States (Idaho, Oregon and Washington).

There are very considerable differences in the composition and, hence, aroma and flavor profiles of the oils from different growing areas. The principal constituents are menthol, menthone, menthyl acetate and other esters, these being common to peppermint oils from all sources. It is their ratio which determines the distinctive profile and flavoring quality of the oil. In all cases the initial impact is that of menthol which tends to smother the finer underlying notes. Considerable care is necessary in evaluating peppermint oils and a snap judgement of odor quality should not be made. It is far better to appraise the full evaporation pattern of the sample on a smelling strip over a period of several hours and after 24 hours. The following are the percentages of the principal constituents of oils from the main producing areas:

Source	Menthol (%)	Menthone (%)	Menthyl Esters (%)
American (midwest)	45–61	14–30	4.6–9.6
American (farwest)	51–65	17–21	3.1–9.5
English "Mitcham"	48–68	9–12	3–21
Italian	about 46	about 27	about 4

The cultivation, harvesting and distillation of peppermint have been extensively reviewed by Virmani and Datta (1970). The following are worthy of note as those factors affecting the quality of the essential oil.

Propagation

The plants are propagated by root-cuttings. In the second year they yield the best quality oil, providing there has been plenty of sunshine and no adverse climatic conditions. The plants continue

Courtesy of Bush Boake Allen, Ltd.

FIG. 2.22. A FIELD OF PEPPERMINT (*Mentha piperita*, L.)

Courtesy of Bush Boake Allen, Ltd.

FIG. 2.23. FLOWERING TOP OF PEPPERMINT AT TIME OF HARVESTING

to thrive for about 5 years after which they are replaced as the yield and quality falls off rapidly.

Harvesting

The herbaceous tops are cut when in full bloom, the exact time being critical to ensure optimum yield and correctly balanced constituents. The oil content decreases rapidly after full bloom due largely to leaf-fall. If tops are cut late, the oil loss is significantly greater than if the cut is made too soon; but, in this case, the menthol/menthone ratio is out of balance and the resulting oil is of poor odor and flavor quality.

Courtesy of Bush Boake Allen, Ltd.

FIG. 2.24. MECHANICAL HARVESTING OF PEPPERMINT HERB IN OREGON

Distillation

The oil is obtained by steam distillation of the partially dried herbaceous tops. The stills vary considerably from simple field stills to very sophisticated operations used in the United States.

In England, the yield of essential oil is as low as 20 lb per acre; but in the United States average yields are upwards of 100 lb per acre (50–60 kg per hectare). Percentage yields depend upon the state of dryness of the herb when distilled; and it is more usual to quote yields as weight of oil per area of ground cut.

Courtesy of Bush Boake Allen, Ltd.

FIG. 2.25. A MOBILE FIELD STILL USED FOR THE PRODUCTION OF PEPPERMINT OIL IN OREGON

Profile of Natural American Far-west Oil of Peppermint

Origin Commercial oil obtained from Yakima valley, Washington.

Odor

Impact:	5—intense.
Initial:	Strongly MENTHOLIC, COOLING.
After 20 min:	SWEET, pleasingly AROMATIC. COOLING, FRESH; clean. MENTHOLIC, MINTY. SMOOTH; creamy, toffee-like. PUNGENT, slightly IRRITATING. GREEN; grassy.
Persistence:	The profile changes rapidly over the first 20 min of airing but then much more slowly. The characteristic notes are clearly identifiable after 12 hr but almost lost after 24 hr airing.

	Dry-out:	BALSAMIC. BUTTERY; oily. GREEN; leafy. Slightly MUSTY, EARTHY.
Flavor (at 0.05% in fondant)		COOLING, REFRESHING. SMOOTH, CREAMY. SWEET with slightly BITTER back-note. Strongly MENTHOLIC. PUNGENT but not irritating.
	Mouth-feel:	Initially WARMING then increasingly and persistently COOLING and FRESH. Slightly NUMBING, anaesthetic.
	After-taste:	COLD, NUMBING. Lingering MINTY. SWEET, slightly SICKLY.

Profile of Natural Italian Oil of Peppermint

Origin	Commercial oil obtained from Po valley, northern Italy.	
Odor	Impact:	5—intense.
	Initial:	RAW, ROUGH, GREEN and strongly MENTHOLIC.
	After 20 min:	COOLING, FRESH. GREEN, grassy. MENTHOLIC but with a marked DIRTY, EARTHY back-note. BALSAMIC.
	Persistence:	The initial rough impact is quickly lost and after 20 min the overall profile is FRESH but this is later replaced by DULL, HEAVY notes becoming distinctly BUTTERY.
	Dry-out:	GREEN, MINTY with a heavy BUTTERY note.
Flavor (at 0.05% in fondant.)		COOLING, FRESH.

		MENTHOLIC but soon losing its impact becoming BALSAMIC and BUTTERY.
	After-taste:	COLD and NUMBING. Lingering GREEN MINTY note with slight BITTER, EARTHY back-note.

Profile of English Oil of Peppermint

Origin	Oil distilled at Long Melford, Suffolk from locally-grown herb.	
Odor	Impact:	5
	Initial:	MENTHOLIC, smoothly COOLING.
	After 20 min:	SWEET, AROMATIC; smooth, rounded. MENTHOLIC. CREAMY; full bodied. GREEN; grassy, herbaceous. Slightly FRUITY.
	Persistence:	The initial fresh mentholic notes are quickly lost and replaced by a rich, creamy, minty profile which is very persistent.
	Dry-out:	DULL, TOFFEE-like. Slightly NUTTY.
Flavor (at 0.05% in fondant)		COOLING, REFRESHING. SMOOTH, CREAMY, TOFFEE-like. SWEET. MENTHOLIC. Slightly GREEN.
	After-taste:	DULL, DRY, NUTTY. COLD TEA-like.

Constituents of Oil of Peppermint (Smith and Levi 1961)

α-pinene	cineole
β-pinene	menthofuran
camphene	3-octanol

limonene
γ-terpinene

linalool
menthone
isomenthone
neomenthone
menthol
menthyl acetate
pulegone
isopulegone
piperitone

Odor Characteristics of the Main Components

Menthol

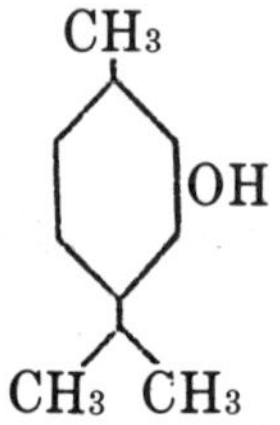

CH_3 OH CH_3 CH_3

Strongly COOLING, slightly IRRITANT.
SWEET (to some SICKLY SWEET).
MINTY, slightly CAMPHORACEOUS.
DRY, MUSTY; woody, heavy.

Menthyl Acetate

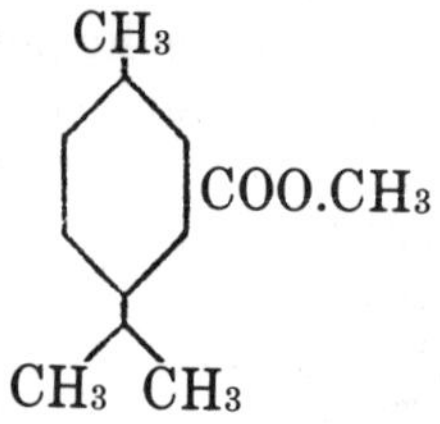

CH_3 $COO.CH_3$ CH_3 CH_3

Mildly SWEET
Pleasantly FRUITY, APPLE-like, WINEY.
Slightly MINTY, REFRESHING, COOL (but not COOLING).
DRY, WOODY; straw-like.
Slight FLORAL back-note.

Menthone

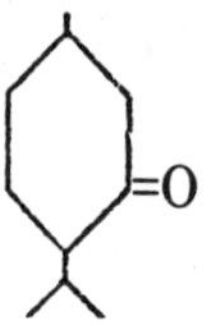

=O

HARSH, IRRITANT; penetrating.
MINTY, slightly COOLING.
HEAVY, LACQUER-like.

Menthofuran

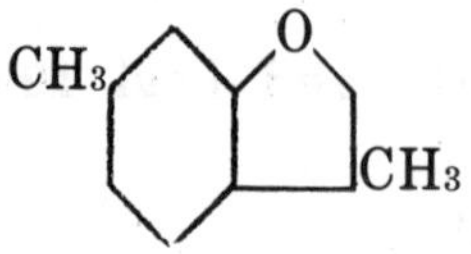

CH_3 O CH_3

HEAVY.
Strongly TAR-like, CREOSOTE-like.
Very slightly MINTY.
The dry-out is CREAMY, NUTTY.

CORN MINT (*Mentha arvensis*, L.)

The essential oil distilled from *Mentha arvensis,* L. var. *piperascens,* is known officially in the United States as "Mint Oil" or "Corn mint Oil" but in commerce it is often referred to merely as "Peppermint Oil." In view of the very considerable difference between this and genuine peppermint oil this nomenclature is misleading. The price differential between these oils also leads to blending to meet price constraints.

Corn mint grows wild in China and parts of Japan but is now extensively cultivated in Japan (particularly around Hokkaido and Hiroshima) in Formosa and in Brazil. Lesser crops are cultivated in Argentina, in India and in South Africa.

When freshly distilled from the cut herb the oil contains 60–80% menthol so that it normally solidifies as it cools to room temperature. This natural oil is usually immediately further processed by freezing to recover 40% of available 1-menthol, leaving a residual oil which still contains 45–55% of menthol. This oil is marketed as "dementholized."

As with oil of peppermint, the geographical source of the essential oil gives distinctly different aromatic profiles due to the ratios of the main components. In this oil it is the total ester content which is of significance: Japanese contains 10–12%; Chinese contains about 2% but up to 10%; Brazilian contains 5–20%.

Profile of Chinese Oil of *Mentha arvensis* (partially dementholized)

Origin Commercial sample.

Odor

Impact: 5

Initial: Strongly COOLING, FRESH; clean.
BITTER-SWEET, to some SICKLY SWEET.
MENTHOLIC.
HARSH, WOODY; sharp.
CREAMY, TOFFEE-like.

Persistence: The strongly MINTY note is quickly replaced by a COOKED note, the odor becoming HARSH. There is a decrease in the rounded CREAMY attribute.

Dry-out: Very delayed but after 24 hours

the residual odor is pleasingly SMOOTH but distinctly HERBACEOUS.

Flavor
(at 0.05% in fondant)

COOLING, REFRESHING.
ROUGH; sharp, coarse, raw.
MENTHOLIC.
GREEN, HERBACEOUS; cooked vegetable.

After-taste: COLD.
MINTY but with marked BITTER note.

Profile of Brazilian Oil of *Mentha arvensis* (partially dementholized)

Origin Commercial sample.

Impact: 5

Initial: TERPENEY; paint-like.
COOLING but UNFRESH, DRY, WOODY.
MENTHOLIC.
MUSTY, STALE; damp, earthy.
RAW; rough, slightly metallic.
TOFFEE-like, CREAMY.

Persistence: After 20 min airing the TERPENEY notes are lost and there is an increase in the HARSH METALLIC and MUSTY notes.

Dry-out: UNFRESH with a lingering GREEN VEGETABLE back-note

Flavor
(at 0.05% in fondant)

Very similar to that of Chinese origin.

Comparison of the Odors of Oils of *Mentha arvensis* and *Mentha piperita*

M. arvensis	*M. piperita*
MUSTY	FRESH
RAW GREEN MENTHOLIC	MENTHOLIC
COOKED VEGETABLE	CREAMY, TOFFEE-like
HARSH	SMOOTH

Components of Oils of *Mentha arvensis*

	Brazilian Natural (%)	Brazilian (Partially Dementholized) (%)	Japanese (Partially Dementholized) (%)	Chinese (Partially Dementholized) (%)
Terpenes	5–15	12–15	12–15	12–15
Menthol	60–80	45–60	35–55	60–68
Menthone/iso-menthone	14–16	25–35	25–30	15–25
Menthyl acetate	3–6	5–7	1–2	5–10
Menthofuran	–	–	–	–

Other constituents include:

- α-pinene
- camphene
- ι-limonene
- ι-menthene
- caryophyllene
- neo-menthol
- menthyl esters of formic, n-caproic and phenylacetic acids
- iso-valeric aldehyde
- furfural
- piperitone
- pulegone
- isopulegone
- isomenthone

The main constituents of the oils from various sources are closely similar but differ significantly in the relative ratios present (Smith and Levi 1961). Menthofuran which is present in *piperita* oils is absent in *arvensis* oils. Several tests have been suggested to distinguish between these two types of oil the best known being that based on the development of a red color due to the presence of menthofuran in *piperita* oil. Unfortunately, as little as 15% of *piperita* oil admixed with 85% of *arvensis* oil still gives a satisfactory color reaction.

By analysis of the oil from plants as they are developing, it has been established that menthone is formed in the plant from piperitone and is then, in turn, converted into menthol which continues to increase at the expense of the menthone. It is the final balance of menthol/menthone which largely dictates the odor and flavor of the finished oil.

SPEARMINT (*Mentha spicata*, **Huds**)

The term "spearmint" comprises numerous species and varieties of the genus *Mentha*, all of which possess the easily recognizable odor and flavor. The plants resemble those of peppermint but do not

FIG. 2.26. SPEARMINT PLANTS (*Mentha spicata*, L.)

FIG. 2.27. CLOSE-UP OF SPEARMINT LEAVES

grow quite as tall; cultivation and harvesting follow the same pattern as for peppermint. The United States accounts for most of the world output of the essential oil with much smaller quantities originating in Japan and Russia. The main species cultivated in North America is *Mentha spicata*, Huds. var. *tenuis*, Mich. (formerly classified as *Mentha viridis*, L.). The original growing regions were Indiana, Michigan and Idaho but, like peppermint, the center has now moved to the far-west in Washington State where yields of up to 81 lb per acre are obtained compared with only 35–40 lb per acre in the mid-west States. There is still a demand for the so-called "Scotch" spearmint from the mid-west; many consider this oil to have a sweeter and more rounded profile.

Profile of Natural American Spearmint Oil

Origin Commercial oil from Washington state.

Odor

Impact: 4

Initial: SHARP, FRESH; clean, bright.
PENETRATING, slightly IRRITATING.
MINTY; mentholic.
GREEN, LEAFY.
SWEET (to some SICKLY SWEET), SMOOTH, CREAMY.
TOFFEE-like, BALSAMIC; slightly BURNT.
Slightly TERPENEY.

Persistence: Airs-off with very little change in profile over several hours. After 12 hours there is a marked increase in BALSAMIC, CARVONE-like and TERPENEY notes and a decrease in the pungent FRESH, GREEN character.

Dry-out: Still distinguishable as "spearmint" after 36 hours though very weak. The GREEN note becomes more reminiscent of garden mint.

Flavor

(at 0.05% in fondant) — WARM, SMOOTH.
Pleasantly AROMATIC.
SHARP; pungent, but not objectionably so.
HERBACEOUS.

Components of Spearmint Oil

ι-limonene
α-phellandrene
α-pinene
ι-carvone (about 56%)
dihydrocuminyl acetate *
dihydrocuminyl valerate*
dihydrocarveyl acetate*
cineole
linalool

* The carriers of the main characteristic spearmint notes.

GARDEN MINT OR CULINARY MINT

The fresh, pleasing character of garden mint—as used domestically to flavor new potatoes, freshly picked peas or as a sauce in vinegar—is derived from very young leaves of various types of spearmint. At the time of picking the leaves contain an essential oil which is still far from complete. The components of the essential oil progressively change as the plant develops achieving an optimum balance and maximum yield over a very short period when the plant is in full flower. It is very difficult to capture the distinctive character of freshly-picked mint leaves and there is no commercial source of this particular essential oil.

BIBLIOGRAPHY

Herbs

ARCTANDER, S. 1960. Perfumery and Flavor Materials of Natural Origin. Stephen Arctander, Elizabeth, N.J.

APPELL, L. 1968. Physical foundations in perfumery. VI. Volatility of the essential oils. Am. Perfum. Cosmet. *83*, No. 11, 37–47.

CLAIR, C. 1961. Of Herbs and Spices. Abelard-Schuman Ltd., London.

FENAROLI 1975. Handbook of Flavor Ingredients, 2nd Edition. T.E. Furia and N. Bellanca (Editors). Chemical Rubber Co., Cleveland.

GUENTHER, E. 1949. The Essential Oils, Vol. I-VI. D. Van Nostrand & Co., New York.

HEATH, H. B. 1973. Herbs and spices for food manufacture. Trop. Sci. *14*, No. 3, 245–259.

HEATH, H.B. 1973–1974. Herbs and spices—a bibliography. Flavour Ind. *4*, Basil, Feb., 65–66; Sweet Bay, Feb., 66; Marjoram, June, 264; Mints, June, 264; Origanum, Aug., 348; Rosemary, Sept., 396; Sage, *Ibid.5*, Mar.–Apr., 79–81; Savory, Mar.–Apr., 81; Tarragon, May–June, 123; Thyme, May–June, 123.

HODGE, W. H. 1975. Survey of flavour producing plants. Int. Flavour Food Additives *6*, No. 4, 244–245.

LAW, D. 1973. The Concise Herbal Encyclopedia. John Bartholomew & Son, Ltd., Edinburgh.

OBERDIEK, R. 1972–1973. Aroma compounds in spices and herbs. Riechst., Aromen, Koerperpflegem. *22*, No. 9, 293–294, 296; *22*, No. 11, 390, 393–395; 22, No. 12, 431–432; *Ibid.*, *23*, No. 1, 3–4,6; *23*, No. 2, 29–30,32; *23*, No. 4, 107–108,110,113. (German)

Additional References to Individual Herbs

Basil

KARAWYA, M. S., HASHIM, F. M., and HIFNAWY, M. S. 1974. Oils of *Ocimum basilicum*, L. and *O. rubrum*, L. grown in Egypt. J. Agric. Food Chem. *22*, No. 3, 520–522.

LAWRENCE, B. M., HOGG, J. W., TERHUNE, S. J. and PICHITAKUL, N. 1972. Essential oils and their constituents. IX. The oils of *Ocimum sanctum* and *Ocimum basilicum* from Thailand. Flavour Ind. *3*, Jan., 47–49.

ZOLA, A. and GARNERO, J. 1973. Contribution to the study of some European-type basil oils. Parfums, Cosmet., Savons Fr. *3*, No. 1, 15–19. (French)

Sweet Bay Laurel

HOGG, J. W., TERHUNE, S. J. and LAWRENCE, B. M. 1974. Dehydro-1,8-cineole: A new monoterpene oxide in *Laurus nobilis* oil. Phytochemistry *13*, No. 5, 868–869.

Marjoram

EL-ANTABLY, H. M. M., AHMED, S. S. and EID, M. N. A. 1975. Effects of some growth hormones on plant vigour and volatile oil of *Origanum majorana*, L. Pharmazie *30*, No. 6, 400–401.

GRANGER, R., PASSET, J., and LAMY, J. 1975. About the essences of marjolaine. Riv. Ital. Essenze, Profumi, Piante Off. *57*, 446–454. (French)

TASKINEN, J. 1974. Composition of the essential oil of sweet marjoram obtained by distillation with steam and by extraction and distillation with alcohol-water mixture. Acta Chem. Scand., Ser. B *28*, No. 10, 1121–1128 (28 refs).

The Mints

CANOVA, L. 1972. The composition of Scotch spearmint oil. Anal. Acad. Brasil. Cienc. *44*, (Supl.), 273–277.

DOW, A. I., NELSON, C. E. and KLOSTERMEYER, E. 1974. Agronomic aspects of mint production in Washington state. Paper No. 6, 6th International Essential Oil Congress, San Francisco.

GREEN, R. J. 1975. Peppermint and spearmint production in the United States—progress and problems. Int. Flavours Food Additives *6*, No. 4, 246–247.

HEFENDEHL, F. W. and MURRAY, M. J. 1973. Relations between biogenesis and genetic data of essential oils in the genus *Mentha*. Riv. Ital. Essenze, Profumi, Piante Off. *55*, No. 11, 791–796.

HEFENDEHL, F. W. and ZIEGLER, E. 1975. Analysis of peppermint oils. Lebensm. Rundsch. *8*, No. 71, 287–290. (German)
LAWRENCE, B. M., HOGG, J. W. and TERHUNE, S. J. 1972. Essential oils and their constituents. X. Some new trace constituents in the oil of *Mentha piperita*, L. Flavour Ind. *3*, No. 11, 467–472.
NAGELL, A. and HEFENDEHL, F. W. 1974. Composition of the essential oil of *Mentha rotundifolia*, Planta Med. *26*, No. 1, 1–8. (German)
PIPER, T. J. and PRICE, M. J. 1975. Atypical oils from *Mentha arvensis*, var. *piperascens* (Japanese Mint) plants grown from seed. Int. Flavours Food Additives *6*, No. 3, 196–198.
QUARRÉ, J. 1975. Cultivation of mint in the U.S.A. Parfums, Cosmet. Arômes *2*, Mar–Apr., 27–30, 33–35. (French)
SACCO, Y. and NANO, G. M. 1973. *Mentha viridis*, L. var. *typica* f. *brevipetiolata* (Rehb) flowers new to the literature of essential oils. Botanical and chemical study. Riv. Ital. Essenze, Profumi, Piante Off. *55*, No. 4, 229–231. (Italian)
SAKATA, I. and MITSUI, T. 1975. Isolation and identification of 1-methyl-β-D-glucoside. Agric. Biol. Chem. (Tokyo) *39*, No. 6, 1329–1330.
SMITH, D. M. and LEVI, L. 1961. Treatment of compositional data for the characterization of essential oils. Determination of geographical origins of peppermint oils by gas chromatographic analysis. J. Agric. Food Chem. *9*, 230–244.
TOORES, O. A. and RETAMAR, J. A. 1975. The oil of *Mentha viridis (Mentha spicata)*. Riv. Ital. Essenze, Profumi, Piante Off. *57*, No. 4, 214–215. (Spanish)
VIRMANI, O. P. and DATTA, S. C. 1970. Oil of *Mentha piperita* (oil of peppermint). Flavour Ind. *1*, No. 1, 59–63; concluded No. 2, 111–113 (684 refs).

Origanum Species

LAWRENCE, B. M., TERHUNE, S. J. and HOGG, J. W. 1974. 4,5-epoxy-*p*-menth-1-ene: A new constituent of *Origanum heracleoticum*. Phytochemistry *13*, No. 6, 1012–1013.
MAARSE, H. 1974. Volatile oil of *Origanum vulgare*, L. ssp. *vulgare*. III. Changes in composition during maturation. Flavour Ind. *5*, No. 11–12, 278–281.
MAARSE, H. and VAN OS, F. H. L. 1973. Volatile oil of *Origanum vulgare*, L. ssp. *vulgare*. I. Qualitative composition of the oil. Flavour Ind. *4*, No. 11, 477–481; II. Oil content and quantitative composition of the oil. *Ibid.*, 481–484.

Rosemary

GRANGER, R., PASSET, J., and ARBOUSSET, G. 1973. The essential oil of *Rosmarinus officinalis*, L. II. Influence of ecological and individual factors. Parfums, Cosmet., Savons Fr. *3*, No. 6, 307–312. (French)
BRIESKORN, C. H., MICHEL, H., and BIECHELE, W. 1973. Flavones in rosemary leaves. Dtsch. Lebensm. Rundsch. *69*, No. 7, 245–246. (German)

Thyme

GRANGER, R. and PASSET, J. 1973. *Thymus vulgaris* native in France: Chemical varieties and chemotaxonomy. Phytochemistry *12*, No. 7, 1683–1691. (French)
GRANGER, R. PASSET, J. and SIBADE, A. 1974. Chemical types of *Thymus herba-barona*, Loiseleur and Deslongchamps, of Corsica. Riv. Ital. Essenze, Profumi, Piante Off. *56*, No. 11, 622–629. (French)
HAZDEN, T. K. and KOUL, G. L. 1975. Composition of wild thyme oils. Riechst., Aromen, Koerperpflegem. *25*, No. 6, 166,168. (German)
RUSSEL, G. F. and OLSON, K. V. 1972. The volatile constituents of oil of thyme. J. Food Sci. *37*, No. 3, 405–407.

3

The Spices

DEFINITION OF SPICES

The soft-stemmed plant materials used in seasoning food are classified as "herbs" and all other aromatic plant products used for a similar purpose are called "spices," although this broad definition admits of several exceptions. Spices are usually only parts of plants and may be either roots, rhizomes, barks, seeds, fruits, flower buds, etc. Unlike the herbs, the spices are very aromatic and may contain large percentages of essential oil as well as other powerful non-volatile flavoring components. They are normally derived from the semi-tropical or tropical regions of the world, are harvested and usually sun-dried to form the spice of commerce.

"Condiments" are seasonings which are added to food after it has been served. In this category the most popular and widely used are salt, mustard, pepper, and ginger.

CLASSIFICATION OF SPICES

In the case of herbs it is possible to classify them into groups having broadly similar sensory attributes based on the prime constituent of the essential oil. Unfortunately, when one is considering the spices such a meaningful grouping is only partially possible as far fewer associations, based on some common organoleptic property, exist; many of the spices are individually distinctive and unlike any other spice. This being so, it is convenient to classify certain of them on a purely botanical basis either by family (e.g., the umbelliferous fruits) or by form (e.g., aromatic fruits). For convenience, this "mixed" classification has been adopted here, although its limitations from a flavor point of view are clearly recognized, the spices are so classified as follows:

Part 1—The pungent spices:

Capsicum, ginger, black and white pepper, mustard, horseradish.

Part 2—The aromatic fruits:
Nutmeg and mace, cardamom, fenugreek.

Part 3—The umbelliferous fruits:
(a) Anise, fennel (cf: star anise, basil, tarragon).
(b) Caraway, dill, Indian dill.
(c) Celery, lovage, parsley (cf: fenugreek).
(d) Cumin.
(e) Coriander.

Part 4—The aromatic barks containing cinnamic aldehyde:
Cinnamon bark, cassia bark.

Part 5—The phenolic spices containing eugenol:
Clove bud, allspice (pimento), cinnamon leaf, clove stem, clove leaf, West Indian bay.

Part 6—The colored spices:
Paprika, saffron, safflower, turmeric.

The most important of the spices commercially is, of course, pepper which is used universally; pepper is followed by cloves, nutmeg and mace, cardamom, cinnamon and cassia, ginger and allspice or pimento. These 9 spices together account for some 90% of the total spice trade and in weight amount to about 120 thousand metric tons per year.

Like herbs, the spices are subject to quite considerable variation both in appearance and flavoring power. There is certainly no guarantee that one shipment of spice will have the same profile as any other shipment, even from the same source.

The profiles evolved in this chapter are the result of very many individual assessments and represent the average pattern. Again the essential oil has been examined as the prime contributor of the spice's odor but the spices themselves were also examined for flavor profile as in many cases the presence of nonvolatile constituents makes a significant difference between the flavor of the essential oil and that of the spice or an extract of the spice.

Part 1: The Pungent Spices

This is a well-defined group of spices which are used universally in food seasonings to add bite or piquancy to the end-product. In the case of pepper and ginger this pungent effect is accompanied by a quite characteristic flavor due to nonpungent aromatic components present in the essential oil of these spices. The familiar bite and flavor of mustard, on the other hand, is due to a very volatile constituent which largely constitutes so-called "oil of mustard."

It is not easy to equate the relative pungency of the spices in this group as the effect of each in the mouth is so very different. Pepper causes a pleasing tingling sensation along the front edge of the tongue with very little effect upon the throat tissues unless it is present far in excess of normal. Capsicums are different in that they can be detected in low concentrations as a sharp pain deep in the throat with almost no effect upon the linings of the mouth and tongue. Ginger differs yet again in having a fuller and more rounded pungency which is evident upon the sides and back of the tongue but not in the throat. Compared with all of these, the distinctive pungency of mustard is in a different class. The pungent principles of pepper, capsicum and ginger are present as nonvolatile constituents of the spice and can be recovered unchanged by solvent extraction. That of mustard is volatile and is created by enzymic action when ground mustard flour is mixed with water. The warm, sharply aromatic pungency of mustard surfuses the whole mouth.

The physiological nature of pungency has been well documented by Moncrieff (1967). One can readily differentiate between the various pungent spices when they are evaluated singly but it is much more difficult to separate out the several effects when they are present together in a seasoning—as is usually the case. With all of these spices, the pungent effect is cumulative so that one feels an increasing response with successive doses of the same level of stimulus until one reaches a threshold of saturation. This makes the evaluation of the spices themselves and products containing them very difficult to achieve with any degree of accuracy as the build-up of pungency detracts from and ultimately overwhelms any flavoring effect. All the pungent spices leave an after-taste which may be of long duration. In the case of both pepper and capsicum there is an additional physiological effect resulting in a feeling of excessive bodily warmth and even perspiration. The degree and intensity of the pungent response is related to the level of the total stimulation. The threshold concentration for detection and the level of its persistence are measures more of individual sensitivity and response; and assessors can show wide variation in this respect.

CAPSICUM

Capsicum, cayenne pepper and chillies are all names given to members of the various *Capsicum* species and varieties used extensively in foods. The names are legion and often very local (particularly in Mexico) so that considerable confusion arises over what is meant in a formulation by the term "cayenne pepper"; to some users it is particularly hot and to others relatively mild, depending upon the regional source of the capsicums used.

The genus *Capsicum* is widely distributed and exists in innumerable sizes, shapes, colors and levels of pungency; but the fruits can be roughly divided into two main groups:

(a) The large fleshy fruits of *Capsicum annuum*, L.
(b) The small fruited *Capsicum minimum*, Roxb. or *C. frutescens*, L.

The least pungent varieties are the large fleshy red and green peppers which are not unlike a tomato and used more as a flavorful vegetable than as a spice. Paprika is also at this end of the scale although some varieties may indeed be slightly pungent. The most pungent are the very small fruits of *C. frutescens* which are also widely known as African Chillies. In between these two extremes there is a whole spectrum of varieties. It would be something of an oversimplification to say that the larger the fruit the more colorful and less pungent it is. Although many of the smaller varieties are certainly less colorful and have a higher level of pungency this is not always the case. The pattern can be presented in the form of a graph which indicates these rough relationships (Fig. 3.5).

The larger and sweeter (less pungent) types of capsicums are generally well known and widely used in food processing and domestically. They are often called "chilli pepper" and this title, particularly in the United States, must not be confused with "chilli powder" which is really a specially blended product consisting of ground capsicums, cumin seed, oregano and garlic which forms a complete piquant seasoning.

The pungency of capsicum is determined by its content of capsaicin, several methods for the determination of which have been published over the past ten years. The most pungent chillies, obtained from Uganda, may contain 0.8 to 1.1% of this intensely pungent component. Most commercial samples of "cayenne pepper" contain less than 0.5% and may indeed test as low as 0.1%. Although these quantitative methods are available it is more usual for the pungency to be expressed in terms of "Scoville units" based on the concentration of a test solution the pungency of which is just detectable. The several problems associated with the quality control of products containing capsicum will be dealt with in Chap. 10.

It is worth reiterating that when considering the use of capsicum

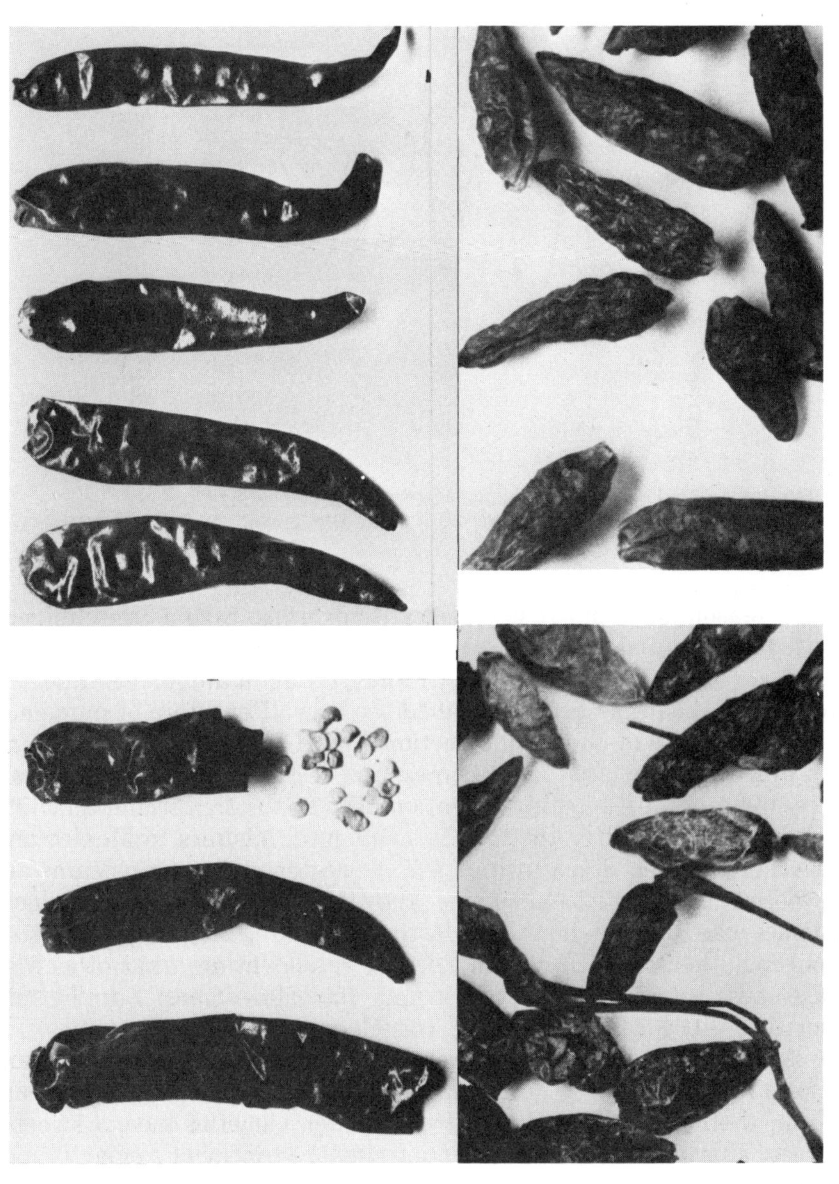

FIG. 3.1. (ABOVE) MEXICAN CAPSICUM (*Capsicum annuum*, L.)

FIG. 3.2. (BELOW) EAST AFRICAN CAPSICUM (*Capsicum annuum*, L.)

FIG. 3.3. (ABOVE) HONTAKA CAPSICUM (*Capsicum annuum*, L.) MAGNIFIED X 2

FIG. 3.4. (BELOW) MOMBASA CHILLIES (*Capsicum minimum*, Roxb.) MAGNIFIED X 2

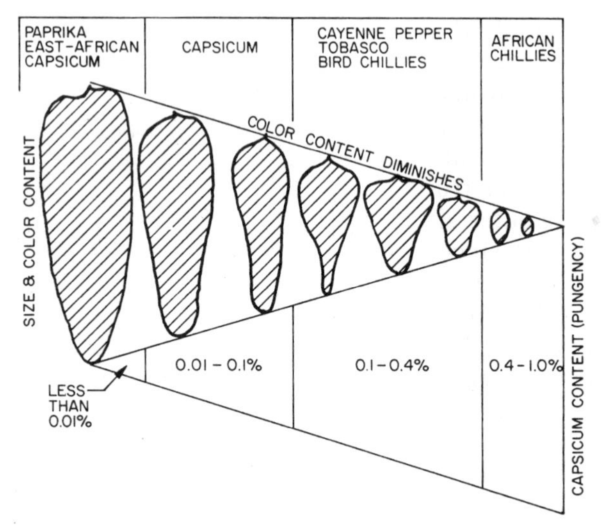

FIG. 3.5. GRAPHICAL RELATIONSHIP BETWEEN SIZE, COLOR CONTENT AND PUNGENCY OF CAPSICUMS

in a seasoning blend it is essential to exercise care as the pungent effect is cumulative. The level of pungency may not appear excessive in one or two mouthsful of the product but by the time the complete meal has been consumed the overall build-up of pungency in the throat can be really objectionable. If the inclusion of cayenne is overdone, the product becomes progressively unpleasant to eat, leaving a painful burning sensation deep in the throat and a feeling of distress long after the food is swallowed. Visiters to Mexico and southern United States, which are the regions where capsicums are used lavishly in the local cuisine, know full well how fiery these local dishes can be and how long-lasting the pungency. It is not surprising, therefore, that many people resist eating what otherwise are most attractively presented and flavorful dishes. For the majority a little cayenne goes a very long way.

Although capsicums do not contain a recoverable amount of essential oil they do, nevertheless, have a distinct flavor profile separate from their pungency. Many of the milder varieties have a sweetly fruity character but the task of defining the profile of so many types is beyond the scope of this book. This is one case where a personal appraisal is the only true guide to the available flavor.

BLACK AND WHITE PEPPER (*Piper nigrum*, L.)

The pepper of commerce is produced from the unripe fruits of the perennial climbing vine and is available in two distinct forms—

black pepper and white pepper. The former consists of the whole dried fruits, picked while still green and sun-dried. During drying they turn to a brownish black color with the individual peppercorns having a much wrinkled outer skin. White pepper is the dried kernels of the fruits which are gathered when they are just turning slightly yellow. The fruits are subsequently soaked in water to soften and loosen the outer skin which is then removed by friction. White peppercorns are smooth surfaced.

Pepper originated in the Western Ghats of India from where it has spread to many parts of tropical Asia, notably Malaysia, Indonesia, Cambodia, Vietnam and Sri Lanka (Ceylon). More recently it has become a commercial crop in parts of West Africa and Brazil. The several grades presently available on the market are listed in Table 3.1. These do not represent discretely different product types but rather are commercially available "blends" of innumerable varieties obtainable in the country of origin.

The annual trade in pepper, both black and white, from all sources is approximately $35 million. As a seasoning and condiment, pepper is second to none, save salt, and its use is ubiquitous. Few plants have a more romantic history or have been the spur to such ruthless exploration and colonization. (There are many popular books on herbs and spices which deal fully with this fascinating subject.) With some considerable variation the total world crop of pepper is about 70,000 tons per annum of which about 22,000 tons are imported into the United States. Actual figures by source are published annually by the American Spice Trade Association.

Although the pungency of pepper is most important and has been the subject of much research (Genest *et al.* 1963) it is the aromatic character of its essential oil which is the subject of this section. The distinctive odor and flavor of pepper overlie its pungency due to its essential oil content which varies both quantitatively and qualitatively between sources and varieties. The chemical composition of the oil is complex (Richards *et al.* 1971) and is present from 1 to 3%. The oil from white pepper contains similar components to that from black pepper but the relative amounts differ. The odor of freshly ground black pepper is markedly different from that of the spice stored in a ground condition as regular users of the domestic peppermill will readily attest. Not only does the ground material soon lose its pleasing freshness but it also develops an obvious and insistent ammoniacal note which detracts from its true peppery character. The profile of the essential oil distilled directly from freshly crushed peppercorns has a most attractive nuance much appreciated in the blending of high quality, spicy fragrances. Oil distilled from previously extracted oleoresin has quite another profile which is com-

Courtesy of Tropical Products Institute

FIG. 3.6. A PEPPER GARDEN IN MALAYSIA

Courtesy of Shell, Ltd.

FIG. 3.7. SARAWAK BLACK PEPPER READY FOR HARVESTING

Courtesy of Bush Boake Allen, Ltd.

FIG. 3.8. HARVESTING BLACK PEPPER IN MALAYSIA

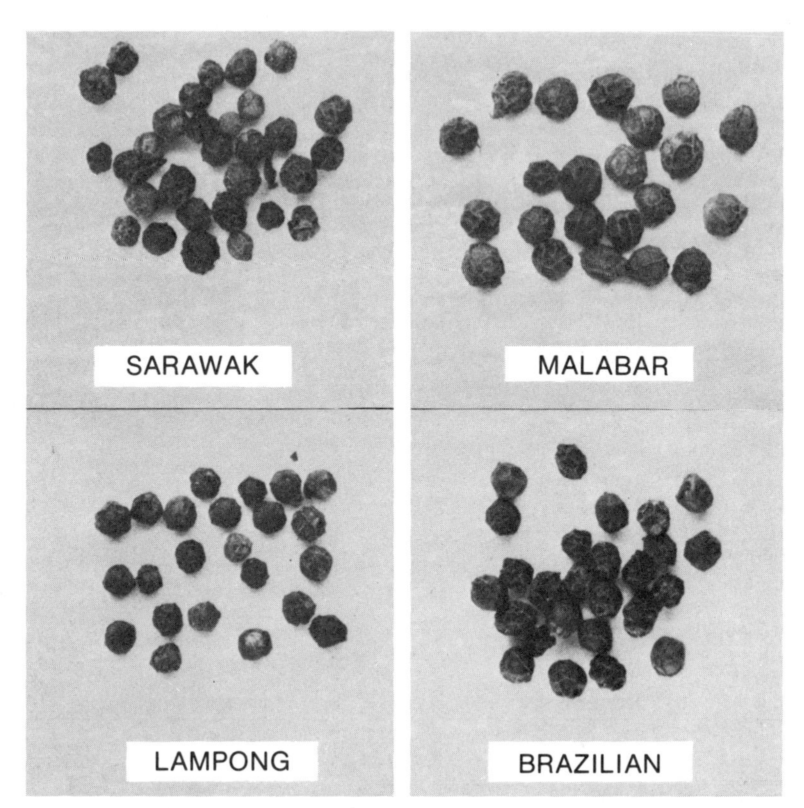

FIG. 3.9. COMPARISON OF BLACK PEPPER (*Piper nigrum*, L.) FROM DIFFERENT SOURCES

FIG. 3.10. CLOSE-UP OF BLACK PEPPER FIG. 3.11. CLOSE-UP OF WHITE PEPPER

TABLE 3.1
COMMERCIAL VARIETIES OF BLACK AND WHITE PEPPER

Commercial Designation	Country of Origin	District	Essential Oil Content (%v/w)
Black pepper			
Malabar	India	Malabar coast,	1.5–4.0–5.0
Mangalore		Mysore, Travancore	
Tellicherry		and the western	
Alleppy		coast of Madras.	
Lampong	Sumatra Indonesia		1.5–3.3–4.0
Singapore	Singapore Malaysia Cambodia	There is a considerable entrepot trade and most grades represent blends from many sources.	2.0–3.0–4.0
Saigon	Vietnam		
Siam	Thailand		
Sarawak	Sarawak		2.3–3.5–4.0
Ceylon	Sri Lanka		2.0–2.8–3.5
Malagasy	Malagasy Republic		
Brazil	Brazil		1.7–2.3–3.0
White pepper			
Muntok	Indonesia	Banga	
Sarawak	Sarawak		
Brazil	Brazil		

Courtesy of Bush Boake Allen, Ltd.

FIG. 3.12. SUN DRYING BLACK PEPPER IN MALAYSIA

paratively flat and lacking in freshness but still decidedly peppery. The profiles of oils distilled from peppers from different geographical regions show a wide spectrum of individual characters by which they can be identified.

The descriptive profiles which follow were derived by the same procedure as those described for the herbs. Additional profiles have been developed for both black and white pepper oil based on an expert evaluation of the effluent gases from a GLC column. These are given in Fig. 3.13 and 3.14. In both cases an agreed vocabulary of descriptive words and phrases was used.

Profile of Oil of Black Pepper (Lampong)

Origin	Laboratory distilled from Lampong black pepper.	
Odor	Impact:	3
	Initial:	FRESH; cool, aromatic. IRRITATING; penetrating. TERPENEY; lemony. SWEET; spicy.
	Persistence:	The profile changes very little as it airs-off but after 24 hours the residual odor is very weak.
	Dry-out:	MUSTY; slightly woody, dry.
Flavor		WARM. Sweetly SPICY. WOODY; musty. Slightly FRUITY.
	After-taste:	Pleasantly WARM.

Profile of Oil of Black Pepper (Malabar)

Origin	Laboratory distilled from Malabar black pepper.	
Odor	Impact:	3
	Initial:	AROMATIC, SWEET. FRESH; cool. TERPENEY; pine-like, lemony, fruity. Slightly IRRITATING; penetrating. WOODY, SPICY.
	Persistence:	The initial fresh impact is quickly lost after which the profile does not change significantly.

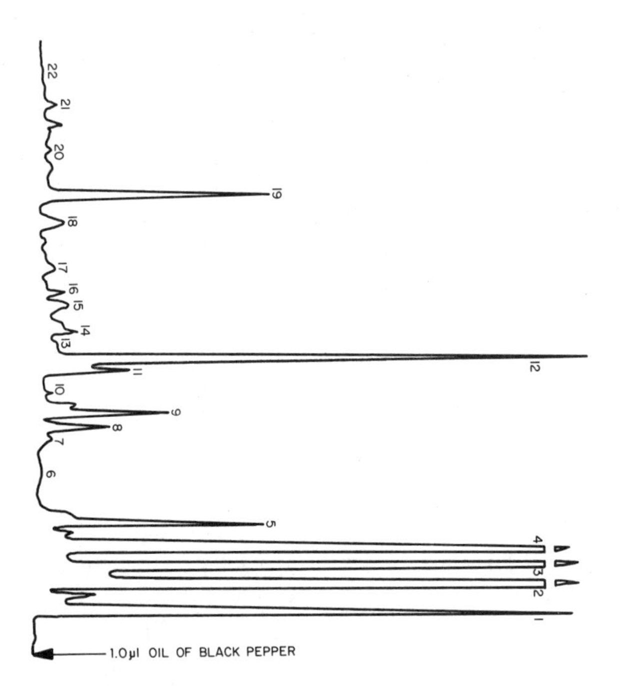

FIG. 3.13. SENSORY ASSESSMENT OF THE GLC EFFLUENT FROM OIL OF SARAWAK BLACK PEPPER

Peak No.	Character
1	DRY, METALLIC, slightly fruity, piny
2	DRY, more fruity than No. 1, piny
3	GREEN, mango-like
4	GREEN, mango-like similar to peak No 3
5	HARSH, BURNT, carvacrol-like
6	HEAVY, MEATY
7	DRY, GRASS-like
8	Strongly MEATY, H.V.P.-like
9	CORIANDER-like
10	VERY DRY, musty
11	WARM, strongly CLOVE-like
12	DRY, WOODY, slightly CLOVE-like
13	GREEN, beany
14	GREEN, grassy
15	No discernable odor
16	HEAVY, DULL
17	GREEN, PEPPERY
18	DRY, HAY-like
19	LIGHT, GREEN
20	WARM, DRYING
21	Very PEPPERY, IRRITATING
22	Very PEPPERY, PUNGENT, GREEN

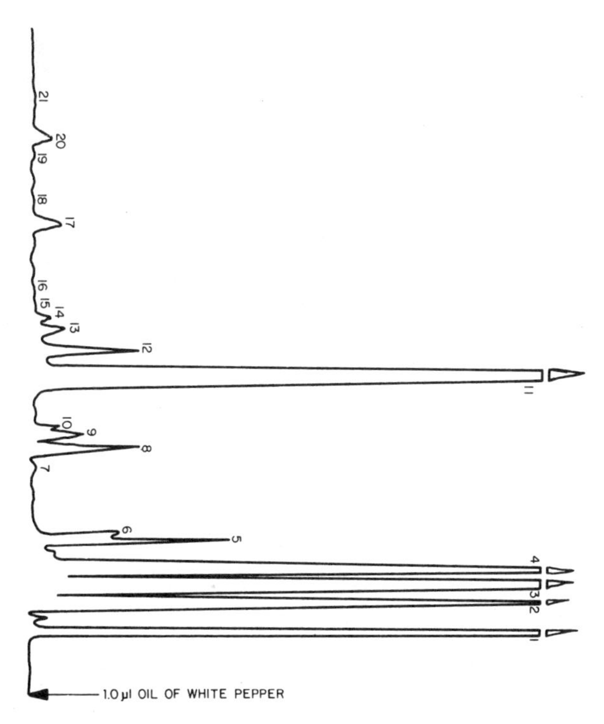

FIG. 3.14. SENSORY ASSESSMENT OF THE GLC EFFLUENT FROM OIL OF MUNTOK WHITE PEPPER

Peak No.	Character
1	METALLIC, slightly fruity
2	TERPENEY, fruity
3	GREEN, fruity, mango-like
4	GREEN, strongly MANGO-like
5	HARSH, BURNT, carvacrol-like
6	MUSHROOM-like, slightly meaty
7	BURNT, MEATY
8	No discernable odor
9	No discernable odor
10	CORIANDER-like
11	SWEETLY FRAGRANT, floral
12	DRY, GREEN, grass-like
13	DRY but nondescript
14	GREEN, SWEET, strongly FLORAL
15	DRY like drying adhesive
16	DRY, WARM, strongly CLOVE-like
17	BURNT
18	CLOVE-like
19	PAINT-like
20	Strongly AROMATIC, sassafras-like
21	Very PEPPERY, IRRITATING

	Dry-out:	MUSTY, WOODY; dry, warm.
Flavor		FLAT; insipid.
		FRUITY; lemony, citrus.
		MUSTY; woody, dry.
		Slightly BITTER.
		Mildly SPICY; warm.
		Slightly BALSAMIC.
		Slightly ASTRINGENT.
	After-taste:	Slightly WARM.

Components of Oil of Black Pepper

Pepper oil is primarily a mixture of hydrocarbons comprising 70 to 80% of monoterpenes, 20 to 30% of sesquiterpenes and with less than 4% of oxygenated compounds to which the main distinguishing notes may be attributed (Pangborn *et al.* 1970).

Richard *et al.* (1971) carried out a very sophisticated GLC computer analysis of the volatile oil obtained from 17 varieties of Indian pepper and reported the presence of the following compounds:

α-thujene	α-cubebene	linalool
α-pinene	α-copaene	1-terpinen-4-ol
camphene	β-elemene	carvone
sabinene	β-caryopyllene	caryophyllene ketone
β-pinene	β-farnescene	
myrcene	humulene	
α-phellandrene	β-selinene	
Δ-3-carene	α-selinene	
α-terpinene	β-bisabolene	
terpinolene		

Lewis *et al.* (1969) also examined a range of locally available varieties of Indian pepper noting marked differences in odor profile. Although specific profiles were not published the authors concluded that:

(a) A high percentage of monoterpenes is necessary before an oil of pepper can be called characteristically "fresh."

(b) Pepper oil which is rich in α- and β-pinenes has a predominantly "terpeney" or "turpentine-like" odor which detracts from the pleasingly "fresh" character.

(c) Pepper oil rich in caryophyllene is strongly "sweetly floral."

The aromatic characters of the compounds present in oil of black pepper are given in Table 3.2.

GINGER (*Zingiber officinale*, Roscoe)

Ginger is the rhizome or rootstock of the herbaceous perennial. It is native to tropical Southeast Asia where it has been cultivated

TABLE 3.2
AROMATIC ATTRIBUTES OF THE COMPONENTS OF OIL OF BLACK PEPPER

Component	Main Character	Subsidiary Attributes
Monoterpenes		
α-Pinene	PINE-like	Warm, resinous, refreshing
Camphene	CAMPHORACEOUS	Oily, terpeney
Sabinene	PEPPERY	Warm, woody, herbaceous
β-Pinene	DRY WOODY	Resinous,pine-like,terpeney
Myrcene	SWEET BALSAMIC	Resinous,lemony,fresh
α-Phellandrene	PEPPERY	Woody,fresh,citrus,minty
Δ-3-Carene	PENETRATING	Sweet,irritating
α-Terpinene	LEMONY	Fresh
β-Phellandrene	PEPPERY	Minty,slightly citrus-like
Limonene	FRESH	Light,orange-like
γ-Terpinene	HERBACEOUS	Warm,lemony
Terpinolene	SWEET PINY	Slightly anisic
Sesquiterpenes		
β-Caryophyllene	WOODY SPICY	Dry,clove-like
β-Farnescene	SWEET	Warm
Humulene	SWEET WOODY CITRUS	Penetrating
β-Selinene	SWEET WOODY	Peppery
α-Selinene	HERBACEOUS	Warm,woody,peppery
β-Bisabolene	WARM SPICY	Balsamic,aromatic
Oxygenated compounds		
Linalool	FLORAL WOODY	Light,refreshing,slightly citrus-like
1-Terpinen-4-ol	WARM PEPPERY	Earthy,woody
Carvone	WARM HERBACEOUS	Spicy,slightly floral
Caryophyllene ketone	FRUITY	minty

both as a fresh vegetable and as a dried spice since time immemorial. Marco Polo was the first to write of finding ginger in China, which country still provides significant commercial crops. This valuable spice is now grown in India, the West Indies (principally in Jamaica), along the west coast of Africa (Sierra Leone and Nigeria), in Mexico and parts of South America and more recently in Queensland, Australia. The annual trade in ginger amounts to about $6.5 million but figures are difficult to establish as the trade is partly indigenous and partly export. The total external market for spice ginger amounts to about 12,000 metric tons per year.

Ginger is harvested at two periods depending upon its ultimate usage:

(a) When the leaves are fully green and the rootstock young and tender and not woody. This is used to make "stem" or preserved ginger and crystallized ginger confections.

(b) When the stems have withered. The woody rootstock after clean-

Courtesy of Tropical Products Institute

FIG. 3.15. HARVESTING GINGER IN JAMAICA

Courtesy of Bush Boake Allen, Ltd.

FIG. 3.16. JAMAICA GINGER AT TIME OF HARVESTING

Courtesy of Bush Boake Allen, Ltd.

FIG. 3.17. HAND PEELING OF GINGER RHIZOMES IN JAMAICA

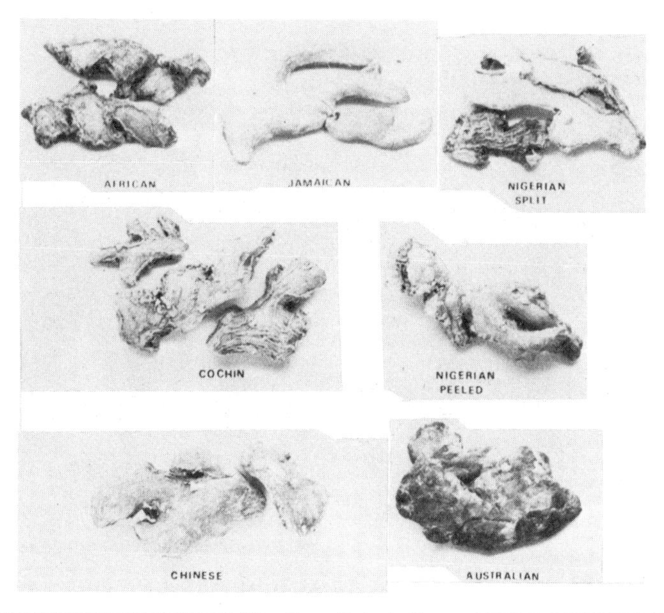

FIG. 3.18. COMPARISON OF GINGER (*Zingiber offinicale*, Roscoe) FROM DIFFERENT SOURCES

Courtesy of Bush Boake Allen, Ltd.

FIG. 3.19. PEELED GINGER RHIZOMES BEING SPREAD OUT TO DRY IN THE SUN

Courtesy of American Spice Trade Association

FIG. 3.20. PEELED JAMACIAN GINGER

Courtesy of Bush Boake Allen, Ltd.

FIG. 3.21. FRESH JAMACIAN GINGER
In this state the spice is widely used as a vegetable in many local dishes.

Courtesy of Albright & Wilson, Ltd.

FIG. 3.22. WORM-INFESTED NIGERIAN SPLIT GINGER
Beetles can cause considerable wastage of spices during storage and are usually controlled by gas disinfestation techniques.

ing, scalding with hot water and sun drying becomes the familiar ginger of commerce.

Ginger from the different growing regions has a distinctly characteristic appearance by which its source can readily be determined. The main features of the commercial varieties presently available are given in Table 3.3, the grades being generally known in the trade by the name of the region of origin. Ginger occurs either unpeeled or peeled. In the former case, the "hands" are left with the rough outer corky, epidermal layers intact; in the latter, these are scraped away from the freshly-dug rhizomes prior to washing and drying. The postharvest treatment of scalding with boiling water, the partial or complete removal of the outer tissues and the often protracted sun drying all have a significant effect upon the aromatic character of this spice. In the case of the peeled varieties, the profile of the essential oil is markedly different from that of the unpeeled gingers. This is due to the presence of additional oxygenated compounds in the oil derived from the epidermal tissues which gives to it a more distinctly "gingery" note. The oil from the inner cortical tissues is richer in terpenes and is more "lemony" in profile. This is clearly reflected in the comparison of the profiles of freshly-ground Jamaican (peeled) and African (unpeeled) gingers.

Ginger, like pepper, is characterized by two major flavoring components:

(a) 1 to 3% of a yellow viscous essential oil which gives the distinctive aroma to the spice.

(b) Nonvolatile pungency due to a highly complex mixture the nature of which is still under investigation (Connell and Sutherland 1969; Connell and McLachlan 1972) but containing zingerone, gingeral and shogaol.

Profile of Ground Jamaican and African Gingers

	Jamaican	African
Odor	SWEET; mildly spicy.	HARSH; strongly spicy.
	WARM; smooth.	WARM; irritating.
	LEMONY.	LEMONY; terpeney.
	Slightly EARTHY; woody.	EARTHY; dirty, woody.
	Slightly irritating.	Strongly MUSTY.
Dry-out:	Persistently characteristic over a long period.	

TABLE 3.3
COMMERCIAL GRADES OF DRIED GINGER

Source	Description	Odor Character	Average Volatile Oil Content (%v/w)
Cochin	Light brown;partially peeled on the flattened sides;cork wrinkled and very light brown.	Strongly LEMONY	2.2
Jamaican	Light buff color;peeled clean,hard.	Delicately AROMATIC SPICY	1.0
African (Sierra Leone)	Very dark greyish-brown; unpeeled,cork tissues wrinkled.	Strongly EARTHY, HARSH	1.6
Nigerian	Light color;usually split and partially peeled.	Strongly aromatic. CAMPHORACEOUS	2.5
Chinese	Pale brown;unpeeled	Mildly AROMATIC, LEMONY	2..5
Japanese	Pale,partially peeled; often limed.	Weakly AROMATIC, LEMONY	2.0
Australian	Light brown;peeled and unpeeled.	Strongly LEMONY	2.5

Flavor	LEMONY; terpeney.	FLAT; dirty.
	Pleasantly WARM.	BITTER; harsh.
	SMOOTH; mildly spicy.	WOODY.
	Slightly WOODY.	EARTHY; green mustiness.
	Slightly BITTER.	DRYING; astringent.
	Slightly IRRITATING.	

The above flavor attributes overlay the strong pungency of these spices.

After-taste	Pleasantly FRESH.	Unpleasantly HARSH.

Profile of Oil of African Ginger

Origin Commercial sample of oil steam-distilled from Sierra Leone Ginger.

Odor	Impact:	5
	Initial:	HEAVY; full-bodied. WARM; spicy. LEMONY; fruity. TERPENEY; turpentine-like, pine-like. Slightly MUSTY. Slightly HARSH, PENETRATING. IRRITATING after a very short time.
	Persistence:	The profile changes hardly at all over several days on a smelling strip.
	Dry-out:	Remains characteristic with slight increase in EARTHY notes.
Flavor		MILD, FLAT. LEMONY; fruity, citrus-like. Slightly PINE-like, terpeney. WARMING. Slightly DRYING; astringent. Slightly BITTER.
	After-taste:	Initially pleasingly SPICY but later BITTER.

Profiles of oils distilled from Nigerian and Cochin gingers and established by an examination of the effluent from a gas chromatograph are given in Fig. 3.23 and 3.24.

Components of Oil of Ginger

Oil of ginger is a mixture of terpenes of which there is a relatively small proportion of sesquiterpenes and oxygenated compounds. Although the oil has a quite distinctive and identifiable profile there is little to indicate that any of the compounds so far identified is unique to or even characteristic of ginger although the sesquiterpene zingiberene and *cis*- and *trans*-sesquiphellandrol are probably the most significant in this respect (Bednarczyk and Kramer 1975).

Although the bulk of ginger used in seasonings is sun dried, there is a considerable trade in fresh ginger (green ginger) throughout the Far East and Pacific where the fresh rhizomes play an important part in the cuisine. The oil distilled from fresh green ginger has been investigated (Krishnamurthy *et al.* 1970). It has a more pronounced spicy odor than normal dried ginger, this being attribbuted to the higher percentage of zingiberene and a lower percentage of sesquiterpene alcohols in the oil.

The following constituents have been identified in oils of ginger:

α-pinene	1,8 cineole (about 7%)
camphene	borneol
β-pinene	zingiberol
sabinene	methyl heptanone
Δ-3-carene	citral
myrcene	n-decylaldehyde
β-phellandrene	n-nonylaldehyde
limonene	
tricyclene	
zingiberene	

The aromatic characters of the principal components are given in Table 3.4.

TABLE 3.4
AROMATIC ATTRIBUTES OF THE COMPONENTS OF OIL OF GINGER

Component	Main Character	Subsidiary Attributes
Monoterpenes		
α-Pinene	PINE-like	Warm, resinous, fresh
Camphene	CAMPHORACEOUS	Oily, terpeney
β-Pinene	DRY WOODY	Resinous, pine-like, terpeney
Sabinene	PEPPERY	Warm, woody, herbaceous
Δ-3-Carene	PENETRATING	Sweet, irritating
Myrcene	SWEET BALSAMIC	Resinous, lemony, fresh
β-Phellandrene	PEPPERY	Minty, slightly citrus-like
Limonene	FRESH	Light, orange-like
Tricyclene	CELLULOID-like	Sickly
Sesquiterpenes		
Zingiberene	WARM, WOODY, SPICY	Sweet
Oxygenated Compounds		
	These represent about 9% of the oil and include:	
1,8 Cineole	CAMPHORACEOUS	Fresh, spicy
Borneol	DRY WOODY	Peppery
Zingiberol		

MUSTARD

Mustard seeds are derived from several species of the genus *Brassica*—the cress family, which also includes horseradish as well as several well-known vegetables such as cabbage, turnip and sweede; many of these have pungent components in their flavor derived by the enzymic breakdown of glucosides present in the plant.

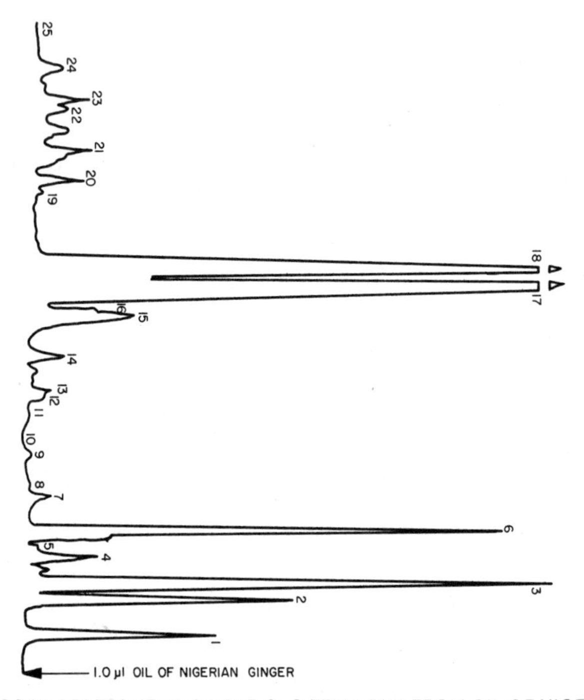

FIG. 3.23. SENSORY ASSESSMENT OF THE GLC EFFLUENT FROM OIL OF NIGERIAN GINGER

Peak No.	Character
1	CITRUS, vaguely gingery
2	METALLIC, slightly fruity, terpeney
3	GREEN, aldehydic
4	GREEN FLORAL, full-bodied
5	HEAVY, powerful
6	COOLING, REFRESHING, disinfectant-like
7	HEAVY, MUSTY, MUSHROOM-like
8	GREEN, sweetly aldehydic
9	GREEN
10	GREEN, *cis*-3-hexenol-like
11	WAXY, candle-like
12	CORIANDER-like
13	MUSTY, slightly burnt
14	LIGHT, GREEN, stimulating
15	CUMINALDEHYDE-like, green back-note
16	EUCALYPTUS-like, cineolic
17	FRUITY, green, orange-peel note
18	FRUITY, green, floral note
19	FRUITY, green, estery
20	DRY but nondescript
21	Strongly GINGERY, woody, earthy
22	Mildly TOFFEE-like
23	Mildly DRY
24	CORIANDER-like, full-bodied
25	SWEET, FRUITY, estery

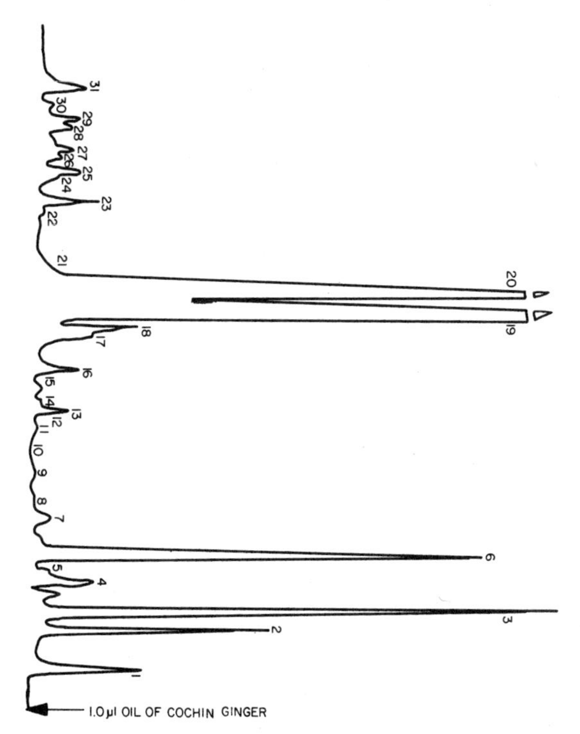

FIG. 3.24. SENSORY ASSESSMENT OF THE GLC EFFLUENT FROM OIL OF COCHIN GINGER

Peak No.	Character
1	Very LIGHT, SWEET, slightly gingery
2	LIGHT, vaguely floral
3	GREEN
4	LIGHT, GREEN, floral, warming
5	Very GREEN, ALDEHYDIC
6	CINEOLIC, very strongly PENETRATING
7	HEAVY, MUSHROOM-like, MUSTY
8	Very SWEET, fruity with floral back-note
9	GREEN
10	GREEN, *cis*-3-hexenol-like
11	CITRUS, orange-like
12	SOAPY
13	CORIANDER-like
14	BURNT, musty
15	HEAVY, MUSTY, stale
16	LIGHT, COOLING, fragrant
17	DRY, WOODY becoming FRUITY
18	FLORAL, light
19	Very strongly LEMONY
20	FRUITY, estery
21	DRY, CLOVE-like
22	DRY but nondescript
23	DRY, mildly TERPENEY
24	DRY, SWEETLY GINGER-like
25	DRY, MUSTY
26	WARM, SWEET, estery
27	LIGHT, SWEET
28	Very DRY, WOODY
29	ETHANOLIC
30	Very FLORAL, SWEET, estery
31	DRY, LINALOOL-like but sweeter

There are two principal types of mustard used as a condiment, in seasonings or as a source of "mustard oil." These are:

(a) *White mustard.*—The seeds of *Brassica (Sinapis) alba*, Boise. The plant is a native of southern Europe and is now widely cultivated throughout Europe and in the United Kingdom. It is an annual which grows only about 2 ft in height and bears a mass of bright yellow flowers. The seed is spherical and pale yellowish in color. The taste of the crushed seeds is at first slightly bitter but an agreeable aromatic pungency soon develops.

(b) *Black mustard.*—Consists of the seeds of *Brassica nigra* (L), Koch, and brown mustard from the seeds of *Brassica juncea* (L), Czerniaew, are so closely similar that both sources are recognized as being "black mustard." The plants are similar to that of white mustard but the flowers are very much smaller. The seeds are smaller and dark reddish to grey-brown in color. Black mustard is widely cultivated in the Netherlands, Italy, the United Kingdom and the United States. The crushed seeds exhale a slight but characteristic odor which becomes intensely strong and irritant when the powder is wetted. The taste is initially slightly bitter but the powerful pungency soon covers this.

The Components of Mustard

Mustard seeds contain much fixed oil together with protein, sugars and mucilage. The oil is obtained by expression and is straw-colored. It is present to about 24 to 26% in white mustard and 24 to 40% in black mustard.

White Mustard.—The most important constituent of white mustard seed is the glycoside SINALBIN. In the presence of water this decomposes by the action of an enzyme MYROSIN yielding *p*-hydroxy benzyl isothiocyanate (sinalbin mustard oil), glucose and sinapine bisulfate.

The sinalbin mustard oil is almost nonvolatile and practically devoid of odor but it has a well-defined pungency and is a vesicant.

Black Mustard.— This variety is characterized by the presence of the glycoside SINIGRIN which decomposes under similar conditions. In this case, the active constituent formed is allyl isothiocyanate (C_3H_5NCS), glucose and potassium hydrogen sulfate.

Allyl isothiocyanate is a very volatile liquid usually called "oil of mustard, artificial." It has a pungent, very irritating odor and a sharply acrid flavor. It, too, is a vesicant.

Some oil is recovered by the maceration of the de-oiled press-cake with water at 70°C followed by distillation to yield 0.5 to 1.2% of an oil containing about 95% allyl isothiocyanate together with further breakdown products such as allyl cyanide and carbon disulfide.

FIG. 3.25. MUSTARD (*Brassica nigra*, L.) IN FLOWER

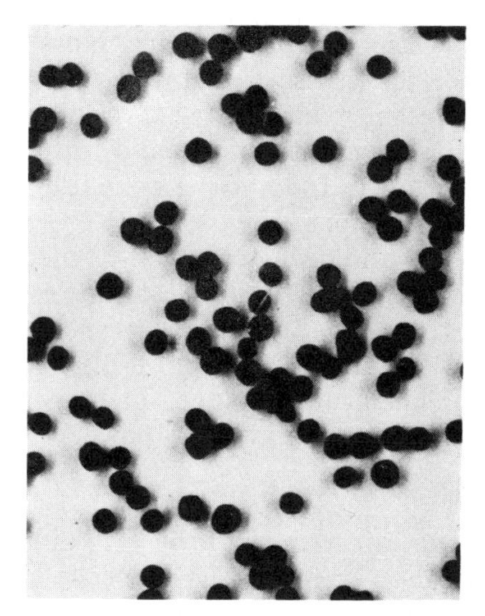

FIG. 3.26. BLACK MUSTARD SEED (*Brassica nigra*, L.) MAGNIFIED X 2

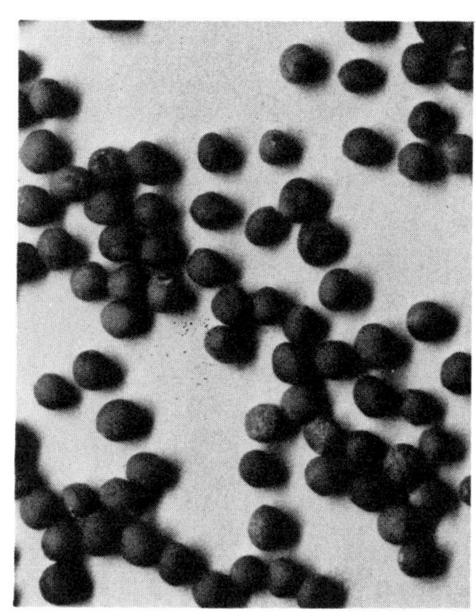

FIG. 3.27. WHITE MUSTARD SEED (*Sinapis alba*, Boise) MAGNIFIED X 2

Profile of Oil of Mustard

Oil of mustard is extremely irritant and lachrymatory. It should never be inhaled directly. An appreciation of the subtle profile of mustard may be obtained by mixing a little mustard powder with water at 70°C and allowing the mixture to stand in a closed vessel for 10 min.

Mustard Powder

Commercial "mustard powder" is a blend of ground black and white mustard in the proportions necessary to produce the level of pungency and flavor required. As mustard seeds contain a fixed oil which tends to go rancid on storage, this is frequently removed from the black mustard used, the final mixture containing about 30% of oil derived from white mustard; this gives to the product its mildly insistent pungency. The final product usually contains added wheat flour or starch to act as an anticaking agent. Ground turmeric has been used as a coloring agent to improve its yellow appearance. In the United Kingdom, there is a regulation requiring that mustard powder shall yield not less than 0.03% of allyl isothiocyanate after maceration with water at 37°C for 2 hr. Added farinaceous matter is limited to 20%.

HORSERADISH

The horseradish (*Armoracia lapthifolia*, Gilib.—formerly classified as *Cochlearia armoracia*, L. Fries) is also a member of the *Cruciferae* and is a perennial herbaceous plant native to eastern Europe. It is now widely cultivated throughout Europe and in the United Kingdom, and forms garden crops in India, Sri Lanka (Ceylon) as well as several other tropical regions.

The plant grows to about 3 ft in height and develops a well-formed tap-root which is the part of the plant which is of most interest as a pungent spice. The freshly-dug roots have the finest flavor and pungency and are best used in this condition. They may also be effectively preserved in the form of horseradish sauce. The roots cannot be dried without considerable loss of flavoring power.

The Components of Horseradish

Horseradish root contains starch, sugars, resins and the glycoside sinigrin which is also present in black mustard. In the presence of water, the sinigrin is decomposed enzymically into allyl isothiocyanate, glucose and potassium hydrogen sulfate in an exactly similar way to mustard. In the fresh root there is sufficient water present in the tissues for this reaction to take place when the root is chopped or finely shredded; it is in this form that it is generally used to make a piquant sauce having a sharp, burning pungency.

The flavor of horseradish is not identical with that of mustard owing to the presence of other aromatics such as diallyl sulfide, phenyl ethyl and phenyl propyl isothiocyanate.

Part 2: The Aromatic Fruits

NUTMEG AND MACE (*Myristica fragrans*, **Houtt.**)

Nutmeg and mace are both derived from the fruit of a large evergreen tree which is native to Indonesia but now cultivated in both the East and West Indies. Propagation is from seeds but as the trees are unisexual it takes several years before the plantations can be thinned to give a preponderance of female trees. Fruiting starts at about 5 years, reaches a peak at 15 years and the trees continue to yield for up to 20 years and longer.

The nutmeg fruit is a peach-like drupe which, when ripe, falls from the tree and splits along a lateral groove revealing the scarlet arillus tightly wrapped around an inner shiny brown shell containing the seed. The arillus becomes MACE and the inner kernel is the NUTMEG of commerce.

Harvesting is simple, the fruit being allowed to fall naturally. Contents of the split fruits are gathered by hand. No use is made of the outer pulp which merely rots around the tree base. Drying of the nutmeg and mace is carried out separately:

Nutmeg: Very slow sun drying until the inner seed (*i.e.*, the nutmeg) is free and rattles when shaken. At this stage the shell is broken and discarded. The nutmegs being collected and graded by hand.

Mace: Flattened between boards and sun-dried during which it changes from the soft leathery scarlet aril to the horny, brittle, pale buff-colored mace of commerce.

World trade in nutmegs and mace is about $14 million per annum, of which 1/5 is as mace. The ratio of nutmegs to mace is approximately 10:1 and this is reflected in their relative cost.

Components of Nutmeg and Mace

The most important constituent of nutmeg and mace is the volatile oil which is present (from 7 to 16% in nutmeg and 4 to 15% in mace) together with about 35% of fat and starch, proteins and fiber.

Oil of nutmeg has been shown to consist of about 80% monoterpenes together with about 4% terpene alcohols and about 11% other

Courtesy of A. H. Taylor

FIG. 3.28. (LEFT) NUTMEGS GROWING IN TRINIDAD

FIG. 3.29. (BELOW) NUTMEG FRUIT SPLIT OPEN SHOWING THE KERNEL (NUTMEG) AND THE SCARLET ARILLUS (MACE)

Courtesy of Tropical Products Institute

Courtesy of A. H. Taylor

FIG. 3.30. SORTING NUTMEGS IN GRANADA, WEST INDIES

Courtesy of A. H. Taylor

FIG. 3.31. HAND PICKING MACE IN GRANADA, WEST INDIES

Courtesy of Tropical Products Institute

FIG. 3.32. DRYING NUTMEGS AND MACE IN INDONESIA

Courtesy of American Spice Trade Association

FIG. 3.33. WEST INDIAN NUTMEGS (*Myristica fragrans*, Houtt.)

FIG. 3.34. EAST INDIAN NUTMEG (*Myristica fragrans*, Houtt)

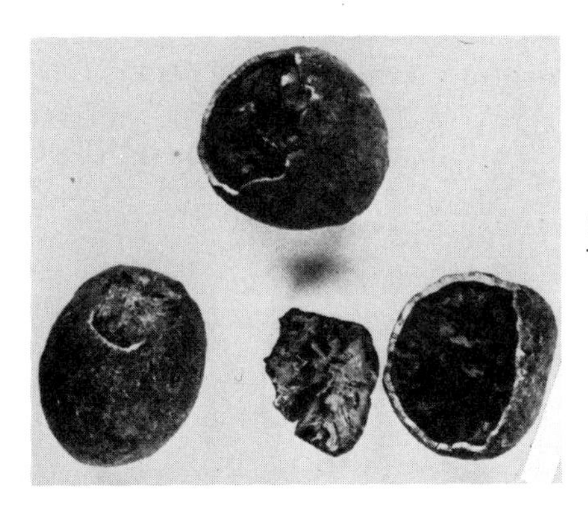

FIG. 3.35. EAST INDIAN NUTMEGS IN SHELL
These are worm-infested and are only used for distillation of nutmeg oil.

aromatics (Baldry *et al.* 1976); whereas oil of mace contains 87.5% monoterpenes, 5.5% monoterpene alcohols and about 7% other aromatics (Forrest and Heacock 1972). The oil from East and West Indian nutmegs differs largely in the percentage of monoterpenes present, that from East Indian nutmegs having less terpenes and, hence, is preferred where a heavier aromatic note is required in a flavor or fragrance.

Profile of Oil of Nutmeg (West Indian)

Origin Distilled commercially in the United Kingdom from selected West Indian nutmegs.

Odor
Impact: 3
Initial: SPICY, WARM.
FRESH, TERPENEY; camphoraceous.
SWEET, slightly fruity, aromatic.

		Mildly PENETRATING; very slightly irritating.
		FULL-BODIED; rounded, smooth.
	Persistence:	The oil airs-off very quickly and loses its fresh top-notes within 1 hour. Thereafter it remains largely unchanged in profile for several days.
	Dry-out:	Very weak, OILY, slightly PHENOLIC.
Flavor		SMOOTH, SWEET, SPICY.
		Slightly BITTER.
		LIGHT, FRESH; terpeney.
		Slightly PUNGENT.
		Very slightly ASTRINGENT; numbing.
	After-taste:	Lingering BITTER.

Profile of Oil of Nutmeg (East Indian)

Origin	Distilled commercially in the United Kingdom from selected East Indian (Padang) nutmegs.	
Odor	Impact:	3
	Initial:	SPICY, WARM.
		PINY, CINEOLIC.
		CAMPHORACEOUS; fresh, terpeney.
		SWEET, very SMOOTH.
		PENETRATING but not irritating.
		MEDIUM-BODIED slightly HARSH.
	Persistence:	The oil displays little change over several hours airing and remains distinguishable over several days.
	Dry-out:	Very HEAVY, OILY, almost SICKLY.
Flavor		HEAVY, SPICY.
		Harshly TERPENEY, PINY.
		Slightly BITTER.
		Slightly PUNGENT.

After-taste: UNFRESH, BITTER.

Profiles of oil of East Indian and West Indian Nutmeg oils were established by an examination of the effluent from a GLC and are given in Fig. 3.36 and 3.37.

Comparison of the Aromas of Nutmeg Oils

West Indian	East Indian
FRESH	HEAVY
LIGHT	CAMPHORACEOUS
SWEETLY SPICY	OILY SPICY

Components of the Essential Oil (Sanford and Heinz 1971)

Monoterpenes	Terpene Alcohols
α-Pinene	*trans*-Sabinene hydrate
α-Thujene	Linalool
Camphene	*cis*-Sabinene hydrate
Sabinene	β-Terpineol
β-Pinene	1-Terpen-4-ol
Myrcene	Borneol
Δ-3-Carene	α-Terpineol
β-Phellandrene	Geraniol
ρ-Cymene	Citronellol
1,4,-*p*-Menthadiene	Other Compounds
Terpinolene	Linalyl acetate
	Bornyl acetate
	Geranyl acetate
	Safrole
	β-Caryophyllene
	Eugenol
	Methyl eugenol
	trans-Isoeugenol
	trans-Methyl isoeugenol
	Myristicin
	Elemicin
	Methoxyeugenol
	trans-Isoelemicin
	Myristic acid

Of these compounds, it is the myristicin which is considered to be toxic and to act as a narcotic if ingested in large quantity.

As nutmegs are frequently stored for very long periods under high ambient temperatures, Sanford and Heinz (1971) also examined the

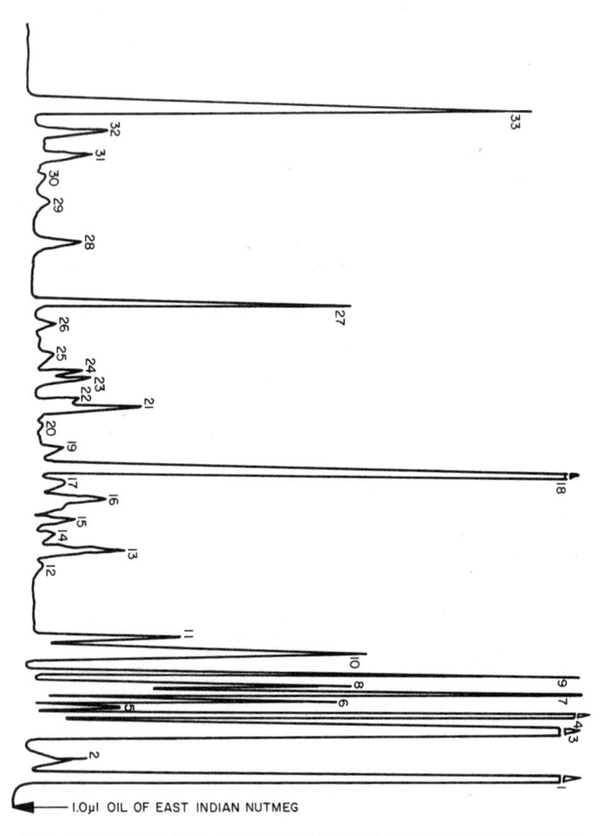

FIG. 3.36. SENSORY ASSESSMENT OF THE GLC EFFLUENT FROM OIL OF EAST INDIAN NUTMEG

Peak Number East Indian	Peak Number West Indian	Character	Tentative Identification
1	1	PINE-like, resinous	α-Pinene
1	2	CAMPHOR-ACEOUS	Camphene
3	3	WOODY; resinous	β-Pinene
4	4	FRESH, CITRUS-like	Myrcene
5	5	PENETRATING, CITRUS-like	Δ-3-Carene
6	6	Strongly LEMONY	α-Terpinene
7	7	ORANGE-like	Limonene
8	8	MINTY	β-Phellan-drene
9	9	COOKED VEGETABLE	ρ-Cymene
10	10	IRRITATING, PAINT-like	
11	11	PINY, SWEET	Terpinolene
12	12	FLORAL, ROSEY	Linalool
13	13	PUNGENT, SPICY	Sabinene
14	14	FRUITY	
15	15	PUNGENT, EARTHY	β-Terpineol
	16	No discernable odor	
16	17	SWEET, FRUITY	

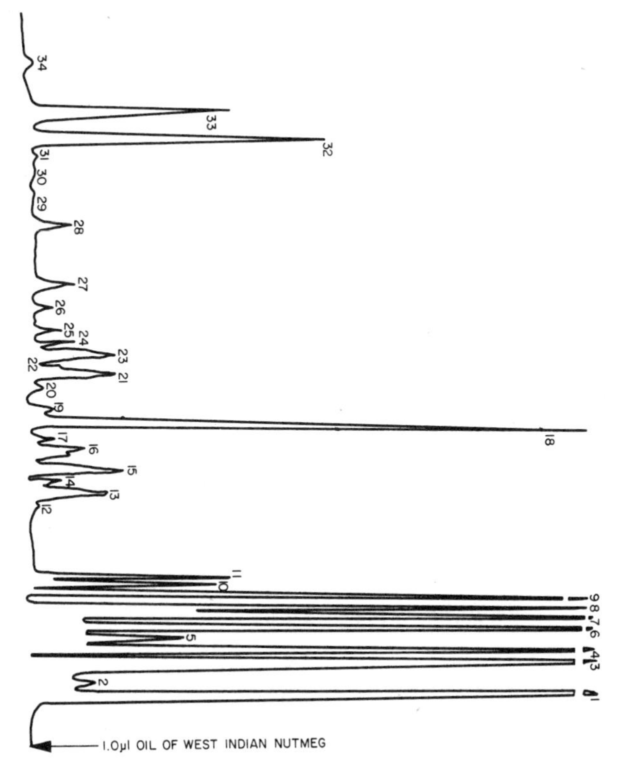

FIG. 3.37. SENSORY ASSESSMENT OF THE GLC EFFLUENT FROM OIL OF WEST INDIAN NUTMEG

Peak Number East Indian	Peak Number West Indian	Character	Tentative Identification
17	18	SWEETLY FRAGRANT	
18	19	MUSTY, PEPPERY	1-Terpen-4-o1
19	20	CAMPHOR-ACEOUS	Borneol
20	21	FLORAL	α-Terpineol
21	22	ROSY	Citronellol
22	23	Strongly FRUITY	Linalyl acetate
23	24	BALSAMIC, GREEN	Bornyl acetate
24	25	FRUITY, FLORAL, GREEN	Geranyl acetate
	26	ROSEY	Geraniol
25	27	DULL, SPICY, CLOVE-like	
	28	SHARP, HERBACEOUS	
26	29	DRY, WOODY	
27		SWEET, WOODY	Safrole
28	30	WOODY, SPICY	β-Caryophyllene
29	31	CLOVE-like	Eugenol
30	32	FLORAL, SPICY	Isoeugenol
31	33	WOODY	
32	34	DULL, SPICY, TEA-like	Methyl eugenol
33	35	WARM, BALSAMIC	Myristicin
	36	DRY, WOODY, GREEN	
	Tail	FAINT WOODY	

effect on the composition of the essential oil and concluded that the variations observed were a function of the relative volatility of the components rather than to any change in their nature.

Nutmeg Butter

Nutmegs yield from 20 to 30% of a fixed oil which can be recovered by expression or by solvent extraction. The essential oil forms part of the product and cannot readily be removed by distillation. To increase the yield the ground nutmeg is generally subjected to heat, the resulting product being an orange-red to red-brown paste having the consistency of butter at room temperature—hence its name "nutmeg butter." The product readily melts to a clear oleoresin when gently heated. A clear fluid oleoresin can also be produced by cold extraction with acetone.

The chief constituent of nutmeg butter is trimyristicin (about 75%) together with essential oil (about 12%) and the glycerides of oleic and linoleic acids.

Profile of Nutmeg Butter

Origin Commercial sample.

Odor

Impact: 2

Initial: WARM, SPICY.
LIGHT, FRESH; terpeney, lemony.
FRUITY; pleasantly aromatic.
Slightly IRRITATING; slightly pungent.
OILY.

Persistence: The presence of the fixed oil makes the aroma very lingering, the profile changing hardly at all over several hours.

Dry-out: HEAVY, OILY with slight FRUITY back-note.

Flavor

(5% dispersed on salt and tasted at 0.5% in a neutral soup)

SMOOTH, WARM, SPICY.
BITTER.
TERPENEY, LEMONY.

After-taste: Lingering BITTER.

MACE

Oil of mace is not a regular commercial article although there are plenty of nutmeg oils offered under this title. For this assessment of the aromatic profile, hand-picked Granada mace was distilled in the laboratory and allowed to mature for one month before examination.

Profile of Oil of West Indian Mace

Odor	Impact:	3
	Initial:	FRESH, LIGHT. PINE-like, TERPENEY, very slightly LEMONY. Sweetly AROMATIC. WARM, SPICY, SMOOTH. BREAD-like. Slightly FRUITY. OILY.
	Persistence:	The initial fresh notes are lost within 10 min, the profile becoming smooth and spicy. After 4 hr it changes to a markedly dry and floral note.
	Dry-out:	OILY with FRUITY back-note.
Flavor		Slightly BITTER. SMOOTH, SWEET, SPICY. FRESH; terpeney. CREAMY, MELLOW. Slightly PUNGENT, slightly NUMBING.
	After-taste:	Lingering SPICY, slightly WARM.

CARDAMOM (*Elettaria cardamomum*, Maton)

Cardamoms are the dried ripe fruits of *Elettaria cardamomum*, Maton, a member of the *Zingiberacae*. The flavoring properties of this spice are confined to the seeds which are normally retained in the natural pericarps (husks) until required as the aroma of the separated seeds is very quickly lost. Cardamom is a perennial bushy herb, bearing fruits in pods, each of which is a capsule containing 8–16 irregular, tightly-packed dark brown seeds. The plants yield an economic crop of fruits from four years after planting.

There are two distinct botanical varieties of this species: var. min-

FIG. 3.38 WEST INDIAN MACE

Courtesy of American Spice Trade Association

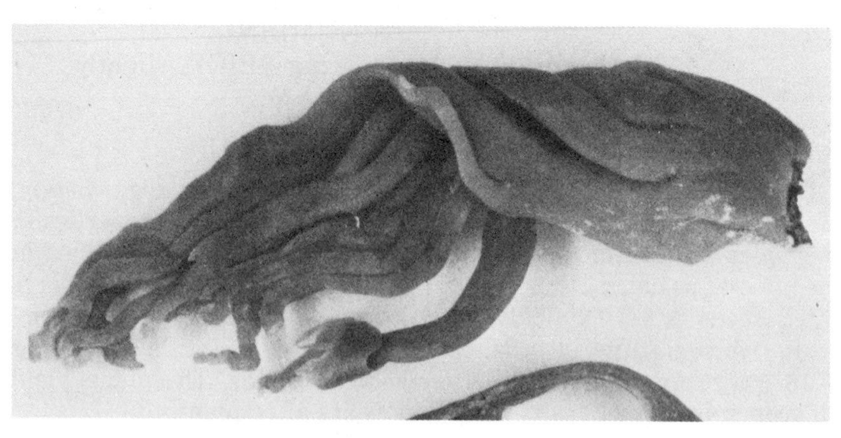

FIG. 3.39. WEST INDIAN MACE MAGNIFIED X 3

uscula—indigenous to Malabar and Mysore; and var. major—a native of Sri Lanka (Ceylon).

The main producing area is India, although smaller crops are produced in Sri Lanka, Laos and Guatemala and, more recently, in El Salvador. The total world crop amounts to about 3000 metric tons per annum.

Although not reflecting the quality of the aromatic seeds, whole cardamoms are marketed more by the appearance of the capsule than on available flavor. In the past, the capsules were often bleached white to improve their value but the major commercial shipments are now in their natural state and are the so-called "Alleppy Green" variety. The fruits vary in length from 5/16 to 9/16 in., being 3-sided and oval-oblong. The surface is generally clean, longitudinally wrinkled and often carrying numerous pale brown scabs.

Constituents of Cardamom Seeds

The reddish-brown angular seeds contain about 6% of essential oil (2-10%), some fixed oil, protein and some sugars. The oil is very volatile and is rapidly lost through evaporation once the seeds are removed from the capsules. After only one week the oil content of the seeds may fall to a half its freshly ground value and be completely lost in as little as 12 weeks.

Profile of Oil of Cardamom

Origin Commercially distilled from Alleppy Green Cardamoms.

Odor

Impact: 5

Initial: PENETRATING, slightly IRRITATING.
CINEOLIC; cooling.
CAMPHORACEOUS; disinfectant-like.
WARM, SPICY.
SWEET, very AROMATIC; pleasing.
FRUITY; lemony, citrus-like.

Persistence: The oil rapidly airs-off on a smelling-strip loosing its freshness and becoming HERBY, WOODY, with a marked MUSTY back-note.

Dry-out: No residual odor after 24 hr.

Courtesy of American Spice Trade Association

FIG. 3.40. ALLEPPY CARDAMOM (*Eletteria cardamomum*, Maton.)

FIG. 3.41. ALLEPPY CARDAMOM (MAGNIFIED X 2) WITH ONE CAPSULE OPENED TO SHOW THE SEEDS

Flavor		SMOOTH, WARM, SPICY. Slightly BITTER. CAMPHORACEOUS. CINEOLIC; sweet. FRUITY; lemony. Slightly HERBACEOUS.
	After-taste:	COOLING; slightly ASTRINGENT.

Components of Oil of Cardamom

α-pinene	cineole (25–40%)*
sabinene*	methyl heptanone (about 1.5%)
myrcene	linalool
limonene*	linalyl acetate
ρ-cymene	β-terpineol*
	α-terpinyl acetate (28–34%)
	borneol*
	neryl acetate
	geraniol
	nerol
	nerolidol

The components marked * are often blended together to make a synthetic "extender" of genuine cardamom oil. These imitation cardamom oils often have a strongly CAMPHORACEOUS top-note with a sickly ROSY, FLORAL overtone which becomes BALSAMIC as it airs-off. They are generally much less sweetly cineolic than the genuine oil.

FENUGREEK (*Trigonella foenum-graecum*, **L.**)

Fenugreek is a small European annual herb of the natural order *Leguminosae* and now widely cultivated in southern Europe, North Africa, Cyprus and India. It is the seeds which constitute the spice. These are oblong-quadrangular somewhat compressed and obliquely truncated at each extremity. They are 3–5 mm long and 2–3 mm broad, brownish-yellow in color with a deep furrow running obliquely from one side.

Ground fenugreek has a strong, pleasant and quite peculiar odor which is very reminiscent of maple due to the presence of a minute quantity of essential oil which is extremely odorous. The quantity of volatile oil is only about 0.01 to 0.02% and is not available commercially. The seeds also contain about 7% of fixed oil, resin, protein and starch. Its marked bitter taste is due to two alkaloids—trigonelline and choline.

FIG. 3.42. FENUGREEK (*Trigonella foenum-graecum*, L.) MAGNIFIED x 3

In flavorings, ground fenugreek is usually extracted with alcohol to give a strongly aromatic essence which, in conjunction with other aromatic chemicals, may be used in the making of an imitation maple syrup flavor.

If the alcoholic extract is processed to remove the solvent the resulting "oleoresin fenugreek" has a quite different profile reminiscent of cooked meat which makes it a useful component in seasonings.

Profile of Oleoresin Fenugreek

Origin	Commercial sample.	
Odor	Impact:	4
	Initial:	MEATY, BROTHY, HYDROLYZED VEGETABLE PROTEIN-like.
		SMOOTH; mellow, soft.
		SWEET, MAPLE-like; slightly sickly.
		HEAVY; dull, warm.
		Slightly SMOKY; woody.
		Very slightly GREEN; lovage-like.
	Persistence:	Very lingering with almost no change in character.

Flavor

(2% dispersed on salt and tasted at 1% in a neutral soup)

Very MILD, SMOOTH.
MEATY, HYDROLYZED VEGETABLE PROTEIN-like.
MAPLE-like.
BITTER/SWEET.

After-taste: Pleasantly MEATY.

Part 3: The Umbelliferous Fruits

The plants in the natural order *Umbelliferae* are mostly biennial or perennial herbs, the name of the order being derived from the fact that the inflorescence is characteristically an umbel. The fruits consist of two fused carpels (cremocarps) which may separate when fully ripe, the half-fruits formed by this being called mericarps. Each contains one seed. The fruits are readily distinguishable by size, shape and color. Examination under a hand lens reveals the presence of five longitudinal ridges or ribs. The oil ducts containing the essential oil lie in the furrows between the ridges; these are called the "vittae." There are usually six vittae in all.

In the spice trade the umbelliferous fruits are more usually referred to as "seeds" (e.g., caraway seed). Those of importance as spices are given in Table 3.5. All of these fruits are highly aromatic and each has its own flavor nuance. In many cases, the fragrant green leaves of these plants are also widely used (particularly on the continent of Europe) to add a delicate and appetizing flavor to salads, soups and casseroles. Where appropriate, the profiles of the green herb will be considered under the heading of the spice. The flavor profiles are very individual but all are mild and sweet making them a perfect flavor adjunct to mildly flavorful dishes such as cheese, fish and eggs. The flavor of the freshly ground fruits is powerful and can be too strong unless used with care. Compared with the equivalent green herb the ground spice is harsh and insistent.

There is a very tenuous relationship between the profiles of the various umbelliferous fruits and these are set out in Table 3.6. The members of this group will, however, be dealt with alphabetically.

ANISE (*Pimpinella anisum*, **L.**)

Anise, or aniseed as it is frequently called, is the dried ripe fruit of *Pimpinella anisum*, L., a plant indigenous to Greece, Egypt and

TABLE 3.5
UMBELLIFEROUS FRUITS OF IMPORTANCE AS SPICES

Spice	Appearance	Essential Oil Content (%v/w)
Anise	Small ovoid fruits having short stems; 3–5 mm long, 1–2 mm wide; 10 well-defined ridges.	1.5–4
Caraway	Dark brown curved mericarps; 4–8 mm long; about 1 mm wide; 5 pale yellowish, prominent ridges.	3–7
Celery	The smallest of the umbelliferous spices; about 1 mm long; 5 clearly-defined ridges visible under hand lens.	2.5–3
Coriander	Globular entire cremocarps which may split if over-ripe; 3–4 mm in diameter; primary ridges are straight and secondary ridges wavy.	0.2–1
Cumin	Oval fruit with a rough surface due to hairs; 5–6 mm long; 5 primary and 4 secondary ridges.	2.5–5
Dill	Broadly oval or eliptical fruit with conspicuous wing-like lateral ridges; compressed and almost flat; five longitudinal ridges.	2.5–4
Fennel	The largest of the umbelliferous spices; 4–10 mm long; easily split into 2 mericarps; surface ridges prominent.	3–4
Parsley	Small, very dark-colored fruits; ridges clearly visible under a hand lens.	1.5–3.5

Asia Minor. It is an annual herb bearing yellowish-white flowers in a loose compound umbel. The upper leaves are narrowly divided, the lower are triangular in outline and deeply toothed. Anise is now cultivated in Russia, Spain, France and Germany and to a lesser extent in other parts of Europe and India.

The fruits of commerce are usually complete cremphores, greyish-green to dull yellowish-brown in color with pedicels attached. They are rough to the touch owing to the presence of surface hairs.

Anise contains 8–11% of fixed oil, some protein and 1.5–4% of a colorless to pale-yellow oil which contains 80–90% of anethole which causes the oil to solidify on cooling.

China Star Anise

The main commercial source of Oil of Anise is not, however, *Pimpinella anisum* but China Star Anise (*Illicium verum*, Hooker) a member of the Magnolia family and botanically unrelated to genuine anise. This is a small evergreen tree indigenous to the southern and southwestern provinces of China. The fruit is star-shaped and is

TABLE 3.6
FLAVOR RELATIONSHIPS WITHIN THE UMBELLIFERAE

	Anise Group	Caraway Group	Celery Group	Cumin Group	Coriander Group
Principal Constituents in Essential Oil	Anethole Methyl chavicol	d-Carvone d-Limonene	d-Limonene	Cumin-aldehyde	d-Linalool Geraniol Borneol
	Anise Fennel Star anise *also* Basil Tarragon	Caraway Dill Indian mill	Celery Parsley *also* Lovage Root	Cumin	Coriander
Odor Character	Anethole	d-Carvone	d-Limonene	Cumin-aldehyde	d-Linalool
	ANISIC SWEET AROMATIC	WARM, then COOLING FRUITY SWEET	ORANGE-LIKE TERPINEY	POWERFUL AROMATIC SPICY PUNGENT	SWEET FLORAL FRUITY

FIG. 3.43. ANISE PLANT (*Pimpinella anisum*, L.)

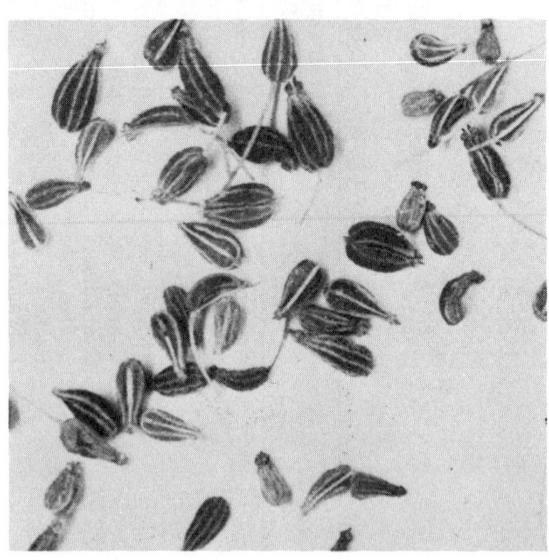

FIG. 3.44. ANISE SEED MAGNIFIED X 3

FIG. 3.45. CHINA STAR ANISE (*Illicium verum*, Hooker) MAGNIFIED X 2

made up of 1-seeded boat-shaped carpels, usually 8 in number, radiating from a central axis. When gathered, the fruits are green but change to dark brown on drying. The paler brown seeds are ovoid, smooth and shiny. The oil is obtained by steam-distillation in Vietnam and the Province of Kwangsi. Only a small proportion of the crop is exported as such.

The odor of genuine oil of anise from *Pimpinella anisum* is considered to be more delicate and superior to that of China Star Anise but as most of the oil available is the latter it is the profile of this oil which will now be described.

Profile of Oil of China Star Anise

Origin	Commerical sample.	
Odor	Impact:	4
	Initial:	SWEET, AROMATIC; pleasant.

		MEDICINAL.
		WARM but also COOLING.
		PENETRATING; irritating.
		FRUITY.
		MELLOW; full-bodied, rounded.
	Persistence:	Almost no change over 24 hr.
	Dry-out:	Very weakly CAMPHORACEOUS.
Flavor		REFRESHING, SWEET.
		MEDICINAL.
		Initially WARM but later very COOLING.
		FRUITY.
	After-taste:	None.

Components of Oil of China Star Anise

α-pinene	anethole (80–90%)
	methyl chavicol
	p-methoxyphenlactone (anisketone)

A synthetic anethole made from methyl chavicol, which is isolated from pine oil fractions, is now very widely used as a flavoring agent in place of Oil of Anise. In terms of cost, neither Oil of Anise can compare with that of synthetic anethole but the odor and flavor of the pure chemical is weak and lacking in the fine bouquet associated with the essential oil.

CARAWAY (*Carum carvi*, L.)

The so-called "caraway seeds" of commerce consist of the dried fruits of *Carum carvi*, L., a biennial plant native to western Asia but now cultivated throughout northern and central Europe, Morocco and parts of the United States. The main commercial source of this spice is The Netherlands where it is extensively cultivated. The plant grows to about 2 ft in height and has very divided leaves. The flowers and fruits are borne in compound umbels.

The spice consists of separated mericarps which are brown in color, curved and tapering towards one end—banana-shaped. They contain about 15% fixed oil and 3–7% of an almost colorless essential oil.

Profile of Oil of Caraway

Origin	Distilled commercially from Dutch Caraway.	
Odor	Impact:	4

FIG. 3.46. CARAWAY PLANT (*Carum carvi*, L.)

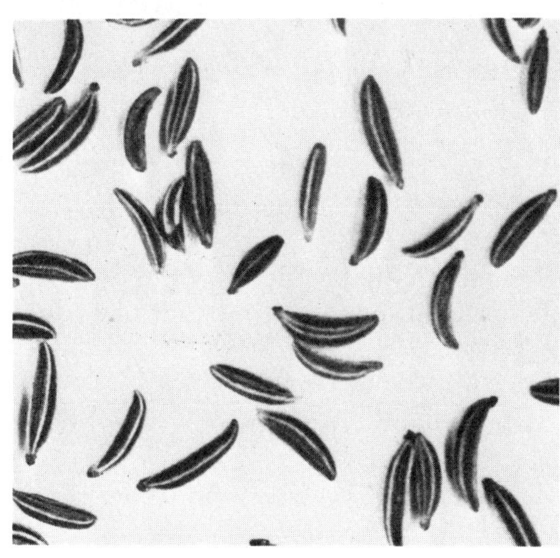

FIG. 3.47. CARAWAY SEED (*Carum carvi*, L.) MAGNIFIED X 3

	Initial:	SWEET, pleasantly FRUITY, AROMATIC.
		FRESH, COOLING.
		MINTY; spearmint-like.
		PENETRATING; slightly irritating, slightly pungent.
		Slightly CINEOLIC; medicinal.
		Slightly ANISIC; piny, light.
	Persistence:	Most of the fresh notes are lost within 1 hr, the profile becoming heavier and somewhat camphoraceous.
	Dry-out:	MINTY, WOODY.
Flavor		SWEET, FRUITY; to some, SICKLY.
		COOLING, MINTY; spearmint-like.
		BITTER.
		ASTRINGENT; drying.
	After-taste:	SPICY with WOODY back-notes.
		Slightly BITTER.

When caraway is chewed, as in caraway cake or bread, the flavor is distinctly BITTER with a NUMBING almost CLOVE-like character and a slightly SOAPY after-taste.

Components of Oil of Caraway

d-limonene	d-carvone (50–60%)
	dihydrocarvone
	dihydrocarveol
	carveol
	eugenol

The aromatic profiles of l- and d-carvone are of particular interest and account for the "spearmint-like" description in the profile of Oil of Caraway.

d-Carvone (in Caraway)	l-Carvone (in Spearmint)
Initially WARM then COOLING.	COOL, LIGHT, FRESH; aromatic.
Thin; very PENETRATING.	PENETRATING; slightly irritating.
FRUITY.	MINTY; spearmint-like.
SWEET; to some, SICKLY.	Pleasingly SWEET.
MUSTY, EARTHY; woody, smoky.	Slightly GREEN; herbaceous.

CELERY (*Apium graveolens*, L.)

So-called "celery seeds" consist of the dried ripe fruits of *Apium graveolens*, L., a native of the United Kingdom and northern Europe. The commerical spice is mostly separated mericarps which are very small, dark brown in color and easily distinguishable.

Celery contains 2.5–3% of a powerfully aromatic essential oil. The world production of celery is in excess of 2000 metric tons most of which is distilled to give the oil. The main producing areas are the south of France, India and in California.

Profile of Oil of Celery Seed

Origin	Distilled commercially from Indian Celery Seed.	
Odor	Impact:	5
	Initial:	WARM. SWEET, SPICY. PENETRATING but not irritating. Slightly FATTY. FRUITY; lemony.
	Persistence:	Although the odor is lingering much of the power of the oil is lost within 1 hr of airing-off; the heavier, more spicy attributes persist for several days.
	Dry-out:	Remains distinguishably CELERY.
Flavor		WARMLY SPICY. BURNING. Very BITTER.
	After-taste:	BITTER.

Components of Oil of Celery Seed

d-limonene (about 60%)	sedanolid
selinene (10–20%)	sedanonic anhydride

Celery Herb

Celery is widely cultivated in most temperate regions as a vegetable crop, the thick fleshy stems being familiar to most for their pleasingly crisp texture and subtle flavor. This part of the plant contains insufficient essential oil to make it commercially recover-

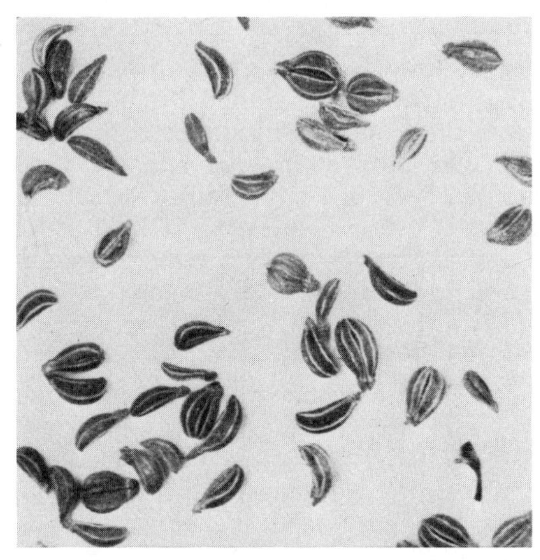

FIG. 3.48. CELERY SEED (*Apium graveolens*, L.) MAGNIFIED X 3

able but the odor of the green stalk and leaves is distinctly different from that of celery seed.

CORIANDER (*Coriandrum sativum*, L.)

Coriander is the dried ripe fruits of *Coriandrum sativum*, L., which, from the literature, appears to be one of the oldest known of the spices. The plant is a hardy annual which grows to about 2 ft in height and bears pinkish-white flowers in a compound umbel. The name is derived from the fact that the whole plant, when freshly bruised, exudes a peculiarly strong, unpleasant odor not unlike that associated with bugs—the name is based on the Greek *kopis* (bug). Fortunately, these unpleasant characters are completely lost as the plant matures and are totally absent from the ripe fruit.

The plant is a native of the Mediterranean region and is now widely naturalized and cultivated. The commercial sources are Russia, Bulgaria, Romania and Morocco with only limited quantities coming from the United Kingdom, India and the United States.

Coriander of commerce consists of firmly-united mericarps in a characteristically spherical appearance. They are yellowish-brown colored often with a purplish-red tinge. In the United Kingdom and Morocco the plant produces large fruits which are lacking in essential oil compared with the much smaller fruited varieties from east-

FIG. 3.49. CORIANDER PLANT (*Coriandrum sativum*, L.) READY FOR HARVESTING

FIG. 3.50. CORIANDER MAGNIFIED X 3

ern Europe. The yield of essential oil is up to 1.0% depending on source. The small fruited Russian coriander gives 0.8–1.0% whereas the large fruited coriander from the United Kingdom and Morocco yields only about 0.2%. There are small differences in profile between the oils from different sources but these are not significant or diagnostic.

Profile of Oil of Coriander

Origin	Commercially distilled from Russian Coriander.	
Odor	Impact:	3
	Initial:	WARM, SWEET, MUSTY. ROSE-like; floral, oriental. PENETRATING but not irritating. FRUITY; lemony, terpeney. Slightly BALSAMIC; woody.
	Persistence:	The oil airs-off with little change in character and is totally evaporated in 24 hr.
	Dry-out:	No residual odor.
Flavor		WARMLY SPICY, AROMATIC. SWEET, FRUITY; citrus-like. Slightly FLORAL; rose-like.
	After-taste:	Pleasantly FRUITY.

Components of Oil of Coriander

α-pinene	d-linalool (60–70%)
β-pinene	n-decylic aldehyde
dipentene	geraniol
ρ-cymene	l-borneol
α-terpinene	acetic esters
β-terpinene	

CUMIN (*Cuminum cyminum*, **L.**)

Cumin is the dried ripe fruits of an annual plant (*Cuminum cyminum*, L.) which is native to Egypt and the Mediterranean region. It is now cultivated in India, Morocco, Sicily and Cyprus.

The plant grows to about 2 feet in height and has very divided string-like leaves. The commercial spice consists of a mixture of whole and divided fruits which are oval and rough to the touch owing to surface hairs. They yield 2.5–5% of an amber-yellow essen-

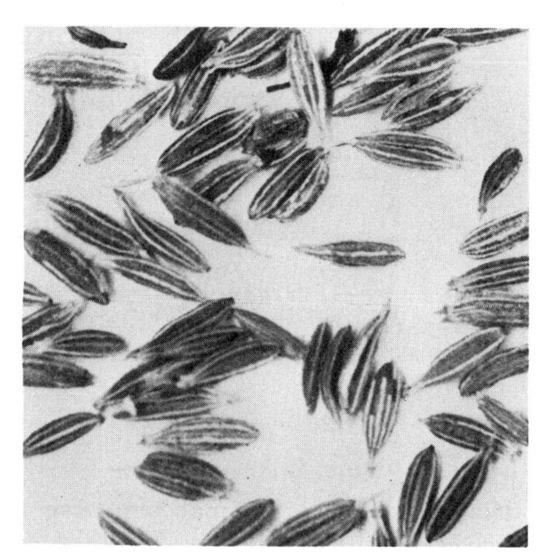

FIG. 3.51. CUMIN (*Cuminum cyminum*, L.) MAGNIFIED X 3

tial oil which is intensely strong and to some even offensive. It is a flavor of particular value in the blending of Indian curry powder. The character, though powerful and dominated by cumin aldhyde, is not unlike that of caraway but is heavy and lacks the pleasing freshness of that spice.

Profile of Oil of Cumin

Origin	Commerically distilled from Indian Cumin.	
Odor	Impact:	5
	Initial:	STRONGLY PENETRATING, IRRITATING. FATTY; oily. OVERPOWERINGLY CUMIN ALDEHYDE-like. SPICY, WARM, HEAVY; curry-like. Slightly PUNGENT.
	Persistence:	Airs-off with little change in profile over 24 hr.
	Dry-out:	Persistently characteristic.
Flavor		WARM, HEAVY, SPICY, very CURRY-like.

	Slightly BITTER.
	PUNGENT.
After-taste:	Pleasantly SPICY, WARM.

Components of Oil of Cumin

α-pinene	cuminyl alcohol
β-pinene	cuminaldehyde (40–65%)
ρ-cymene	
β-phellandrene	

Tassan and Russell (1975) determined that there are four aldehydes influencing the main odor impact of cumin. One aldehyde in particular, 3-*p*-menthen-7-al, was found to be responsible for the characteristic odor of heated cumin.

DILL (*Anethum graveolens*, **L.**)

Dill is the dried ripe fruits of a small annual or biennial plant (*Anethum graveolens*, L.) a native of the Mediterranean region and Russia. It is now cultivated in Germany, Hungary, The Netherlands, India, Pakistan and in the United States. The plant grows up to 3 ft in height and has distinctive, deeply-divided leaves. The flowers and fruits are borne in a compound umbel.

The commercial spice is usually separated mericarps which are light brown in color, oval and very compressed so as to be almost flat. The fruits yield 2.5–4% of a yellowish essential oil rich in d-carvone. The odor of the oil is somewhat reminiscent of caraway but is flatter and less agreeable.

Dill Weed

The whole plant is very aromatic and the green herbaceous tops together with the formed but unripe seeds are harvested and distilled to yield a pale yellow, strongly aromatic essential oil which is extensively used in the flavoring of pickled cucumbers (dill pickles).

The profile of the seed and weed oil is quite different and these are compared below. As with most of the herbs, the quality of the oil depends on the correct time of harvesting and distillation for, as the fruits ripen, the character of the oil more nearly approaches that of the normal seed oil. The carvone content of the weed oil is much lower than that of the seed oil (20–40%) so that the terpenoid fraction is higher and the oil is, in consequence, milder and more pleasingly fresh.

Profile of Oil of Dill Seed

Origin Commercial sample from German Dill.

Courtesy of Bush Boake Allen, Ltd.

FIG. 3.52. A FIELD OF DILL READY FOR HARVESTING

FIG. 3.53. DILL (*Anethum graveolens*, L.) MAGNIFIED X 3

Odor	Impact:	3
	Initial:	SWEET, pleasingly FRUITY. FRESH. CARAWAY-like. MEDICINAL. AROMATIC, slightly GREEN.
	Persistence:	The sweet freshness is quickly lost and replaced by a fuller but flatter warm spicy note with a suggestion of anise.
	Dry-out:	Very slight FLAT, HERBACEOUS note.
Flavor		Pleasantly AROMATIC, CARAWAY-like. MEDICINAL; carminative, smooth. WARM, SPICY. Slightly FRUITY. Slightly MINTY.
	After-taste:	None.

Profile of Oil of Dill Weed

Origin	Commercial sample of American distilled Oil of Dill Weed.	
Odor	Impact:	3
	Initial:	FRESH, GREEN, HERBACEOUS. AROMATIC. Strongly WARM, SPICY. Slightly MEDICINAL. Reminiscent of CARAWAY.
	Persistence:	The fresh green top-notes are lost after about 2 hr airing being replaced by a more mellow herbaceous minty character.
	Dry-out:	Smoothly HERBACEOUS with MINTY back-note.
Flavor		Pleasantly AROMATIC. FRESH, GREEN, HERBACEOUS. Slightly BURNING.
	After-taste:	Freshly HERBACEOUS.

Components of Oil of Dill

d-limonene	d-carvone
α-phellandrene	(seed oil, 42–60%)
α-pinene	(weed oil, 20–40%)
dipentene	dihydrocarvone

The typical odor and flavor of the weed oil are chiefly due to its content of α-phellandrene and the higher the content of this constituent the more nearly the oil resembles the freshly-cut herb. Weed oils with 20% or less of carvone are preferred as having the finer and more appetizing herby flavor which is so appreciated in dill pickle.

INDIAN DILL

Indian Dill is derived from *Anethum sowa*, Roxb., a plant related to *Anethum graveolens* but native to northern India and now widely cultivated as a fresh green herb throughout southeast Asia and Japan. The essential oil of Indian dill is characterized by containing dillapiole and a much lower percentage of carvone. The fruits yield 1.2–3.5% of essential oil (Fenaroli 1975).

Components of Oil of Indian Dill (Baslas and Gupta 1971)

α-pinene (about 5%)	dihydrocarvone (7–14%)
d-limonene (9–34%)	d-carvone (20–46%)
d-phellandrene (about 11%)	myristicin (about 1%)
α-terpinene (about 4%)	eugenol (about 3%)
caryophyllene (about 4%)	apiole (about 6%)
	dillapiole (9–39%)

The figures reveal a very wide variation in the aromatic quality of this spice depending on cultural and harvesting conditions.

Traces of anisic aldehyde, anethole, thymol, and eugenol have also been reported as present.

FENNEL (*Foeniculum vulgare*, Miller)

Fennel is the dried ripe fruits of cultivated varieties of *Foeniculum vulgare*, Miller, a biennial or perennial plant which grows to some 5 ft in height. Sweet Fennel (var. *dulce*) is a smaller variety which is widely cultivated throughout the Mediterranean countries where it is highly regarded as a vegetable rather than for its yield of fruits. Bitter Fennel (var. *amara*) is also cultivated but to a more limited extent as it is not of such culinary interest as the sweet variety. The dried seeds are steam distilled to give 3–4% of a pale yellow essential oil rich in anethole; in consequence, the flavor of the seeds resembles that of anise but with a more camphoraceous herby character.

FIG. 3.54. FENNEL PLANTS (*Foeniculum vulgare*, Mill.)

FIG. 3.55. FENNEL SEED MAGNIFIED X 2

Profile of Oil of Fennel

Origin	Commercial sample.	
Odor	Impact:	4
	Initial:	AROMATIC, pleasingly FRESH. WARM, SPICY; soft, mild. ANISE-like. Slightly CAMPHORACEOUS. Slightly GREEN; herbaceous.
	Persistence:	The profile slowly changes with the loss of the fresh top-notes and a marked increase in the heavier camphoraceous character.
	Dry-out:	ANETHOLE-like with GREEN back-note.
Flavor		WARM, SPICY; aromatic. ANISE-like. GREEN, HERBACEOUS. Initially BITTER but then SWEET.
	After-taste:	FRESH with trace of BITTERNESS.

Components of Oil of Fennel

α-pinene	anethole (50–60%)
camphene	fenchone (10–15%)
d-α-phellandrene	ρ-hydroxy phenyl acetone
dipentene	methyl chavicol

PARSLEY (*Petrosilinum sativum*, **Hoffm.**)

There is very little culinary use for the fruits of the hardy biennial plant *Petrosilinum sativum*, Hoffm., or *P. hortense* as the more common garden variety is classified; although small quantities are distilled as a source of the essential oil. By far the greatest use for parsley is as an attractive green garnish or in the form of dried parsley flakes as a seasoning in a variety of food products.

The fruits yield 1.5–3.5% of essential oil whereas the freshly harvested herb yields as little as 0.06%. The profiles of the two types of oil are quite different, the most preferred oil being obtained from the first-year, young leaves. As with dill, the distillation of the second-year herb when the fruit is set gives an oil having a profile nearer to that of the seed oil. Commercially, the latter option is the

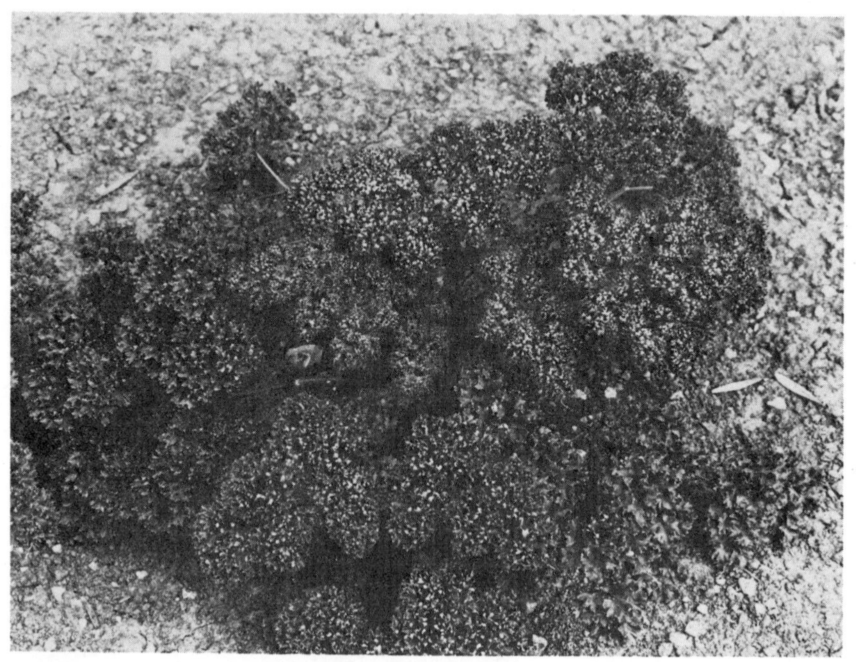

FIG. 3.56. PARSLEY HERB PLANT (*Petrosilinum sativum,* Hoffm.)

FIG. 3.57. PARSLEY SEED MAGNIFIED X 2

more attractive as the yield of oil is so much higher; but the aromatic quality of the oil is far removed from that of the freshly picked, green parsley leaves.

Comparison of the Profiles of Oil of Parsley Seed and Parsley Herb

	Parsley Seed	Parsley Herb
Odor	WARM	FRESH
	SPICY	HERBACEOUS
	TURPENTINE-like	SWEET
Flavor	WARM	WARM but FRESH
	AROMATIC, SPICY	HERBACEOUS
	BITTER	Slightly BURNING
After-taste	BITTER	CLEAN, FRESH with a little BITTERNESS

Part 4: The Aromatic Barks Containing Cinnamic Aldehyde

Cinnamon and the related bark spices occupy a particular place in the armament of the flavorist as they are compatable with both sweet and savory products. Although there are very many varieties available commercially, these spices may be classified either as cinnamon or cassia depending on their source and nature. All belong to the natural order *Lauraceae.* Some are thick-stemmed bushes growing 5–10 ft in height; others are large trees growing up to 50 ft. The bark is stripped off the trunk and branches; then scraped to remove surface cork and rolled into quills the thickness of which is characteristic of the type of cinnamon. The main distinguishing features are given in Table 3.7.

CINNAMON (*Cinnamomum zeylanicum*, **Nees.**)

In the United Kingdom, the term "cinnamon" is taken to mean exclusively the bark of the shoots of coppiced trees of *Cinnamomum zeylanicum,* many different varieties of which exist in Sri Lanka and southern India. The most esteemed bark is derived from cultivated trees in Sri Lanka but an increasingly significant crop is being obtained from plantations in the Seychelles and in Brazil, where the production of this spice has taken a new impetus.

TABLE 3.7
COMMERCIAL VARIETIES OF CINNAMON/CASSIA

Botanical Name	Common Names	Country of Origin	Description and Commercial Grades	Volatile Oil Content
True Cinnamon *Cinnamomum zeylanicum*, Nees.	Ceylon cinnamon Seychelles cinnamon	Sri Lanka Seychelles	Single or double closely packed compound quills, up to 1 m long, 1 cm in diameter and 0.5–1 mm thick. Surface dull yellowish-brown, longitudinally ridged. Also as quillings, featherings and chips. Nominated grades (quills): 00000 Special (highest grade) to 0 (coarsest grade). Other gradings are sometimes used.	1–2%
Cassia-type Cinnamon *Cinnamomum cassia*, Blume.	Chinese cassia China Junk cassia Cassia lignea The following local names may also be used: Canton Kwantung Kwansi Yunnan Honan	China	Kwantung: Single whole quills or channelled pieces, 20–40 cm long, 12–18 mm in diameter and 1–3 mm thick. Outer surface dark brown with greyish cork patches. Nominated grades: Selected quills Selected broken Kwansi/Honan: Similar to the above but graded: thin quills medium quills thick broken	1–2%

Cinnamomum loureirii, Nees.	Saigon cinnamon The following local names may also be used: Vietnam Danong Annam Tonkin	Laos Cambodia Vietnam	Quills about 30 cm long, 5 cm in diameter and 0.5–0.7 mm thick. Outer surface light greyish-brown to dark reddish-brown, surface warty and longitudinally ridged with crustose lichen patches. May occur in flattened pieces up to 10 mm thick. Nominated grades: thin medium thick broken	2.5–4%
Cinnamomum burmanii, Blume.	Batavia cassia Padang cassia Korintje cinnamon The following local names may also be used: Indonesia Java Macassar Timor	Indonesia Padang–the lowlands of Sumatra. Korintje–the uplands of Sumatra.	Smooth regular single quills up to 30 cm long, 12 mm in diameter and 1 mm thick. Light reddish-brown in color. Graded on thickness and color of quills: Padang: AA, A, B, C (broken) Korintje: A, B, C (broken)	Padang: 0.5–1.8% Korintje: 1–1.5%

The tree is a bushy evergreen usually grown from seed. The first crop of bark is available after 2–3 years. Under cultivation conditions, the trees are not allowed to grow above 8 ft in height and are cut back or coppiced each year. The whole of the outer bark is peeled from the cut branches and the inner wood and outer corky layers very carefully removed by scraping. The thin strips of inner bark are laid out in the sun to dry and are then packed together to make "compound quills" which may be anything up to 1 m in length. In the Seychelles, the bark is taken every two years from older trees and hence the quills are often thicker and usually single, not compound.

The total trade in cinnamon bark is about $9 million and in cassia bark about $6.5 million. The exportation of cinnamon from Sri Lanka is of prime importance to that country's economy and the average figure for exports of this spice is 2.7 thousand tons per annum. Imports of this type of cinnamon into the United States account for less than 10% of the total annual cinnamon imports much of it being re-exported to Mexico.

Cinnamon bark contains 0.9 to 2.3% essential oil, depending on the variety.

Profile of Oil of Cinnamon (Ceylon-type)

Origin	Commercially distilled in the U.K. from bark imported from Sri Lanka.	
Odor	Impact:	4
	Initial:	SWEET, pleasantly AROMATIC. WARM, SPICY. IRRITATING, PUNGENT. Slightly WOODY. Slightly CLOVE-like.
	Persistence:	The oil airs-off very slowly with little change in profile under 24 hr.
	Dry-out:	TERPENEY, PINY with slight FLORAL back-note.
Flavor:		WARM, SPICY. Pleasantly SWEET, AROMATIC. Slightly PUNGENT, BURNING. Very slightly BITTER.

After-taste: WARM, SPICY and very pleasing.

Components of Oil of Ceylon Cinnamon

caryophyllene	cinnamic aldehyde (65–75%)
l-phellandrene	l-linalool
ρ-cymene	furfural
l-α-pinene	methyl amyl ketone
	nonyl aldehyde
	benzaldehyde
	hydrocinnamic aldehyde
	cumin aldehyde

Wijesekera *et al.* (1974) carried out a comparative study of the essential oils from the bark, leaves and roots of *C. zeylanicum* and confirmed that the same terpenes were present in each oil but in different proportions. They also reported the presence of benzyl benzoate in the bark oil.

CASSIA

The classification of the several species of *Cinnamomum* comprising the so-called cassia-type cinnamon is difficult owing to the problems of establishing the exact source of the spice. The cassia group is native to China, Vietnam and Indonesia and the commercial spice is available from all of these regions.

Cassia barks are known under several different names in the spice trade but throughout North America are called simply "cinnamon" without qualification. The following are the major sources:

Saigon Cinnamon (*Cinnamomum loureirii*, Nees.)
Cultivated in Vietnam and exported via Saigon from which port it takes its name.

Korintje Cinnamon
From the uplands around Korintje in Sumatra.

Padang or Batavia Cinnamon (*Cinnamomum burmanii*, Blume)
Cultivated in the lowland areas along the west coast of Sumatra and exported via the port of Padang.

Although there is a generic likeness in the aromatic profiles of the various species and varieties of cassia and cinnamon they do differ considerably and it is not uncommon for spice traders to make blends of the ground material to suit particular requirements. The names given commercially to cinnamon do not mean very much as they only loosely indicate the port of export and not the growing region from which the bark originated.

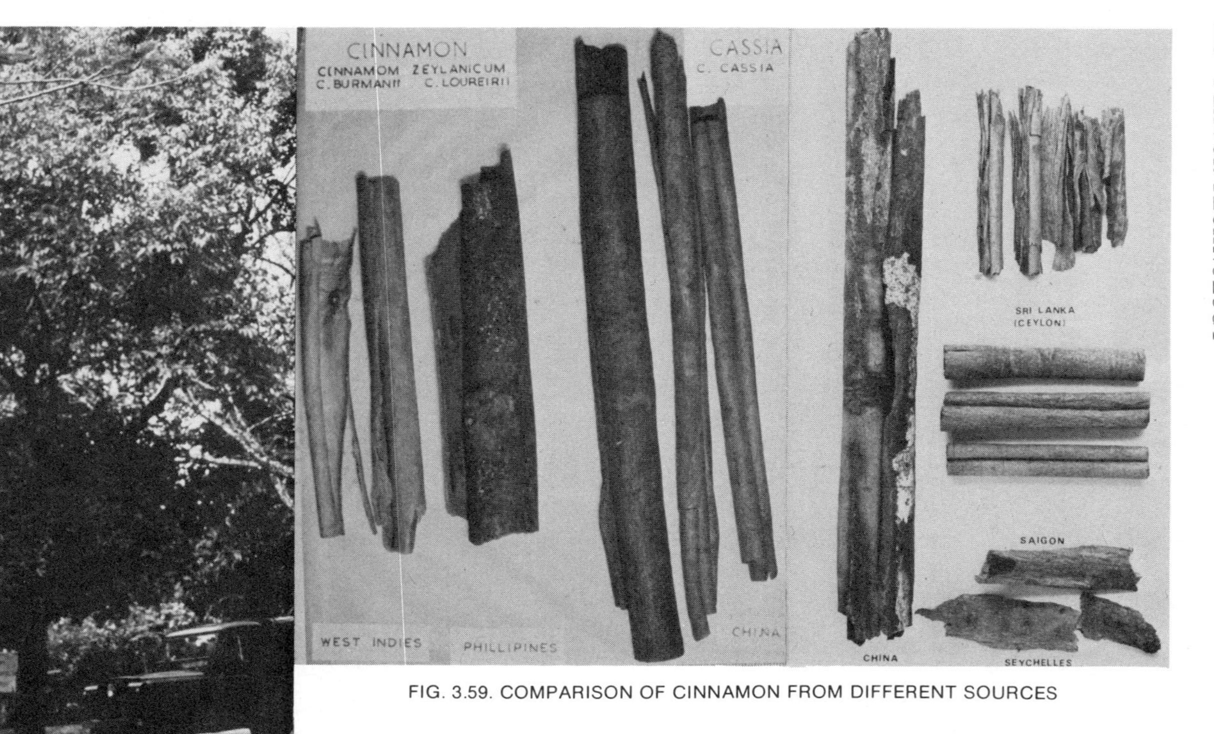

Courtesy of Dr. Wm. Mitchell

FIG. 3.58. CINNAMON TREES (*Cinnamomum zeylanicum, Nees.*)

FIG. 3.59. COMPARISON OF CINNAMON FROM DIFFERENT SOURCES

Courtesy of Bush Boake Allen, Ltd.

FIG. 3.60. HAND PEELING CINNAMON STICKS TO FORM QUILLS

Courtesy of American Spice Trade Association

FIG. 3.61. PREPARATION OF CINNAMON QUILLS FOR SHIPMENT IN SUMATRA

CHINA CASSIA

Chinese Cinnamon or Cassia is the bark of *Cinnamomum cassia*, Nees. obtained from the provinces of Kwangsi, Kweichow and Kwantung, most of the bark being distilled locally to give the essential oil. The yield of oil has been variously reported to be as low as 0.3% and as high as 2% depending on the ratio of leaves to twigs used. As the oil is stored and transported in lead-soldered tins, the lead content of such imported oils can be higher than is acceptable for food flavorings. It is usual for such oils to be re-distilled to remove any lead contamination, a process which also improves the color of the oil and its aroma and flavor. Unfortunately, most cassia oils are contaminated in one way or another and it is not unusual to find kerosene and other extenders used to cheapen the price. As the oil contains 70–90% of cinnamic aldehyde, some commercially available oils may be adjusted with synthetic cinnamic aldehyde.

Profile of Oil of Cassia—Re-distilled and "Lead-free"

Origin	Re-distilled in the United Kingdom from imported Cassia Oil.	
Odor	Impact:	4
	Initial:	Pleasantly SWEET; to some, slightly SICKLY. SPICY, WARM; smooth. Strongly PUNGENT, IRRITATING. DRY, WOODY; powdery. Slightly FRUITY; almond-like.
	Persistence:	Very little change in profile over 24 hr.
	Dry-out:	FATTY, slightly GREEN.
Flavor		Very SWEET. WARM, BALSAMIC. HARSH, slightly PUNGENT; burning. POWDERY. Slightly CLOVE-like. Slightly BITTER back-note.
	After-taste:	SWEETLY WARM and SPICY.

Components of Oil of Cassia

terpenes similar to those in Ceylon Cinnamon.	cinnamic aldehyde (80–95%) cinnamyl acetate cinnamic acid benzaldehyde

methyl salicylate
methyl *ortho* coumaraldehyde
salicylaldehyde
coumarin

Comparison of the Odors of Oil of Ceylon Cinnamon and Oil of Cassia (Re-distilled)

Ceylon Cinnamon	Cassia	Cassia Oil
TERPENEY LIGHT, FRESH WARM, SPICY	FLAT but SPICY *Less* WARM *More* FULL-BODIED, HEAVY, POWDERY	(laboratory distilled from genuine cassia bark) SMOOTH LIGHT, TERPENEY SWEETLY SPICY WOODY

CINNAMON LEAVES

The leaves of the cinnamon tree are also aromatic and yield an essential oil rich in eugenol. Such oils have a similar flavoring effect as that of Oil of Clove, particularly with oils obtained from the leaves and stems of the clove tree. Cinnamon Leaf Oil will be considered in the following group so that these aromatic relationships may be better compared.

Part 5: The Phenolic Spices Containing Eugenol

The distinctive note of clove buds and allspice (pimento) berries and the essential oils derived from clove leaves, clove stems, pimento leaves, cinnamon leaves and West Indian bay leaves is that of eugenol, the principal constituent in the essential oils.

CLOVE

Cloves are the dried unopened buds of an evergreen tree (*Eugenia caryophyllata,* Thunb., formerly classified as *Caryophyllus aromatis,* L.). By far the biggest clove producing country is Tanzania (formerly Zanzibar and the island of Pemba) which provides about 80% of the world crop. Other countries also producing this spice are the Malagasy Republic, Indonesia and Sri Lanka.

Clove trees are slow growing and do not produce the first crop of flowers until 6 years after planting, the maximum yield not being

Courtesy of Shell, Ltd.

FIG. 3.62. SORTING AND SPREADING CLOVE BUDS FOR SUN DRYING IN TANZANIA

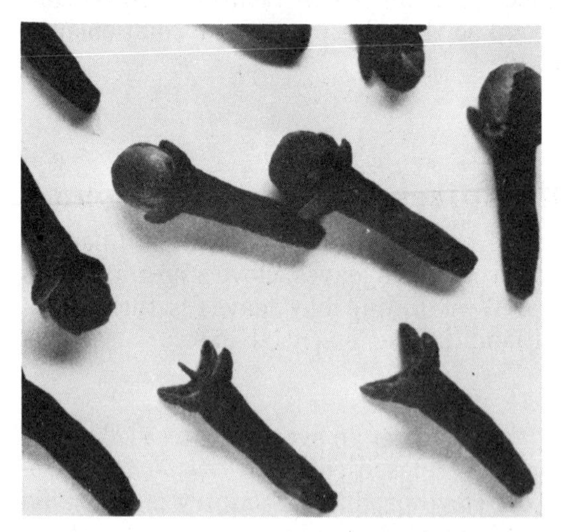

FIG. 3.63. CLOVE (*Eugenia caryophyllata*, Thunb.) SHOWING BOTH CLOVE BUDS AND BLOWN CLOVES (MAGNIFIED X 2

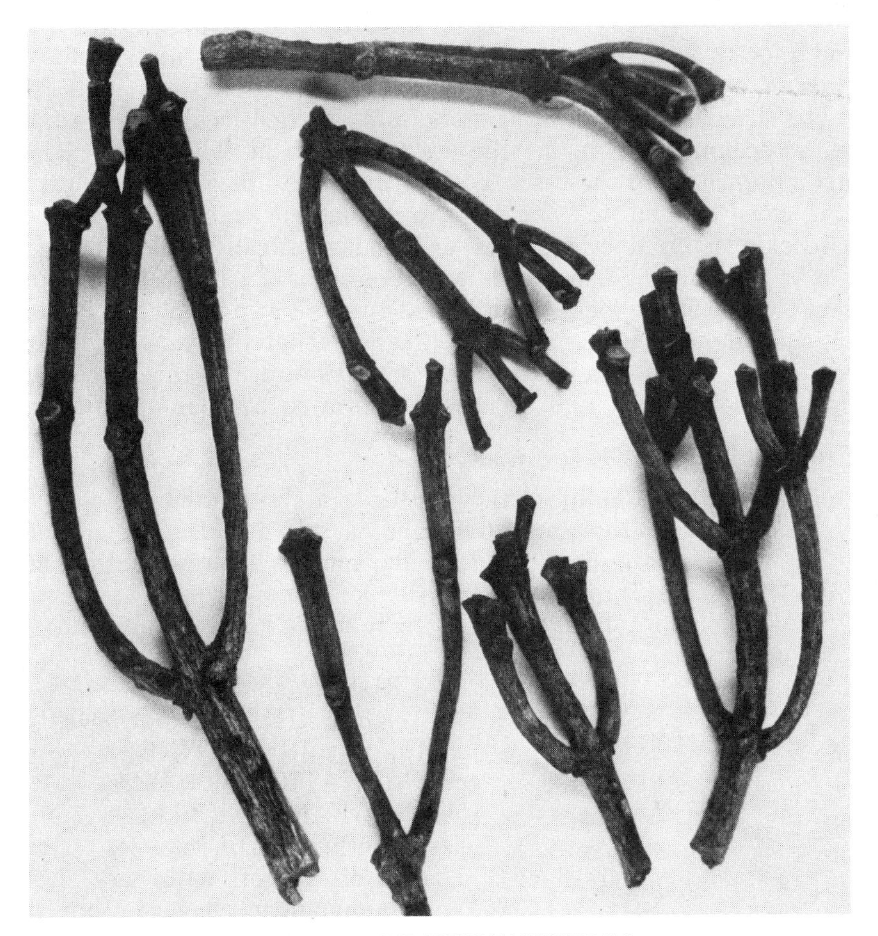

FIG. 3.64. CLOVE STEMS MAGNIFIED X 2

obtained until the trees are some 20 years old. The unopened buds, borne in clusters of 10 to 15, are picked by hand and the buds separated from the short stems prior to sun drying. During the drying process the buds change from green to a deep brown and assume the familar characteristic appearance. The average yield of cloves is only about 3 kg per tree. The total world crop is variable and cyclic as the clove tree exhibits a periodicity of production which is not yet fully understood. On average, the total crop amounts to some 20,000 metric tons per annum.

Clove buds, the most esteemed source of Clove Oil, contain 16–17% of essential oil, the nature of which depends to some extent upon the method of distillation. Normally, the buds are steam distilled and the oil then contains 91–95% of eugenol; oils having a finer aroma and a lower eugenol content are obtained by water distillation of the buds.

The dried stems or flower stalks are distilled locally as are the leaves accumulated during the harvesting of the flower buds. The oils obtained from these sources are much less pleasant than genuine clove bud oil and are also very much cheaper. Such oils are used extensively to create cheaper blends of so-called "Clove Oil" to suit particular end-user needs or are used as a source of pure eugenol. Clove stems yield about 6% essential oil containing 80–85% of eugenol and the leaves 2–3% oil having about the same phenolic content. It is usual to rectify these oils before use as this increases the eugenol content and markedly improves the odor and flavor.

Profile of Oil of Clove Bud

Origin	Commercially distilled in the United Kingdom from imported Zanzibar clove buds. The oil had been matured for four months before examination.	
Odor	Impact:	3
	Initial:	SPICY, PEPPERY; pungent. SWEET. FRUITY; apple-like. WARM, HEAVY, PHENOLIC. Slightly IRRITATING. CARNATION-like, FRAGRANT, FLORAL. Slightly ACIDIC; acetic.
	Persistence:	The oil airs-off with very slow change in character but with the floral notes becoming pronounced. Even after 5 days the residual odor is moderately strong and identifiable.
	Dry-out:	PHENOLIC.
Flavor		BURNING, very WARMING. PHENOLIC, MEDICINAL; dental. SPICY, FRUITY back-note. Slightly ASTRINGENT.
	After-taste:	Slightly BITTER and very WARM leaving the mouth DRY.

The flavor of ground clove differs from that of the oil owing to the presence of tannins. The initial warm, spicy notes are quickly overtaken by a marked astringency and an unpleasant bitterness which tends to persist. The after-effect of eating clove is a warm, numbing accompanied by bitterness.

Components of Oil of Clove Bud

caryophyllene (5–10%)

eugenol (70–90%)
eugenyl acetate (10–17%)
methyl-n-amyl ketone
caryophyllene oxide
methyl salicylate

Profile of Rectified Clove Leaf Oil

Origin Clove Leaf Oil imported from the Malagasy Republic and rectified in the United Kingdom.

Odor

Impact: 3

Initial: SPICY, PEPPERY; pungent.
SWEET.
DRIED INK-like, slightly METALLIC.
WARM, HEAVY, PHENOLIC.
Slightly CARNATION-like.
GREEN, LEAFY.
WOODY, MUSTY, DRY; damp earth-like.
BREAD-like.

Persistence: The initially pleasing character is lost after about 1 hr to be replaced by an INKY note which persists with an increasing WOODINESS.

Dry-out: Thinly PHENOLIC and METALLIC.

Flavor

Similar to that of Clove Bud Oil but FLATTER and much HARSHER with an absence of FRUITY notes.

After-taste: WARM, strongly ASTRINGENT.

Components of Clove Stem and Leaf Oils

The percentage of free eugenol is in the same order as that in Clove Bud Oil but the stem and leaf oils contain only very small percentages of eugenyl acetate. This results in a less sweet and fruity odor; in consequence, these oils are harsh and coarse by comparison. Napthalene has been reported as present in the stem and leaf oils but has not been identified in the bud oil.

Comparison of the Aroma of Oil of Clove Bud and Clove Leaf with Eugenol

Clove Bud Oil	Clove Leaf Oil	Eugenol
FRUITY	PHENOLIC	ACIDIC
SWEETLY SPICY	WOODY	PHENOLIC
PHENOLIC	DAMP	SMOKY

CINNAMON LEAF

In the preparation of cinnamon bark, the freshly-cut branches of the cinnamon tree (*Cinnamomum* sps) are first stripped of leaf material. If the market price is right these are distilled in local field stills to give a pale yellow oil having a strong smell of eugenol. If the price is not attractive the leaves are merely ploughed in as a green fertilizer. In Sri Lanka the producing season is all the year round but the best oil is obtained after the monsoon period (July-December). The whole process is very primitive but, even so, the total production is in the order of 100,000 lb of oil per annum.

Profile of Rectified Oil of Cinnamon Leaf

Origin	Cinnamon Leaf Oil imported from Sri Lanka and rectified in the United Kingdom.	
Odor	Impact:	3
	Initial:	Strongly SPICY, HEAVY, WARM. PHENOLIC. SWEET, slightly FRUITY. LEATHER-like. EARTHY, MUSTY; stale, dirty. TERPENEY; paint-like. Slightly PUNGENT; irritating.
	Persistence:	The initial harsh character is only slowly lost and the profile becomes more clove-like.

	Dry-out:	FLAT with EARTHY, PHENOLIC note.
Flavor		Harshly PHENOLIC, SHARP.
		WARM, SPICY.
		SWEET but with definite suggestion of BITTERNESS.
		ASTRINGENT.
	After-taste:	WARM, strongly ASTRINGENT.

Comparison of the Aromas of Ceylon and Seychelles Cinnamon Leaf Oils

Ceylon	Seychelles
LEATHER-like	*More* CLOVE-like, SPICY
TERPENEY, PAINT-like	*More* ACIDIC, HARSH
Slightly FRUITY	*More* MUSTY, DRY, MALT-like
	Less LEATHER-like
	Less FRUITY

Components of Cinnamon Leaf Oil

Cinnamon leaf oils are frequently adulterated or "extended" with kerosene or other diluents. In a freshly distilled oil the chief components are:

terpenes (about 7.5%)	eugenol (75–95%)
α-pinene	safrole (about 0.5%)
β-phellandrene	cinnamic alcohol and esters (about 7%)
α-caryophyllene	
β-caryophyllene	benzyl benzoate (about 27%)

ALLSPICE (PIMENTO) (*Pimenta dioica*, L.)

Allspice or pimento is the dried unripe berries of a large evergreen tree (*Pimenta dioica*, L., formerly classified as *Pimenta officinalis*, Lindl) which is native to the West Indies and parts of Central America. The bulk of the world's crop now comes from Jamaica; indeed the name often used for this spice is "Jamaican Pepper." Most of the crop is sun-dried and exported as such, the distillation of the essential oil taking place in the United Kingdom, the United States and Germany. Only a very small quantity of the oil is distilled locally. The leaves of the pimento tree are also aromatic and these are collected for distillation.

The berries yield about 3–4% of a yellow oil which, like clove oil, is heavier than water owing to its high eugenol content.

Courtesy of Tropical Products Institute

FIG. 3.65. ALLSPICE (PIMENTO) BERRIES IN JAMAICA

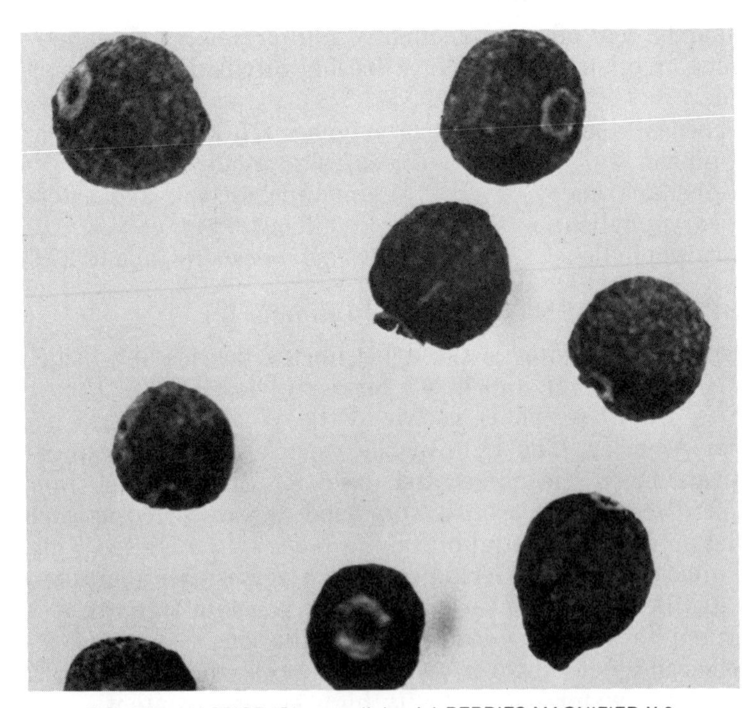

FIG. 3.66. ALLSPICE (*Pimenta dioica*, L.) BERRIES MAGNIFIED X 3

Profile of Oil of Pimento Berries

Origin	Distilled commercially in the United Kingdom from imported Jamaican pimento berries.	
Odor	Impact:	3
	Initial:	SWEET, SPICY, WARM.
		CINEOLIC, LIGHT; clean, ethereal.
		PHENOLIC, CLOVE-like; medicinal.
		MENTHOLIC, COOLING; camphoraceous.
		DAMP, WOODY; stale, musty.
		Slightly FLORAL; scented, tea-like.
		TOFFEE-like; sickly.
	Persistence:	The light top-notes are unusually persistent but after about 5 hr the profile becomes duller and more heavily phenolic and clove-like.
Flavor		Strongly ASTRINGENT, DRYING.
		WARM, SPICY.
		Slightly PUNGENT, PEPPERY.
		Very slightly FRUITY.
	After-taste:	Persistent ASTRINGENCY leaving mouth unpleasantly DRY.

Comparison of the Aromas of Oil of Pimento Berry and Leaf.

Pimento Berry	Pimento Leaf
WARM, SPICY	*Less* WARM, CLOVE-like
Strongly PHENOLIC	*More* CINNAMON-like, SPICY
Full-bodied	*More* WOODY, DRY
	Much thinner

Components of Oil of Pimento Berry and Pimento Leaf

Nabney and Robinson (1972) reported the following components in the berry oil:

α-pinene 1,8 cineole

thujene
β-pinene
Δ-3-carene
myrcene
α-phelladrene
γ-terpinene
limonene
ρ-cymene
isocaryophyllene
β-caryophyllene
β-elemene
alloaromadendrene
α-terpineol
γ-muurolene
α-selinene
β-selinene
curcumene
calamenene
terpinolene-4,8-oxide
linalool
terpinen-4-ol
methyl eugenol (8.8%)
caryophyllene oxide
β-caryophyllene alcohol
eugenol (68.6%)
chavicol

Veek and Russell (1973) carried out an extensive examination of the leaf oil reporting the presence of many of the same compounds. They also carried out a sensory assessment of the effluent from the gas chromatograph using an agreed list of descriptive terms similar to those employed in sections of this chapter. They, too, note the difference in sensitivity between the chromatographic detector and the human nose, obtaining responses in areas where only trace peaks or no peaks are recorded.

WEST INDIAN BAY (*Pimenta racemosa*, Miller)

The Bay tree, formerly classified as *Myrcia acris*, D.C., grows wild throughout the West Indies but is also cultivated both in Dominica and Puerto Rico. The leaves are rarely offered as such but are distilled locally to produce some 35,000 to 50,000 lb of a dark yellowish-brown essential oil per annum. In view of the nomenclature, care should be taken not to confuse this oil with that of Sweet Bay Laurel (*Laurus nobilis*, L.) which has a quite different profile.

In Puerto Rico the leaves are reported to yield 1–2.75% oil but figures of up to 3.5% have been obtained. The oil is often very dark owing to iron contamination; and commercial lots are often wet and dirty. A much better oil is obtained by re-distillation.

Profile of West Indian Bay Oil

Origin Commercial sample.
Odor Impact: 5

	Initial:	HARSH, PUNGENT, SHARP, very PENETRATING.
		SPICY, WARM.
		GREEN, COOLING; fresh, leafy.
		PHENOLIC, CRESOL-like; medicinal, tar-like.
		SWEET to some pleasant, to others SICKLY.
		Slightly FRUITY; estery.
	Persistence:	Little change in profile over 5 hr, the initial very penetrating character is slowly lost to be replaced by a sweet balsamic character.
	Dry-out:	GREASY, DAMP, EARTHY.
	Hedonic reaction:	The panel showed a very marked difference of opinion on this oil ranging from pleasant to obnoxious.
Flavor		BITTER.
		PHENOLIC, CRESOL-like.
		Slightly PUNGENT; tingling.
		ASTRINGENT; drying.
		Slightly FRUITY; lemony.
	After-taste:	Persistently ASTRINGENT and BITTER.

Components of West Indian Bay Oil

α-pinene	eugenol	total phenols 50–65% (eugenol + chavicol)
myrcene	chavicol	
α-phellandrene	cineole	
limonene	methyl chavicol	
	methyl eugenol	

Some samples of West Indian Bay Oil have a phenolic content far in excess of 65%. These should be suspect as they may have been adjusted by the addition of clove oil.

Part 6: The Colored Spices

Although possessing aromatic qualities the members of this group are used more for their coloring power than for their flavoring effects.

Courtesy of Bush Boake Allen, Ltd.

FIG. 3.67. HUNGARIAN PAPRIKA: (LEFT) HARVESTING; (CENTER) SORTING; (RIGHT) PACKING AND SHIPPING

PAPRIKA

Paprika is another member of the large *Capsicum* family and is the large fleshy fruits of *Capsicum annuum*, L., cultivated in Hungary, Spain, the United States, Mexico and in several other parts of the world. There are several different varieties available in the form of a moderately fine powder. These are classified as sweet, semisweet, mild and pungent depending on quality of the start material and the proportion of various parts of the fruit which are included in the final grinding. The outer fleshy pericarp carries the most color, the inner tissues and seeds carry the pungent principle—capsaicin. The color, too, shows marked graduation depending upon how much of the washed seeds are incorporated into the mix. The most esteemed grade of paprika is "extra" and is derived entirely from the pericarp, this grade has no pungency, a very delicate flavor and a high level of color. Paprika containing the powdered seeds may go rancid on storage owing to the presence of fixed oil and for this reason paprika should always be stored in well-closed containers in a cool place away from direct heat or sunlight.

The various varieties and grades are well described by Parry (1969).

SAFFRON (*Crocus sativus*, **L.**)

Saffron is the dried golden stigmas of a small crocus which is a member of the iris family. The cultivation of this plant dates back to antiquity for saffron has always occupied a very special place for medicinal and culinary use as well as being an attractive yellow dye. Today, the main commercial crops are obtained from Spain, France, northern India and China with smaller quantities coming from several other European countries.

The reddish-lilac flowers appear in October and each has a long bright red stigma. Usually the whole flowers are gathered and the stigmas removed by hand; these are then dried with the aid of gentle heat. Harvesting is difficult and has to be carried out speedily as the life of the flowers is short. The yield of stigmas is small and it is not surprising that this is the most expensive of the spices when it takes 30,000 to 35,000 hand-picked blooms to give just 1 lb of dried saffron.

Constituents of Saffron

Saffron contains three prime constituents:

(a) A volatile oil (0.5–1%) having an intense odor of the spice.

FIG. 3.68. SAFFRON (*Crocus sativus*, L.) MAGNIFIED X 2

(b) A colorless, bitter glycoside, picrocrocin.
(c) A highly-colored glycoside, crocin.

It is this latter constituent which gives to the spice its characteristic color. Crocin is a yellowish-brown powder readily soluble in hot water; its coloring power is intense and only very low levels of usage are necessary in food products.

The Flavor of Saffron

Saffron is not to everyone's taste. The flavor is delicate with a sharply bitter back-note and in certain areas is particularly esteemed; but to most, it is an acquired taste. The odor is faint but somewhat reminiscent of iodine. The two Spanish varieties commercially available are considered to have the finer aroma and flavor. Saffron is particularly popular in European dishes such as bouillabaisse and Spanish fried rice.

MEXICAN SAFFRON, AMERICAN SAFFRON, SAFFLOWER

The tubular florets of the annual herb *Carthamus tinctorius*, L. have for long been known as "Bastard Saffron," but are now more

FIG. 3.69. AMERICAN SAFFLOWER (*Cartbamus tinctorius*, L.)

generally known by the names given in the above heading. Mexican saffron bears no relationship to genuine saffron as it is the flowers of a thistle-like plant belonging to the natural order *Compositae* and growing in the central highlands of Mexico.

The dried florets are deep orange in color and have a strong aromatic and rather unpleasant odor with a bitter taste. They are widely used as a source of natural yellow color but they lack the fine delicate flavor associated with genuine saffron. The colored constituent is safflor-yellow which is present at up to 30%. A red glucoside, safflor-red, is also present but at only about 0.5%.

TURMERIC (*Curcuma longa*, L.)

Turmeric is the dried rhizome of an herbaceous perennial, a member of the natural order *Zingiberaceae* and closely related to ginger. This spice has a well-defined aroma and flavor but is used more for its strong tinctoral power. The color is bright yellow but varies from pure yellow to deep orange depending upon the source of the spice and the acidity of the medium in which it is used. The color of turmeric is very pH sensitive and is a good indicator. The clear yellow color is only developed in acid media.

Like ginger, turmeric is widely cultivated throughout the tropics,

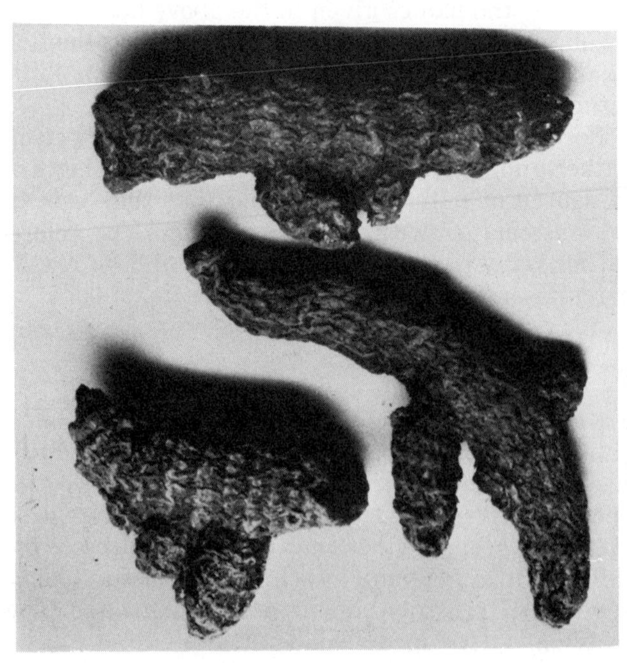

FIG. 3.70. ALLEPPY TURMERIC (*Curcuma longa*, L.) MAGNIFIED X 2

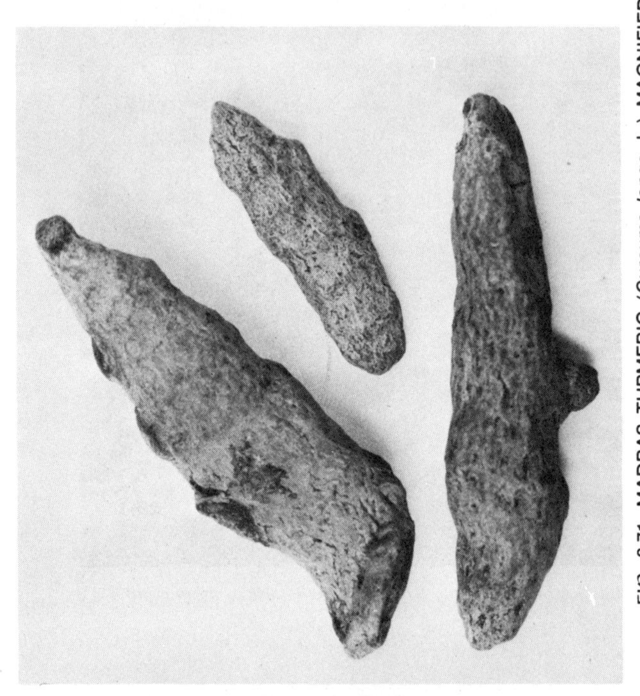

FIG. 3.71. MADRAS TURMERIC (*Curcuma longa*, L.) MAGNIFIED X 2

the main commercial crops being obtained from India (Madras and Alleppy), Sri Lanka and the East Indies with lesser quantities coming from China and Jamaica. The rhizomes are dug up when the leaf has died down, they are cleaned of fibrous roots, washed and then either boiled or steamed to inactivate them. They are then sun-dried.

Constituents of Turmeric

Turmeric has two main constituents of value in food:
(a) 1.5–5% of an orange-yellow, highly aromatic essential oil.
(b) An orange-yellow compound—curcumin.

The rhizomes are also rich in starch, much of which is gelled by the postharvest treatment; they also contain some fixed oil, protein and resins.

Profile of Oil of Turmeric

Although this essential oil is not particularly liked, the aroma and flavor of this spice is of significance in the profile of most curry powders. There is no great demand for the essential oil although it is available as a by-product of the deodorizing of turmeric oleoresins (q.v.).

Odor	Impact:	4
	Initial:	FLAT, METALLIC.
		SPICY, WARM.
		Strongly PENETRATING, IRRITATING.
		FRESH, ORANGE-like; terpeney.
		GREEN with flat PAINT-like back-note.
		HEAVY, ORIENTAL.
		Slightly GINGER-like.
	Persistence:	The oil airs-off very slowly with little change in profile.
	Dry-out:	HEAVY, EARTHY.
Flavor		BITTER.
		PUNGENT.
		AROMATIC, SPICY: curry-like.
		METALLIC, GREEN back-notes.
	After-taste:	Warmly SPICY, BITTER.

STABILITY OF NATURAL COLORS

It is worthy of note that none of the natural colors included in this group of spices has anything like the stability to light as do the synthetic permitted food colors. It is not unusual to see products containing these natural materials very badly spoiled through exposure to sunlight.

BIBLIOGRAPHY

Spices

ANON. 1974. Flavouring substances. II. Br. Food J. *76*, No. 860, 133–135, 143.

FENAROLI, G. 1975. Handbook of Flavour Ingredients, 2nd Edition, T. E. FURIA and N. BELLANCA (Editors). Chemical Rubber Co., Cleveland, Ohio.

GUENTHER, E. 1949. The Essential Oils, Vol. I-VI. D. Van Nostrand & Co., New York.

HEATH, H. B. 1972. Tropical Products Institute conference on spices. Flavour Ind. *3*, 294–303. (Note: The above is a detailed review of Tropical Prod. Inst. Conf., London, Apr. 10–14, 1972. Complete conference papers available from Tropical Products Institute, Grays Inn Road, London, England.)

HEATH, H. B. 1973–1974. Herbs and spices—a bibliography. Flavour Ind. *4*: Anise and star anise, Feb., 65; Bay (W. I.), Feb., 66; Capsicum, Feb., 68–69; Caraway, Apr., 169; Cardamom, Apr., 169; Celery, Apr., 170; Cinnamon/Cassia, Apr., 171; Clove, Apr. 171; Coriander, Apr., 172; Cumin, Apr., 172; Dill, May, 217; Fennel, May, 217–218; Fenugreek, May, 218; Ginger, June, 263–264; Lovage, June, 264; Mustard, June 265–266; Nutmeg/Mace, Aug., 346; Parsley, Sept., 394; Paprika, Aug., 348; Pepper, Sept., 394; Pimento (allspice), Sept., 396. Flavour Ind. *5*: Saffron, No. 3/4, 79; Turmeric, No. 5/6, 124.

INTERNATIONAL TRADE CENTRE UNCTAD/GATT. 1970. Markets for Spices in North America, Western Europe and Japan. International Trade Centre, Geneva, Switzerland.

LEWIS, Y. S. 1973. The importance of selectng the proper variety of a spice for oil and oleoresin extraction. Tropical Prod. Inst. Conf. Papers, London, 183–185.

MACLEOD, A. J. 1973. Spices. Chem. Ind. No. 19, 778–780.

MELCHOIR, H. and KASTNER, H. 1974 Spices. A Botanical and Chemical Study. Verlag Paul Parey, Hamburg, Germany.

MONCRIEFF, R. W. 1967. The Chemical Senses, 3rd Edition. Leondard Hill Books, London, England.

OBERDIECK, R. 1973–1974. Essences, aromatic substances and raw materials. I. Aromatic substances in essences from spices, herbs and drugs. Alkohol-Ind. *86*, No. 10, 204–205, 208; *87*, No. 16, 356–358; *87*, No. 17, 381–385. (German)

PARRY, J. W. 1969. Spices, Vol. 1 and 2, 2nd Edition. Chemical Publishing Co., New York.

STRATEN, S. van and VRIJER, F. de. 1973. Lists of Volatile Compounds in Food, 3rd Edition. Rapport *4030*. Central Institute for Nutrition and Food Research, Zeist, The Netherlands.

ZURCHER, K. and HADORN, H. 1974. Apparatus for the isolation of essential oils from spices, fitted with sampling device for gas chromatographic analysis. Mitt. Geb. Lebensmittel-unters. Hyg. *64*, No. 4, 466–469. (German)

Additional References to Individual Spices

Allspice

VEEK, M. E. and RUSSELL, G. F. 1973. Chemical and sensory properties of pimento leaf oil. J. Food Sci. *38*, 1028–1031.

Anise and Star Anise

TABACCHI, R., GARNERO, J. and BUIL, P. 1974. Contribution to the study of the chemical composition of Turkish anise seed oil. Riv. Ital. Essenze, Profumi, Piante Off. *56*, No. 11, 683–698. (French)

Bay, West Indian

MARTINEZ NADAL, N. G., MONTALVO, A. E. and SEDA, M. 1973. Antimicrobial properties of bay and other phenolic essential oils. Cosmet. Perfum. *88*, No. 10, 37–38.

Capsicum

MAGRA, J. A. 1975. Capsicum. Crit. Rev. Food Sci. Nutr. *6*, No. 2, 177–199.

Caraway

ROTHBACHER, H. and SUTEU, F. 1974. Origin and formation of carvenone in caraway oil (*Carum carvi*). Planta Med. *26*, No. 3, 283–288.

Cardamom

BARUAH, A. K. S., BHAGAT, S. D. and SAIKIA, B. K. 1973. Chemical composition of Alleppy cardamom oil by gas chromatography. Analyst *98*, No. 1164, 168–171.
MUKHERJI, D. K. 1973. Large cardamom. World Crops *25*, No. 1, 31–33.
MITRA, C. R. 1975. Important Indian Spices. IV. Cardamom (*Elettaria cardamomum, Amomum subulatum, A. aromaticum, A. xanthioides, Zingiberaceae*) Riechst., Aromen, Koerperpflegem. *25*, No. 11, 326. (German)
MIYAZAWA, M. and KAMEOKA, H. 1975. Composition of the essential and fixed oil from cardamom seed. J. Japan Oil Chemists' Soc. *24*, No. 1, 22–26.
SHANKARACHARYA, N. B. and NATARAJAN, C. P. 1971. Cardamom. Chemistry, technology and uses. Indian Food Packer *25*, No. 5, 28–36.

Celery

KNIGHT, D. W. and PATTENDEN, G. 1975. Synthesis of alkylidenephthalide constituents of celery odor and models for freelingyne. J. Chem. Soc., Perkin Trans. *1*, No. 7, 635–640.

Cinnamon/Cassia

TER HEIDE, R. 1972. Qualitative analysis of the essential oil of cassia (*C. cassia*, Blume). J. Agric. Food Chem. *20*, No. 4, 747–751.

WIJESEKERA, R. O. B., JAYEWARDENE, A. L. and RAJAPAKSE, L. S. 1974. Volatile constituents of leaf, stem and root oils from cinnamon (*Cinnamomum zeylanicum*). J. Sci. Food Agric. *25*, No. 10, 1211–1220.

ZURCHER, K., HADORN, H. and STRACK, C. 1974. Evaluation of the quality of cinnamon by gas chromatographic analysis of the essential oil. Mitt. Geb. Lebensmittelunters. Hyg. *65*, No. 4, 440–452. (German)

Coriander

TASKINEN, J. and NYKANEN, L. 1975. Volatile constituents obtained by the extraction with alcohol-water mixture and by steam distillation of coriander fruit. Acta Chem. Scand. Ser. B *29*, No. 4, 425–429.

Cumin

TASSAN, C. G. and RUSSELL, G. F. 1975. Chemical and sensory studies on cumin. J. Food Sci. *40*, No. 6, 1185–1188.

Dill

BASLAS, B. K. and BASLAS, R. K. 1972. Chemical examination of the oil from *Anethum graveolens* herb. Riechst., Aromen, Koerperpflegem. *22*, No. 5, 155–156, 158. (German)

BASLAS, B. K. and BASLAS, R. K. 1972. Biogenetic relationship between terpenic and non-terpenic constituents of the essential oils of *Anethum graveolens* and *A. sowa*. Riechst., Aromen, Koerperpflegem. *22*, No. 8, 270–271.

BASLAS, B. K. and GUPTA, R. 1971. Chemical examination of essential oils from plants of genus *Anethum* (Umbelliferae)—oil from seeds of East Indian dill. Part II. Flavour Ind. *2*, 363–366.

BELAFI-RETHY, K. KERENYI, E. and KOLTA, R. 1974. Composition of domestic and foreign essential oils. III. Components of dill oil. Acta Chim. Acad. Sci. Hung. *83*, No. 1, 1–13. (Hungarian)

MIYAZAWA, M. and KAMEOKA, H. 1974. The constitution of the essential oil from dill seed. J. Japan Oil Chem. Soc. *23*, No. 11, 746–749 (Japanese)

ZLATEV, S. K. and FEDIN, P. I. 1975. Influence of presowing irradiation by gamma-rays of dill seeds on the yield of essential oil. Riv. Ital. Essenze, Profumi, Piante Off. *57*, No. 4, 248–252.

Ginger

BEDNARCZYK, A. A. 1974. Identification and evaluation of the flavor-significant components of ginger essential oil. Diss. Abstr. Int. B *35*, No. 1, 306. (Available from University of Maryland, College Park, Maryland, U.S.A.).

BEDANRCZYK, A. A. and KRAMER, A. 1975. Identification and evaluation of the flavor-significant components of ginger essential oil. Chem. Senses Flavor *1*, No. 4, 377–386.

BROWN, B. I. 1975. Further studies on ginger storage in salt brine. J. Food Technol. *10*, No. 4, 393–405.

CONNELL, D. W. and McLACHLAN, B. 1972. Natural pungent components. IV. Examination of the gingerols, shogaols, paradols and related compounds by thin-layer and gas chromatography. J. Chromatogr. *67*, 29–35.

CONNELL, D. W. and SUTHERLAND, M. D. 1969. A re-examination of gingerol, shogaol and zingerone, the pungent principles of ginger (*Zingiber offinale*, Roscoe). Aust. J. Chem. *22*, 1033–1043.

KAMI, T., NAKAYAMA, M. and HAYASHI, S. 1972. Volatile constituents of *Zingiber officinale.* Phytochemistry *11,* No. 11, 3377–3381.

KRISHNAMURTHY, N. *et al.* 1970. Essential oil of ginger. Indian Perfum. *14,* Part 1, 1–3.

MASADA, Y. *et al.* 1974. Studies on the constituents of ginger (*Zingiber officinale,* Roscoe) by GC-MS. J. Pharm. Soc. Japan *94,* No. 6, 735–738. (Japanese) *See also* Proc. 4th Int. Congr. Food Sci. Technol. 1a, 84–86.

MATHEW, A. G. *et al.* 1973. Oil of ginger. Flavour Ind. *4,* No. 5, 226–228.

MITRA, C. R. 1975. Important spices of India. III. Ginger (*Zingiberaceae*). Riechst., Aromen, Koerperpflegem. *25,* No. 6, 170. (German)

NATARAJAN, C. P. *et al.* 1972. Chemical composition of ginger varieties and dehydration studies on ginger. J. Food Sci. Technol., India, *9,* No. 3, 120–124.

THOMPSON, E. H., WOLF, I. D. and ALLEN, C. E. 1973. Ginger rhizome; a new source of proteolytic enzyme. J. Food Sci. *38,* No. 4, 652–655.

Horseradish

GILBERT, J. and NURSTEN, H. E. 1972. Volatile constituents of horseradish roots. J. Sci. Food Agric. *23,* No. 4, 527–539.

HANSEN, H. 1974. Glucosinolates in horseradish. Tidsskr. Planteavl. *78,* No. 3, 408–410.

KOJIMA, M. 1973. Evaluation of the quality of Japanese horseradish powder by gas chromatography. J. Food Sci. Technol. Japan, *20,* No. 7, 316–320.

LU, A. T. and WHITAKER, J. R. 1974. Some factors affecting rates of heat inactivation and reactivation of horseradish peroxidase. J. Food Sci. *39,* No. 6, 1173–1178.

Mustard

ROY, U. and DAS GUPTA, S. K. 1975. Note on the changes in some qualitative characters during the last phase of maturation of Indian mustard seed (*Brassica juncea,* Coss). Curr. Sci. *44,* No. 1, 18–19.

KARIG, F. 1975. Procedure for the analysis of mustard seed constituents. J. Chromatogr. *106,* No. 2, 477–480.

Nutmeg/Mace

BALDRY, J. *et al.* 1974. Chemical composition and flavour of nutmegs of different geographical origins. Conf. Proc. 4th Int. Congr. Food Sci. Technol. *1a,* 38–40.

BALDRY, J. *et al.* 1976. Composition and flavour of nutmeg oils. Int. Flavours Food Additives *7,* 28–30.

FORREST, J. E. and HEACOCK, R. A. 1972. Identification of the major components of the essential oil of mace. J. Chromatogr. *69,* 115–121.

SANFORD, K. J. and HEINZ, D. E. 1971. Effects of storage on the volatile composition of nutmeg. Phytochemistry *10,* 1245–1250.

Parsley

FRANZ, C. and GLASL, H. 1974. TLC and GLC of the essential oil in the leaves of some varieties of parsley. Qual. Plant.—Plant Foods Hum. Nutr. *24,* No. 1–2, 175–182. (German)

FREEMAN, G. G. *et al.* 1975. Volatile flavor components of parsley leaves (*Petroselinum crispum* (Mill, Nyman). J. Sci. Food Agric. *26,* No. 4, 465–470.

Paprika

SULC, D. *et al.* 1974. Development of technological procedures for production of new spice concentrates and extracts from paprika. Conf. Proc. 4th Int. Congr. Food Sci. Technol. *8b*, 34–36.

Pepper

DEBRAUWERE, J. and VERZELE, M. 1975. New constituents of the oxygenated fraction of pepper essential oil. J. Sci. Food Agric. *26*, No. 12, 1887–1894.

GENEST, G. *et al.* 1963. A critical study of two procedures for the determination of piperine in black and white pepper. J. Agric. Food Chem. *11*, 509–512.

HARPER, R. S. 1974. Pepper in Indonesia. World Crops *26*, No. 3, 130–133.

JENNINGS, W. G. and BERNHARD, R. A. 1975. Identification of components of the essential oil from the California pepper tree (*Schinus molle*, L.) Chem., Mikrobiol. Technol. Lebensm. *4*, No. 3, 95–96.

LEWIS, Y. S., NAMBUDIRI, E. S. and KRISHNAMURTHY, N. 1969. Composition of pepper oil. Perfum. Essen. Oil Rec. *60*, No. 7, 259–263.

PANGBORN, R. M. *et al.* 1970. Preliminary examination of odour quality of black pepper oil. Flavour Ind. *1*, 763–767.

PAULOSE, T. T. 1973. Pepper cultivation in India. Tropical Prod. Inst., London, Conf. Proc. 91–94.

RICHARDS, H. M., RUSSELL, G. F., and JENNINGS, E. G. 1971. Volatile components of black pepper varieties. J. Chromatogr. Sci. *9*, No. 9, 560–566.

RUSSELL, G. F. and ELSE, J. 1973. Volatile compositional differences between cultivars of black pepper (*Piper nigrum*). J. Assoc. Off. Anal. Chem. *56*, No. 2, 344–351.

Pimento

NABNAY, J. and ROBINSON, F. V. 1972. Constituents of Pimento berry oil (*Pimenta dioica*). Flavour Ind. *3*, 50–51.

Turmeric

MITRA, C. R. 1975. Important Indian spices. I. *Curcuma longa* (*Zingiberaceae*). Riechst., Aromen, Koerperpflegem. *25*, No. 1, 15. (German)

PEROTTI, A. G. 1975. Curcumin—a little known useful vegetable colorant. Ind. Aliment. *14*, No. 6, 66–68. (Italian)

SHOLTO-DOUGLAS, J. 1973. Commercial Scitamineae. III. Profitable turmeric cultivation. Flavour Ind. *4*, No. 9, 387–388.

Saffron

AMELOTTI, G. and MANNINO, S. 1977. Analytical evaluation of the commercial quality of saffron. Riv. Soc. Italiana Sci. Allimentazione *6*, No. 1, 17–20. (Italian)

4

Aromatic Vegetables and Mushrooms

The appeal of many dishes depends upon the use of one or more of the aromatic vegetables or the edible fungi, the nature and availability of which varies considerably from country to country. This group of food materials includes onions, garlic, leeks and tomatoes as well as many less flavorful vegetables and a wide variety of edible fungi or mushrooms. All of these contribute both to the texture and to the flavor of the product, some being of value nutritionally while others are of use mainly because of their strongly aromatic character. The replacement of the latter by alternatives in the form of extracts or imitation flavorings or in the dehydrated form may fundamentally alter the nature of the end-product and can lead to problems during product development.

From the point of view of flavor profiles only two groups of these materials will be considered here:

(a) Part 1: The alliaceous vegetables.
(b) Part 2: Edible fungi (mushrooms).

Part 1: The Alliaceous Vegetables

These include:

Onion (*Allium cepa*, L.)
Garlic (*Allium sativum*, L.)
Chives (*Allium schoenoprasum*, L.)
Leeks (*Allium porrum*, L.)

Whereas several other species are used domestically throughout the Far East (Jones and Mann 1963), the first two are of major worldwide importance in processed foods.

Onions and garlic have been used domestically since earliest times and in spite of the problems of handling and preparation in bulk these fresh vegetables are still widely used in factory-produced products such as pickles, sauces, relishes and canned or frozen ready-to-

Courtesy of Bush Boake Allen, Ltd.

FIG. 4.1. THE AROMATIC VEGETABLES: ONIONS, GARLIC, MUSHROOMS

serve meals because of their wide popularity. The advent of dehydrated and convenience foods has led to a considerable increase in the demand for onion and garlic chips and powder. Unfortunately the dehydration process results in a loss of the distinctively fresh top-notes as well as the lachrymatory pungency which is so characteristic of many varieties of freshly-cut onion. This is due to the partial inactivation of the enzyme systems responsible for the development of the complex mixture of aromatic components. Not all of the enzymes are inactivated if the drying is carried out under carefully controlled conditions but sufficient to make a considerable difference to the flavor profile when the dried product is re-hydrated. In addition to any losses, the powder also acquires a background of boiled onion notes which it contributes to any product in which it is used.

ONION (*Allium cepa*, L.)

The Flavor of Onion

In the natural products so far considered, the aromatic components exist per se in the plant material and can be recovered by simple distillation or extraction. The flavor of onions, on the other hand, is an excellent example of the action of enzymes in flavor development as the characteristic notes are created spontaneously only when the onion is cut and the cells disrupted. Research into the chemistry of the complex reactions involved (reviewed by Johnson *et al.* 1971) has established that much of the odor and, hence, flavor arise as a result of the conversion of odorless, nonvolatile, S-substituted cysteine sulfoxide derivatives, first to an unstable alkyl sulfenic acid intermediate by the action of an alliinase enzyme system. The sulfenic acid moiety is very unstable and the reaction proceeds directly to a mixture of mercaptans, di-, tri- and poly-sulfides as well as thiosulfonates. All possible combinations of methyl and propyl derivatives have been isolated but little or no allyl compounds are present in onion. A similar system in garlic results in the formation of mainly allyl compounds, the difference being due to the nature of the precursor present (Freeman and Whenham 1975).

Most onion flavor components are the result of the degradation of the thiosulfonates, resulting in a complex mixture of sulfides plus ammonia and pyruvic acid which is produced in the first stage of the reaction. A determination of pyruvic acid has been suggested as a measure of the flavoring strength of onions and onion products even though the acid itself does not contribute to the flavor profile (Schwimmer and Weston 1961).

Dehydrated Onion

The domestic onion is one of the world's oldest crops. The many advances in the formulation and manufacture of dehydrated meals have created a demand for onions in a powder form and considerable tonnages are now dehydrated annually to satisfy this need. As pointed out above, there is some loss in the fresh odor and flavor character during the processing but, in spite of this, the dehydrated form has the advantage that it saves valuable preparation time on the factory floor, there is little or no waste or spoilage and, within acceptable limits, it gives a consistent flavor to the end-product. The nature of this flavor is different from that of fresh onions so that a direct equivalence cannot be established. However, allowance is made for this during product formulation and an acceptable usage level established by trial and error. For guidance, initial usage levels assume an 8:1 fortification of the flavor in the dehydrated form and the missing top-notes may be added in the form of a liquid flavor made from onion oil.

Onions grown for dehydration are not generally the same as those used domestically. The prime consideration is, of course, that of a high dry-solids content coupled with a good level of flavor and pun-

Courtesy of Gentry International Sales, Ltd.

FIG. 4.2. MECHANICAL HARVESTING OF ONIONS

Courtesy of Gentry International Sales, Ltd.

FIG. 4.3. HARVESTED ONIONS ARE INSPECTED IN THE FIELD WHERE ALL DAMAGED AND DECAYED ONIONS AND OTHER EXTRANEOUS MATTER ARE REMOVED

gency. In California, special strains of white globe onions have been adopted for the production of high quality onion powder, flakes and chips. The United States is by far the largest producer of dehydrated onion products followed by Japan, Egypt and the central European states of Czechoslovakia and Hungary. Although care is

Courtesy of Gentry International Sales, Ltd.

FIG. 4.4. A CONTINUOUS BELT DEHYDRATOR IN GILROY, CALIFORNIA
Onions begin the dehydration with 20% solids and leave with approximately 95% solids.

taken during all stages of manufacture, onion powder is still hygenically suspect and unless special handling techniques are adopted the product may show very high total bacterial and spore counts. However, these counts are of far less significance to the food processor than is the presence of pathogens and Salmonella which clearly indicate faulty processing. Most specifications for onion powder define a limit for total count but this is relatively meaningless. Of far greater significance is a limit of Coliform organisms (specifically *E. coli*) and a requirement that pathogens and Salmonella be completely absent.

Onion Oil

The distillation of so-called "onion oil" is a very involved operation as yields are small and differ greatly depending on the variety of

Courtesy of Gentry International Sales, Ltd.

FIG. 4.5. A FIELD OF ONIONS IN BLOOM FOR SEEDING

onion used and the distillation technique employed as the components are to some extent water-soluble. The average yield is 0.015% of a dark brown oil which may crystallize on standing.

The comparative flavoring strength of the oil is in the order of 4000 times that of fresh onion. In this form it is much too powerful to incorporate directly into foods and diluted versions, both liquid and dry, are available for this purpose. It is necessary to establish with the supplier the relative flavoring strength as compared with either fresh or dehydrated onion. It has to be appreciated that, though these products may have the flavor character of fresh onion, they do not contribute any of the textural quality given by the raw vegetable.

In addition to their characteristic flavor notes, it is well recognized that onions when freshly cut have a marked pungency and lachrymatory property. This is an important factor in the flavor profile but these attributes are difficult to define, let alone assess accurately. The pungency level is somewhat related to flavor strength and aromatic character but its nature is not yet well understood. Matikkala and Virtanen (1967) proposed that thiopronanal-S-oxide, derived by the enzymic degradation of S(1-propenyl)-L-cysteine sulfoxide might be the chemical responsible. This compound is highly unstable which is consistent with the fact that the lachrymatory character of onion is destroyed by even gentle heat and is not present in either onion powder or onion oil.

Profile of Freshly-cut Onion

Origin	English onion, as purchased for domestic use, chopped very fine immediately prior to the evaluation.	
Odor	Initial:	slightly LACHRYMATORY. LIGHT, GREEN, FRESH. SWEET. PENETRATING, slightly PUNGENT. Slightly SOAPY.
	Persistence:	The full-bodied fresh character is retained for several hours.
Flavor: (determined at 4% in a neutral soup)		Slightly PUNGENT, PERSISTENT. Light GREEN, VEGETABLE-like. Slightly SWEET, FULL, ROUNDED. SULFUROUS.
	After-taste:	Pleasantly lingering, WARM.

Profile of Californian Onion Powder

Origin	Commercial sample of highest available quality.	
Odor (determined by evaluation of the head-space over a sample at room temperature)		
	Initial:	DULL, COOKED ONION. GREEN, VEGETABLE-like. Slightly PUNGENT but not penetrating. SWEET.
(determined at 0.5% in a neutral soup at 60°C)		
	Initial:	LIGHT, FRESH, over-marked COOKED note. SWEET, ROUNDED; full-bodied. GREEN, VEGETABLE-like. Slightly PUNGENT, slightly PENETRATING.
	Persistence:	The initial fresh notes are quickly

over-ridden by the BOILED ONION note which is very persistent.

Flavor
(determined at 0.5% in a neutral soup)

Slightly PUNGENT.
GREEN, VEGETABLE-like, BEANY.
SWEET, BREAD-like, full-bodied, round.
COOKED ONION.

After-taste: Pleasantly SWEET.

Profile of Onion Oil

Origin Commercial sample distilled in the United Kingdom from English onions.

Odor
Impact: 3
Initial: PUNGENT, very PENETRATING.
Strongly SULFUROUS.
FRESH, GREEN.
OILY, SOAPY.
Persistence: The aromatic character persists with little change over 24 hr.

Flavor: PUNGENT.
(determined at 0.0004% in a neutral soup)

Strongly SULFUROUS.
Pleasantly FULL-BODIED, VEGETABLE-like.
Slightly SWEET, slightly GREEN.
OILY.

After-taste: Pleasantly SWEET.

Components of Onion Oil

Oil of Onion obtained by steam distillation or by solvent extraction of the expressed juice differs in detailed profile though each has been shown to contain a mixture of similar di- and trisulfides (Carson and Wong 1961). The following components have been identified:

ethanol
1-propanol

hydrogen sulfide
1-propanethiol

2-propanol	methyl disulfide
methanol	methyl 1-propyl disulfide
propanal	1-propyl disulfide
1-butanal	methyl trisulfide
acetone	methyl 1-propyl trisulfide
methyl ethyl ketone	1-propyl trisulfide

Sulfides are not the only odorous components but they are so powerful as to predominate.

The lachrymatory character of fresh onions (thiopropanal-S-oxide) does not exist in the essential oil and is not really part of the onion flavor. American wild onions are very flavorful but do not have this lachrymatory effect.

GARLIC (*Allium sativum*, L.)

Because of their attractive flavor and acknowledged medicinal properties the bulbs or "cloves" of garlic have been used in the cuisine of most Mediterranean countries since the dawn of history. Like onions, the entire cloves are almost without odor but once cut or bruised they produce an intensely strong and characteristic odor which to many is obnoxious.

The chemistry of the compounds responsible for the garlic profile is similar to that found in onion (Freeman and Whenham 1975A). The differences are attributed to qualitative and quantitative differences in the precursors present; the active ingredients being primarily allyl (2-propenyl) sulfides together with much smaller amounts of methyl and 1-propyl compounds.

The flavor of onion and garlic is complementary, the former being mild and sweet whereas the latter is harsh and insistent. Because of its relatively high flavoring power, garlic is frequently blended with onion in order to increase the initial impact of the onion but this can only be done to a very limited extent as garlic is quickly recognizable as such and its flavor associations are not always acceptable. If garlic is incorporated into an end-product which is to be distributed in a container such as a screw-capped bottle or jar, the head-space above the product nearly always has a higher proportion of the garlic odor. This may be detectable as such and detract from the product even though the product itself may not contain a sufficient level of garlic to be noticeable when the product is consumed.

Garlic Oil

The essential oil of garlic can be recovered by steam distillation of the freshly crushed cloves, the yield being 0.1 to 0.2%. Garlic oil is a powerful flavoring agent and is widely used in seasonings either as a liquid flavor or dispersed as a dry-carrier.

Many problems arise when fresh garlic is included in a food product. The commercially-available vegetable occurs as a compound bulb made up of 10 to 14 small "cloves" encased in a tough outer skin. The bulbs must first be cleaned and sorted, the outer tissues removed and the cloves separated. The garlic is then ready for mincing, cutting or crushing. The whole process is tedious and the manufacturing department becomes permeated with the smell. This can give rise to the problem of cross-contamination unless great care is taken. Like all other natural products, the flavoring effect of fresh garlic is variable whereas that of garlic oil is relatively consistent. It is not surprising, therefore, that garlic oil is now widely used in place of the fresh vegetable. To overcome the objectionable odor associated with both fresh garlic and garlic oil, the use of an encapsulated garlic oil is strongly recommended. This dry powder product is almost free of odor and does not release its contents until the capsule is broken down by admixture with water.

Profile of Fresh Garlic

Origin Egyptian garlic, as purchased for domestic use, chopped very fine immediately before use.

Odor	Initial:	POWERFUL; to many, strongly objectionable.
		SHARP, THIN, PENETRATING; harsh.
		GREEN, LEEK-like.
		SULFUROUS.
		Slightly PUNGENT; burning.
		SWEET back-note.
	Persistence:	Very lingering with little change, in character.

Flavor
(determined at 0.5% in a neutral soup)

		SWEET but FLAT.
		PUNGENT (at back of mouth).
		SULFUROUS and slightly METALLIC.
		GREEN, LEAFY.
	After-taste:	Very lingering on the breath, BURNING.
		TINGLING with slight BITTERNESS.

Profile of French Garlic Powder

Origin Commercial sample of highest available quality.

Odor
(determined at 0.5% in hot neutral soup)

Initial: POWERFUL, objectionably STRONG.
GREEN, COOKED VEGETABLE: heavy.
Slightly PUNGENT, PENETRATING but not irritating, sharp.
SULFUROUS, slightly BURNT; slightly TOASTED.
SMOOTH; rounded, full-bodied.

Persistence: Lingering with almost no change in profile.

Flavor
Sharply PUNGENT (at back of mouth); burning.
BITTER/SWEET, WARM; full-bodied.
SULFUROUS, slightly METALLIC.
BOILED VEGETABLE-like, BREAD-like.
Slightly TOASTED.

After-taste: Very lingering with a distinctive BITTERNESS.
WARM, some DRYNESS.

Profile of Garlic Oil

Origin Commercial sample distilled in France from French garlic.

Odor
Initial: POWERFUL, OBJECTIONABLE.
FRESH, GREEN, LIGHT.
Mildly PUNGENT, PENETRATING; sharp, irritating.
Strongly SULFUROUS, slightly METALLIC.

Persistence: Very lingering with no change in profile over 24 hr.

Flavor
(determined at 0.00125% in a neutral soup)
Strongly SULFUROUS.
WARM; full-bodied, smooth.

	SWEET with slightly BITTER back-note.
	Slightly GREEN.
	Very slightly PUNGENT.
After-taste:	Lingering with slight BITTERNESS, slightly METALLIC.

Components of Garlic Oil

Garlic oil consists mainly of di- and tri-sulfides having an allyl (2-propenyl) radical. Freeman and Whenham (1975A) carried out a study of the volatile components of garlic oil and concluded that the radicals were present in the following proportions:

1-propyl	2.00%
methyl	13.4%
2-propenyl	84.6% (mostly as 2-propenyl disulfide).

ASAFETIDA (*Ferula foetida*, **Regel**)

Although totally unrelated to the alliaceous vegetables there is one other aromatic plant material which has a profile not unlike that of garlic. Asafetida (or Devil's Dung) is an oleo-gum-resin obtained from the thick fleshy roots of an umbelliferous plant, *Ferula foetida*, and allied species which are found throughout Afganistan and Iran. The latex is collected by incising the plant at the top of the main root.

The dried gum when steam-distilled yields 5–15% of an orange-brown oil having a pungent, strongly garlic-like odor. The components responsible for this character have yet to be studied but are mainly di- and tri-sulfides.

Profile of Oil of Asafetida.

Origin	Oil distilled in the United Kingdom from imported Asafetida.	
Odor	Impact:	4
	Initial:	Strongly PUNGENT, very PENETRATING; acrid.
		Sharply ALLIACEOUS; garlic-like.
	Persistence:	Very lingering, the profile changing very slowly on evaporation becoming smoother and less pungent.

Flavor

(determined at 0.0004% in a neutral soup)

HARSH, strongly GARLIC-like.
BITTER, slightly ACRID.
Vaguely GREEN.
Slightly PUNGENT.

After-taste: Lingering BITTERNESS and unpleasantly GARLIC-like.

Part 2: The Fungi

This is a big and important group of raw materials comprising many thousands of nonpoisonous species of fungi which are used widely in the preparation of food. The flavor of many species is strong and peculiar so that their acceptance is an acquired taste. Others are mild and are used as much for their textural effect as for flavor. Food processors are mainly interested in only two types:

Common field mushroom (*Psalliota campestris*), (*Agaricus campestris*, L.), (*A. bisporus*)
Honeycomb mushroom (*Boletus edulis*, Bull)

Three other species are also cultivated in significant quantities (Gray 1974), namely:

Padi Straw mushroom (*Volvariella volvacea*) (Orient)
Shiitake (*Lenzites elodes*) (Japan)
Truffle (*Tuber melanospermum*) (Europe)

The Flavor of Mushrooms

Almost all edible mushrooms develop an earthy, woody character as they mature to the point where the moldy note dominates and becomes truly objectionable. There is such a wide range of varieties available that it is impossible to generalize, it being necessary to study each in order to establish its profile under the conditions of use.

The common field mushroom, whether it is grown out-of-doors or is cultivated under the special conditions of intensive farming, has only a slight taste. The immature "button" mushroom has a marked nutty flavor which is lost as the mushroom develops. Once the mushroom is cooked a quite different flavor profile develops, the nature depending on the method and degree of cooking employed. The delicate flavor of a freshly picked mushroom cannot be captured by extraction and is largely lost during dehydration. Attempts to make mushrooms by deep-culture methods have been successful but the resulting product, although of considerable nutritional value, is almost flavorless.

The *Boletus* mushroom, derived from Poland, is used as a source of "mushroom extract." This is a thick dark-brown extract prepared by the alcoholic extraction of selected *Boletus* caps. The flavor of this product is strongly meaty with a pleasant but dull earthy background reminiscent of mature mushrooms. The relative flavoring strength of such extracts should be ascertained from the manufacturer.

BIBLIOGRAPHY

Asafetida

ABRAHAM, K. O. *et al.* 1973. Asafoetida, I. Oxidimetric determination of the volatile oil. Flavour Ind. *4*, 301–302.

RAGHAVAN, B. *et al.* 1974. Asafoetida, II. Chemical composition and physicochemical properties. Flavour Ind. *5*, 184–185.

Garlic

CHARANDAEV, M. G. 1974. The mechanical cleaning of garlic bulbs. Konserv. Ovoshchesush. Promst. No. 11, 21. (Food Sci. Technol. Abstr., 1975. *7*, 10T 506.)

CHIVADZE, M. O. *et al.* 1975. The effect of storage conditions on the hygroscopicity of powdered garlic. Konserv. Ovoshchesush. Promst. No. 5, 31–32. (Food Sci. Technol. Abstr., 1976, *8*, 7T 344.)

DU, C. T. and FRANCIS, F. J. 1975. Anthocyanins of garlic (*Allium sativum*, L.). J. Food Sci. *40*, 1101–1102.

HEATH, H. B. 1973. Herbs and spices—a bibliography. IV. Flavour Ind. *4*, 218–219.

HEIKAL, H. A. *et al.* 1972. A study on the dehydration of garlic slices. Agric. Res. Rev. *50*, 243–253.

RAGHEB, M. S. *et al.* 1972. Seasonal changes in garlic and its effect on bulbs during storage. Agric. Res. Rev. *50*, 157–165.

Mushrooms

GRAY, W. D. 1974. Fungi as food. *In* Encyclopedia of Food Technology. A. H. Johnson and M. S. Peterson (Editors). AVI Publishing Co., Westport, Conn.

HAYES, W. A. 1976. A new look at mushrooms. Nutr. Food Sci. *42*, 2–6.

HEATH, H. B. 1973. Herbs and spices—a bibliography. V. Flavour Ind. *4*, 264–265.

IGOLEN, G. 1973. The truffles. Rev. Ital. Essenze, Profumi, Piante Off. *55*, 631–641.

PICCARDI, S. M. and ISSENBURG, P. 1973. Investigation of some volatile constituents of mushrooms, *Agaricus bisporus*: Changes which occur during heating. J. Agric. Food Chem. *21*, 959–962.

THOMAS, A. F. 1973. An analysis of the flavor of the dried mushroom, *Boletus edulis*. J. Agric. Food Chem. *21*, 955–958.

TOYO SEIKAN Co. 1975. Mushroom-like Flavor. Japanese Pat. 5,033,147, Oct. 28, 1975. (Japanese)

Onion

ANON. 1975. Functional and price advantages with artificial onion and garlic flavors. Food Process. *36*, 57.

BRODNITZ, M. H. and PASCALE, J. V. 1971. Thiopronanal S-oxide: a lachrymatory factor in onions. J. Agric. Food Chem. *19*, 269–272.

BRODNITZ, M. H. and PASCALE, J. V. 1973. Alliaceous flavors. U. S. Pat. 3,723,135, Mar. 27.

BUSH BOAKE ALLEN, Ltd., 1975. Onion-like flavour. Bri. Pat. 1,386,236, Mar. 5.

CARSON, J. F. and WONG, F. F. 1961. The volatile flavor components of onions. J. Agric. Food Chem. 9, 140–143.

FREEMAN, G. C. 1975. Distribution of flavor components in onion (*Allium cepa*, L.), leek (*A. porrum*, L.) and garlic (*A. sativum*, L.). J. Sci. Food Agric. *26*, 471–481.

FREEMAN, G. C. and MOSSADEGHI, N. 1971. Influence of sulfate nutrition on the flavor components of garlic (*Allium sativum*, L.) and wild onion (*A. vineale*, L.). J. Food Sci. *21*, 1529.

FREEMAN, G. C. and WHENHAM, R. J. 1974A. Changes in onion (*Allium cepa*, L.) flavor components resulting from post-harvest process. J. Sci. Food Agric. *25*, 499–515.

FREEMAN, G. C. and WHENHAM, R. J. 1974B. Flavor changes in dry bulb onions during winter storage at ambient temperatures. J. Sci. Food Agric. *25*, 517–520.

FREEMAN, G. C. and WHENHAM, R. J. 1975A. The use of synthetic (±)-S-1-propyl-L-cysteine sulfoxide and of alliinase preparations in studies of flavor changes resulting from processing of onion (*Allium cepa*, L.). J. Sci. Food Agric. *26*, 1333–1346.

FREEMAN, G. V. and WHENHAM, R. J. 1975B. Survey of volatile components of some *Allium* species in terms of S-alk(en)yl-L-cysteine sulfoxide present as flavor precursor. J. Sci. Food Agric. *26*, 1869–1886, (35 refs).

HEATH, H. B. 1973. Herbs and spices—a bibliography. VI Flavour Ind. *4*, 346–348, (40 refs).

JONES, H. A. and MANN, L. K. 1963. Onions and Their Allies. Leonard Hill, London.

JOHNSON, A. E., NURSTON, H. E. and WILLIAMS, A. A. 1971. Vegetable volatiles: A survey of components identified. Part I. Chem. Ind. No. 21, 556–565.

JOHNSON, A. E., NURSTON, H. E. and WILLIAMS, A. A. 1971. Vegetable volatiles: A survey of components identified. Part II. Chem. Ind. No. 43, 1212–1224.

LEDL, F. 1975. Analysis of a synthetic onion aroma. A. Lebensm. Unters. Forsch. *157*, 28–33. (German)

MITIKKALA, E. J. and VITANEN, A. I. 1967. The quantitative determination of the amino acids and gamma glutamyl peptides of onion. Acta Chem. Scand. *21*, 2891.

SCHULTZ, H. W. (Editor). 1967. Onion flavor. *In* The Chemistry and Physiology of Flavors. AVI Publishing Co., Westport, Conn.

SCHWIMMER, S. 1973. Flavor enhancement of *Allium* product. U.S. Pat. 3,725,085, Apr. 3.

SCHWIMMER, S. and WESTON, W. J. 1961. Enzymic development of pyruvic acid in onion as a measure of pungency. J. Agric. Food Chem. *9*, 301–304.

5

The Citrus Fruits

This genus of the family *Rutaceae* includes lemon, orange (both sweet and bitter), mandarin (or tangerine), lime and grapefruit in addition to very many other fruits which are pleasant to eat but which are not of great importance commercially. Throughout the world the citrus fruits are appreciated and in many cases form a valuable part of the diet as a source of vitamin C. Those grades which are not suitable for domestic use are generally processed in the growing regions to yield either the juice, which may be further processed to give concentrates, or for the essential oil which is obtained from the peels. In this chapter only the profiles of the highly aromatic essential oils will be discussed as the juices are not used so much as flavorings but as raw materials contributing far more than just flavor. The use of these products in food processing will be discussed in a later chapter.

THE CITRUS OILS

In all citrus fruits the essential oil is contained in numerous, oval, balloon-shaped oil sacs or glands situated irregularly just below the surface of the colored portion of the peel—the flavedo. The white mesocarp or albedo does not contain any oil sacs but carries the bitter glycosides such as hesperidin in lemons, oranges and tangerines or naringin in grapefruit.

The essential oils are generally mechanically extracted from the peels by methods which will be briefly noted under each particular fruit. It is not only the citrus fruits but also the flowers and leaves of citrus plants which are aromatic and yield essential oils of distinctive profiles which are of considerable value in the compounding of fragrances and to a lesser extent in the formulation of imitation flavors. The essential oil distilled from the leaves is known as oil of petitgrain and that from the flowers, oil of neroli. The particular source of the oil is designated in its name (e.g., Oil of Neroli Bigarade from the flowers of the bitter orange *Citrus aurantium*, L.).

Part 1: Lemon

Lemons are the fruits of the tree *Citrus limon* (L) Burm. which is extensively cultivated in Sicily, Italy (Calabria) and California and to a limited extent in Florida, Spain, Brazil and Argentina although the tree will grow in many other areas having a typical Mediterranean type of climate. The ovoid fruits are from 2½ to 4 in. in length, the yellow rind enclosing 8 to 10 fleshy segments, the cells of which contain a very acidic juice.

FIG. 5.1. LEMONS (*Citrus limon* (L) Burm.)

The essential oil is recovered from the peel by cold expression or by distillation, depending upon the region of production. Full details of the methods employed are given in the literature (Guenther 1949–1952; Nagy *et al.* 1977).

Sicilian Lemon Oil

The methods used for oil extraction of Sicilian lemon oil are: (a) Hand pressing. Two methods were widely used in the past but both now are obsolete: (1) sponge; (2) ecuelle. (b) Machine processing. Two methods are currently in use: (1) Sfumatrici. The fruit is halved and the juice first expressed by reeming. The peels are then individually pressed to release the oil. (2) Pellatrice. The whole fruit is rasped in a fine stream of water, the oil and cellular detritus being separated and the water recycled. The oil is obtained by centrifuging the liquor and pressing the solid matter.

The oxygenated constituents of lemon oil, to which the oil owes most of its odor and flavor, are the more soluble in water and for the highest quality oil it is desirable to separate it from any aqueous phase as rapidly as possible. Although widely used, centrifuging is not an entirely satisfactory method of achieving separation owing to the formation of persistent emulsions.

Courtesy of Albright & Wilson, Ltd.

FIG. 5.2. THE CONTINUOUS PRODUCTION OF LEMON OIL ON AN FMC MACHINE IN GREECE

Lemon trees tend to produce fruit continuously but the age of the tree and growing conditions result in a more-or-less seasonal cropping. In Sicily the harvest times are:

September-October	10–20% of crop	"winter" lemons
Mid-November-December	40% of crop	
January-February	30% of crop	
August	Special picking of "verdelli" lemons	

The total Sicilian production of lemon oil is about 750,000 lb per annum, the yield being in the order of 6–7 lb of oil per ton of fruit processed.

Californian Lemon Oil

Because of much higher labor costs and the enormous quantities to be processed, none of the Sicilian type machines are used in California where a special high-capacity plant has been developed.

The processing scheme is as follows: (a) The fruit is washed on arrival at the factory. (b) The fruit is then crushed in heavy stainless steel rollers. (c) The coarse mass is reduced to give: (1) solid matter, which is continuously removed by scraper blades; and (2) a thin emulsion of oil and watery juice. (d) The emulsion is immediately broken in a high-speed centrifuge to give juice, essential oil and fine cell detritus. (e) The press residues are collected and steam-distilled to give a further quantity of a lower grade oil.

As in Sicily, the crop is to some extent seasonal, the essential oil content of the peels being lowest in the winter months and spring and highest in the autumn.

The annual production of lemon oil in California is variable but averages to some 600,000 lb, the yield being up to 14 lb per ton of fruit processed.

Profiles of Lemon Oil

Origin Commercial samples of recently extracted oil.

		Sicilian Lemon Oil	Californian Lemon Oil
Odor	Impact:	3	4
	Initial:	FRESH, LIGHT.	FRESH, LIGHT; airy.
		SMOOTH.	SHARP, PENETRATING.
		FRUITY, PEELY.	FRUITY, PEELY, FULL-BODIED.
		Slightly GREEN.	Slightly FLORAL; fragrant.
			Slightly OILY; fatty.

Persistence: Initial fresh notes are quickly lost and after 1 hr the profile is richly fruity with a marked juicy note. The green note is not perceptible in the Californian oil until after 2 hr airing. Older oils demonstrate a marked rancidity and terpeney character after 2 hr on a smelling strip.

Dry-out: Good quality lemon oils air-off almost completely leaving the very faintest of residual peely note, or no residual odor.

Flavor REFRESHING, SWEET. FRUITY but FLAT.

Note: Lemon flavor is always associated with an acidic character so that its evaluation per se is not meaningful. A suitable medium for tasting is a 12% solution of sucrose with 0.05% citric acid.

Components of Lemon Oil

Terpenes constitute about 90% of the essential oil, mainly:
- α-pinene
- β-pinene
- α-limonene
- β-ocimene

Oxygenated compounds include:
- citral
  - (Sicilian oil: 3.5–5.5%)
  - (Californian oil: 2.5–3.5%)
- citronellal
- linalool
- terpen-4-ol
- α-terpineol
- geranyl acetate

Scora (1975) has used the variation and compositional pattern of components as a basis for a taxonomical study of citrus fruits.

Part 2: Orange

The fruits of the common sweet orange (*Citrus sinensis*, Osbeck.) are spherical, 2 to 4 in. in diameter, the smooth orange colored skin enclosing a juicy cellular pulp divided into loosely joined segments. The essential oil is contained in the peel.

Two types of orange are used as a source of the essential oil:

Sweet Orange	*Citrus sinensis*, Osbeck.
	Citrus aurantium, L., var. *dulcis.*
Bitter Orange	*Citrus aurantium*, L. subsp. *amara.*

Only the sweet orange is used commercially as a source of the juice; bitter oranges are grown mainly for the production of orange conserves such as marmalade.

FIG. 5.3. ORANGES (*Citrus sinensis*, Osbeck)
The flowers and leaves also produce distinctive essential oils of value in flavors and fragrances.

Other parts of the orange tree are also aromatic and two very important essential oils are derived from the leaves and flowers:

Oil of Petitgrain—from sweet and bitter orange leaves.

Oil of Neroli—from sweet and bitter orange flowers.

In the case of oils from the bitter orange the term "Bigarade" is used to distinguish it from similar oils from the sweet orange. These oils are of particular value in the compounding of fragrances.

Sweet Orange

The orange is a native of Asia but is now widely cultivated in very many countries including: United States (California and Florida), Guinea, Brazil, Cyprus, Italy (Sicily and Calabria), West Indies

(Puerto Rico and Jamaica), Spain, South Africa, Israel and Japan (which produces a local variety, *natsudaidai*).

Californian and Florida Sweet Orange Oil.—Two types of orange are processed to give the essential oil: Washington Navel and Valencia.

Methods Used—Modern large-scale processing consists of the following stages: (a) The fruit is halved. (b) The juice is extracted by reeming in a rotary juice extractor. (c) The residual peel is submitted to high pressure in a Pipkin Peeloil Press to yield best oil. (d) The press-cake is steam-distilled to yield a further quality of lower quality oil.

The throughput of this technique is about 7 tons of fruit an hour yielding, in the case of California oranges, up to 6 lb of oil per ton and for Florida oranges about 3–4 lb per ton.

If the oil recovered by steam-distillation is taken into account, the total yield is between 11 and 15 lb per ton of fruit, but this extra oil is very inferior in profile and keeping qualities to the cold-pressed oil and can only be used in cheaper blends for applications where the finer profile is not called for.

Kesterson and Braddock (1975) claim that most citrus processors recover only 10–20% of the total oil available from the fruit and ignore the possible increased yields based on a study of 12 different citrus cultivars over a 3–5 year period.

The total orange oil production of California is about 500,000 lb and of Florida, 350,000 lb per annum. The United States, Spain, Brazil and Guinea together account for some 80% of the total world production.

Profile of Sweet Orange Oil

The aromatic profiles of sweet orange oils from different growing regions can be distinguished but generally reveal a remarkable degree of uniformity being sufficiently similar to enable oils from different sources to be interchanged in a fragrance or flavor formulation. The profile described below was derived from Florida sweet orange oil.

Origin	Commercial sample of recently produced oil.	
Odor	Impact:	2
	Initial:	SWEET. LIGHT, FRESH; terpeney. FRUITY, ALDEHYDIC. Smoothly PENETRATING but not irritant. PEELY.

	Persistence:	The odor of sweet orange is not very persistent and airs-off very quickly but with little change in profile.
	Dry-out:	The residual note is citrus but slightly GREASY.
Flavor		FRESH, SWEET, REFRESHING, FRUITY. Very slight SHARPNESS. Very slight BITTERNESS.
	After-taste:	Cleanly FRUITY.

Note Good quality sweet orange oil should not have any turpentine-like notes nor should it be harsh or irritating.

Components of Sweet Orange Oil

Terpenes constitute about 90% of the essential oil, mainly:
- α-limonene.

Oxygenated compounds include:
- n-decylic aldehyde (0.9–3%)
- citral
- linalool

Shaw and Coleman (1975), in studying the components necessary for natural orange flavor, identified two components, isoprene and 3-methyl-1-butene which contribute significantly to the juicy character of the profile.

Bitter Orange

Several varieties of bitter orange (*Citrus aurantium*, L., subsps. *amara*) are grown throughout southern Europe. This type of orange is used mainly as a component of conserves such as marmalade. The principal growing regions are Spain (Analusia) and Sicily but commercial crops are raised in Guinea, the West Indies (Jamaica, Haiti, Cuba, and Puerto Rico), Brazil and East Africa. Much smaller crops are produced in Mexico, the U.S.S.R. and China.

Methods Used —The production of essential oil from Seville or Bigarde bitter oranges is only about 30 tons per annum and is small in comparison with that from sweet oranges. Expression is often by hand but some machine-pressed oil is produced on a Ramino machine.

Profile of Bitter Orange Oil

The hand-pressed oils are generally of high quality, having a much finer aroma than oil obtained by steam distillation of the pressed peels. The profile of commercial oils is very variable, depending on source and age.

Origin	Commercial sample of Spanish Bitter Orange Oil.	
Odor	Impact:	2
	Initial:	FRESH but DRY and somewhat FLAT. SICKLY SWEET; rich, full-bodied. Slightly FLORAL; perfumed. Slightly PEELY; terpeney.
	Persistence:	Very lingering with little change in profile over several hours.
	Dry-out:	OILY with persistent SICKLY SWEET back-note.
Flavor		FRUITY, PEELY. BITTER; sour. Slightly ACIDIC, SHARP. Slightly FLORAL, fragrant.
	After-taste:	BITTER but pleasingly FRESH.

Components of Bitter Orange Oil

Terpenes constitute about 90–92% of the essential oil, mainly:	Oxygenated compounds include:
d-limonene	n-decyl aldehyde (0.8–1%) linalool (about 0.3%) linalyl acetate (2–2.5%) acetic, capric and caprylic acids have been reported present.

According to Guenther (1949) the substance causing the bitter taste of the oil is contained in the nonvolatile residue.

Part 3: Tangerine, Mandarin, Satsuma

The designation "tangerine" and "mandarin" is used synonymously for the same fruit and its products but some authorities contend that there are distinct differences between the varieties and that the group should be subdivided to account for these. The tree *Citrus reticulata*, Blanco, is cultivated throughout the Mediterranean region but principally in Sicily and Calabria. The fruit is characterized by having a very loose skin, segments which readily separate and is flattened at the ends.

Tangerines are cultivated in Italy (where they are called Mandarin) and Florida (where they are called Tangerine) and in Spain, Algeria and Cyprus. The Brazilian (Mexerica) Mandarin produces a distinctly different fruit and essential oil which should not be confused with either the Italian or Florida products.

The fruits are washed and machine-pressed during October-November to give about 5 lb of oil per ton of fruit processed.

Profile of Sicilian Tangerine (Mandarin) Oil

Origin	Commercial sample of freshly pressed oil.	
Odor	Impact:	3
	Initial:	SWEET, FRUITY. UNFRESH, somewhat SICKLY. Slightly FISHY, terpeney. FLORAL; perfumed, fragrant. DRY.
	Persistence:	Little change in profile over 24 hr.
	Dry-out:	Distinctly of tangerine but less sweet.
Flavor		SWEET, FRUITY. SMOOTH, FLAT. Slightly FLORAL, fragrant. Slightly BITTER.
	After-taste:	Pleasantly FRUITY.

Components of Tangerine Oil

d-limonene
β-phellandrene
ρ-cymene
methyl-N-methyl anthranilate (about 1%)
n-decyl aldehyde
linalool
terpineol
nerol
linalyl acetate
terpinyl acetate

Part 4: Lime

Limes are widespread throughout the Caribbean islands and Central America, Mexico and Polynesia. Apart from these main areas, limes are also grown to a limited extent in Florida and in southern Italy. There are two main types of lime and very many varieties:

Acid lime (*Citrus aurantifolia,* Swingle)

(a) The Mexican, West Indian or Key lime being small bushy trees having small thin-skinned fruit.

(b) The Tahitian or Persian lime being a large tree having thin-skinned ovoid fruit which is seedless.

Sweet lime (*Citrus limetta,* L. or perhaps another variety of *Citrus aurantifolia,* Swingle)

(a) The tree grows wild in Brazil and Mexico; the fruit is sweet and bergamot-like.

Lime oil is used extensively as a source of flavor and two types are commercially available from the West Indies which is the principal source:

(a) Expressed: The so-called "cold pressed" oil is prepared by hand although small quantities are machine expressed.

(b) Distilled: As part of the lime juice operation.

Distilled Lime Oil

Methods Used.—Lime oil is obtained in two ways: (a) by steam distillation of the crushed or comminuted fruits; and (b) by steam distillation, usually under vacuum, of the separated top and bottom layers of the crude expressed acid juice. Distillation time is critical to the quality of the oil and must be continued sufficiently long to ensure the maximum content of high boiling oxygenated components.

At one time lime oil was the by-product of lime juice production; today it is the other way round and the oil is now the prime product. The annual production of distilled lime oil is in excess of 400 metric tons. The treatment, so different from that used to recover other citrus oils, does result in considerable changes in the flavor profile. The "cold pressed" oil has a much finer fresh fruity character. Most users, however, prefer the steam-distilled oil for its consistency of flavor and its cheaper price. The chemical changes which take place during distillation are discussed by Azzouz and Reineccius (1976).

Profile of West Indian Distilled Lime Oil

Origin	Commercial sample from Jamaica.	
Odor	Impact:	3
	Initial:	SHARP, HARSH; pungent, penetrating.

		SWEET, FRESH.
		TERPENEY; paint-like.
		PERFUMED; fragrant.
		CITRUS with a slight tendency to ORANGE-like.
	Persistence:	Lime oil evaporates very quickly and much of the impact is lost within one hour of airing on a smelling strip. The profile remains unchanged.
	Dry-out:	Almost no residual note.
Flavor		Refreshingly SHARP.
		CLEAN, ASTRINGENT.
		GREEN, CITRUS; lemony/ orange-like.
		Slightly FRAGRANT.
	After-taste:	Pleasingly FRESH and COOL.

Components of Distilled Lime Oil (Azzouz and Reineccius 1976)

There have been reported 31 volatile components representing 98.2% of the volatile fraction of Mexican cold-pressed lime oil and 37 components representing 91.2% of distilled Mexican lime oil. The major components are:

d-limonene	citral (about 8–10%)
α-pinene	linalool
β-pinene	geraniol
dipentene	terpineol
β-bisabolene	borneol
terpinolene	terpinene-4-ol

Part 5: Grapefruit

Grapefruit (*Citrus paradisi*, Macf.) is a very popular citrus fruit which is eaten as such, provides a juice which is widely taken at breakfast and is also an essential oil of considerable interest as a flavoring agent. Grapefruit is extensively grown in Florida, Texas and California with much smaller amounts coming from the West Indies, Brazil and Israel.

Florida produces about 40,000 lb of cold-pressed grapefruit oil per annum but this represents only a very small fraction of the oil which could be produced if the demand warranted it. The oil is located deep in the peel which overlays a thick sponge-like albedo which tends to soak up the oil during the extraction. According to Guenther (1949) the yield of oil is only about 5% of what can be recovered by steam-distillation.

Profile of Florida Grapefruit Oil

Origin	Commercial sample.	
Odor	Impact:	2
	Initial:	FRESH; light, pleasing. SWEET; delicate. SMOOTHLY CITRUS with a suggestion of ORANGE. FRAGRANT; slightly perfumed.
	Persistence:	The odor is quickly lost on exposure on a smelling strip.
	Dry-out:	Almost no residual odor.
Flavor		BITTER. FRESH; clean. CITRUS with a suggestion of both LEMON and ORANGE.
	After-taste:	BITTER.

Components of Grapefruit Oil

Terpenes constitute about 90% of the essential oil, mainly:	Oxygenated compounds include:
d-limonene	n-decyl aldehyde
α-pinene	linalool
	geraniol
	citral
	dimethyl anthranilate
	linalyl and geranyl esters

The components of grapefruit oil are the subject of papers by Moshonas (1971) and Coleman *et al.* (1972).

BIBLIOGRAPHY

Citrus Oils (General)

BALBAA, S. I. *et al.* 1971. Study of the peel oils of lemon, lime and mandarin growing in Egypt. Am. Perfum. Cosmet. *86*, No. 6, 53–56.

DI GIACOMO, A. 1971. Citrus oils. 9. The presence of individual constituents in various oils. Riv. Ital. Essenze, Profumi Piante Off. *53*, 117–119. (Italian)

DI GIACOMO, A. 1972. Non-volatile residues of citrus oils: Constitution, characteristics, analytical importance. Riv. Ital. Essenze, Profumi Piante Off. *54*, 159–165. (Italian)

DI GIACOMO, A. 1973. Estimates of the quality of cold-pressed citrus oils at the experimental station in Reggio Calabria. Riechst. Aromen, Koerperpflegem. *23*, 318,320,323. (German)

GUENTHER, E. 1949–1952. The Essential Oils, Vols. I to VI. D. Van Nostrand Co., New York.

KESTERSON, J. W. *et al.* 1971. Florida citrus oils. Agric. Exp. Sta., Gainesville, Florida, Bull. *749*.

KESTERSON, J. W. and BRADDOCK, R. J. 1975. Total peel oil content of the major Florida citrus cultivars. J. Food Sci. *40*, 931–933.

KESTERSON, J. W. and BRADDOCK, R. J. 1976. Method for determining the α-tocopherol content of citrus essential oils. J. Food Sci. *41*, 370–371.

MASDEN, B. C. and LATZ, H. W. 1970. Qualitative and quantitative *in situ* fluorimetry of citrus oil thin-layer chromatograms. J. Chromatog. *50*, 288–303.

MISITANO, F. 1971. Sicilian citrus, a survey. Riv. Ital. Essenze, Profumi Piante Off. *53*, 85–88. (Italian)

NAGY, S., SHAW, P. E. and VELDHUIS, M. K. 1977. Citrus Science and Technology, Vol. 1 and 2. AVI Publishing Co., Westport, Conn.

SCORA, R. 1975. Volatile oil components in citrus taxonomy. Int. Flavours Food Additives *6*, 343–346.

THIEBAUT, P. H. 1973. Production of citrus essential oils on the Ivory Coast. Riv. Ital. Essenze, Profumi Piante Off. *55*, 494–497. (French)

UBERTIS BOCIA, B. 1974. Relation between some I.R. bands of compounds present in citrus essential oils and their structure. Riv. Ital. Essenze, Profumi Piante Off. *56*, 747–748. (Italian)

UBERTIS BOCIA, B. *et al.* 1975. Minimum and C.D. % as quality evaluation criteria in citrus essential oils. Riv. Ital. Essenze, Profumi Piante Off. *57*, 19–24. (Italian)

VELDHUIS, M. K. *et al.* 1972. Oil- and water-soluble aromatics distilled from citrus fruit and processing waste. J. Food Sci. *37*, 108-112.

ZEIGLER, E. 1971. The examination of citrus oils. Flavour Ind. *2*, 647–653.

Grapefruit

COLEMAN, R. L. *et al.* 1972. Analysis of grapefruit essence and aroma oils. J. Agric. Food Chem. *20*, 100–103.

KESTERSON, J. W. *et al.* 1970. Spectrofluorometric and ultraviolet comparisons of Florida, California and Arizona citrus oils. II. Orange and Grapefruit. Am. Perfum. Cosmet. *85*, No. 10, 41–44, 46.

MOSHONAS, M. G. 1971. Analysis of carbonyl flavor constituents from grapefruit oil. J. Agric. Food Chem. *19*, 769–770.

SULSER, H. *et al.* 1971. The structure of paradisiol, a new sesquiterpene alcohol from grapefruit oil. J. Org. Chem. *36*, 2422–2426.

Lemon

DI GIACOMO, A. 1971. The yield and quality of "Femminello Commune" lemon oil in relation to conditions of cultivation and nutritional state of the trees. Riechst. Aromen, Koerperpflegem. *21*, 361–362, 364,366,368. (German)

DI GIACOMO, A. and CALVARANO, M. 1973. U.V. examination of lemon oil and control of its purity. Riv. Ital. Essenze, Profumi Piante Off. *55*, 310–311. (Italian)

EISERLE, R. J. 1970. Lemon: A new fragrance favorite. Soap Chem. Spec. *46*, 52, 54,70,72.
LUND, E. D. and BRYAN, W. L. 1976. Comparison of lemon oil distiled from comminuted mill waste. J. Food Sci. *41*, 1194–1197.
MOSHONAS, M. G. *et al.* 1972A. Analysis of volatile constituents of Meyer lemon oil. J. Agric. Food Chem. *20*, 751–752.
MOSHONAS, M. G. and SHAW, P. E. 1972B. Analysis of flavor constituents from lemon and lime essence. J. Agric. Food Chem. *20*, 1029–1030.
UBERTIS BOCIA, B. *et al* 1973. Analysis of lemon oil. Riv. Ital. Essenze, Profumi Piante Off. *55*, 662–671. (Italian)

Lime

AZZOUZ, M. A. and REINECCIUS, G. A. 1976. Comparison between cold-pressed and distilled lime oils through the application of gas chromatography and mass spectrometry. J. Food Sci. *41*, 324–328.
MOSHONAS, M. G. and SHAW, P. E. 1972. Analysis of flavor constituents from lemon and lime essence. J. Agric. Food Chem. *20*, 1029–1030.

Mandarin / Tangerine

COLEMAN, R. L. and SHAW, P. E. 1972. Analysis of tangerine essence oil and aroma oil. J. Agric. Food Chem. *20*, 1290–1292.
MOSHONAS, M. G. and SHAW, P. E. 1972. Analysis of volatile flavor constituents from tangerine essence. J. Agric. Food Chem. *20*, 96–99.
MOSHONAS, M. G. and SHAW, P. E. 1974. Quantitative and qualitative analysis of tangerine peel oil. J. Agric. Food Chem. *22*, 282–284.

Orange

COLEMAN, R. L. and MOSHONAS, M. G. 1970. Nootkatene from *Citrus sinensis* (orange peel oil). Phytochemistry *9*, 2419–2422.
COLEMAN, R. L. and SHAW, P. E. 1971. Analysis of Valencia orange essence and aroma oils. J. Agric. Food Chem. *19*, 520–523.
DI GIACOMO, A. and MAMIDELLO, M. 1973. On the U.V. spectrophotometric behavior of Italian sweet orange essential oil. Riv. Ital. Essenze, Profumi Piante Off. *55*, 593–598. (Italian)
KARAWYA, M. S. *et al.* 1971. Peel oils of different types of *C. sinensis* and *C. aurantium*, L. growing in Egypt. J. Phar. Sci. *60*, 381–386.
KESTERSON, J. W. *et al.* 1970. Spectrofluorimetric and ultraviolet comparisons of Florida, California and Arizona citrus oils. II. Orange and grapefruit. Am. Perfum. Cosmet. *85*, No. 10, 41–44,46.
LIFSHITZ, A. *et al.* 1970. Comparison of Valencia essential oil from California, Florida and Israel. J. Food Sci. *35*, 547–548.
LUND, E. D. and MOSHONAS, M. G. 1969. Composition of orange essential oil. J. Food Sci. *34*, 610–611.
LUND, E. D. *et al.* 1972A. Distribution of aqueous aroma components in the orange. J. Agric. Food Chem. *20*, 688–690.
LUND, E. D. *et al.* 1972B. Quantitative composition studies of water-soluble aromatics from orange peel. J. Agric. Food Chem. *20*, 685–687.
MOSHONAS, M. G. and LUND, E. D. 1969. Aldehydes, ketones and esters in Valencia orange peel oil. J. Food Sci. *34*, 502–503.

SHAW, P. E. and COLEMAN, R. L. 1971. Quantitative analysis of a highly volatile fraction from Valencia orange essential oil. J. Agric. Food Chem. *19*, 1276–1278.

SHAW, P. E. and COLEMAN, R. L. 1974. Quantitative composition of cold-pressed orange oils. J. Agric. Food Chem. *22*, 785–787, 796–800.

SHAW, P. E. and COLEMAN, R. L. 1975. Composition and flavor evaluation of a volatile fraction from cold-pressed Valencia orange oil. Int. Flavours Food Additives *6*, 190.

6

Vanilla

Vanilla and the various extracts and essences made from it is one of the most important natural flavorants used throughout the food, beverage and confectionery industries as well as domestically. The vanilla beans of commerce consist of the cured, fully grown but unripe fruits of *Vanilla planifolia,* Andrews, a climbing orchid indigenous to the moist forests of Central America, particularly Mexico; the best quality beans still come from Mexico. The better-known "Bourbon" vanilla beans are produced in the Malagasy Republic where crops of up to 1000 metric tons account for some 80% of the world's total annual production. Réunion and Comoro Islands to the north of Malagasy produce a further 150–200 tons and Mexico about another 100 tons annually. Other smaller producing areas include Guadaloupe, the Seychelles and Mauritius.

The vanilla which grows in Tahiti and Java is of a different species, *Vanilla tahitensis,* Moore, the cured beans of which have a totally different profile from those of *V. planifolia.*

The vanilla plant grows in the form of a vine, clinging to other trees and posts which are necessary for its support. The lemon-yellow flowers form in the second year of growth; these are short-lived and wither after only a few hours. The shape of the flower is tubular with an intricate petal design so formed that self-pollination is impossible. Pollination, in nature, is probably carried out by insects or by the humming birds which frequent the plantations. To ensure an optimum yield of beans, the flowers of the cultivated plants are pollinated by hand by skilled native workers on the plantation using a needle or a sharp pointed stick. The flowers occur in clusters and it is usual to limit the number pollinated in any one of these to ensure adequate growth room for the developing pod.

After the flower has dried off, approximately seven months elapse before the bean will be at the right stage for picking. This must be

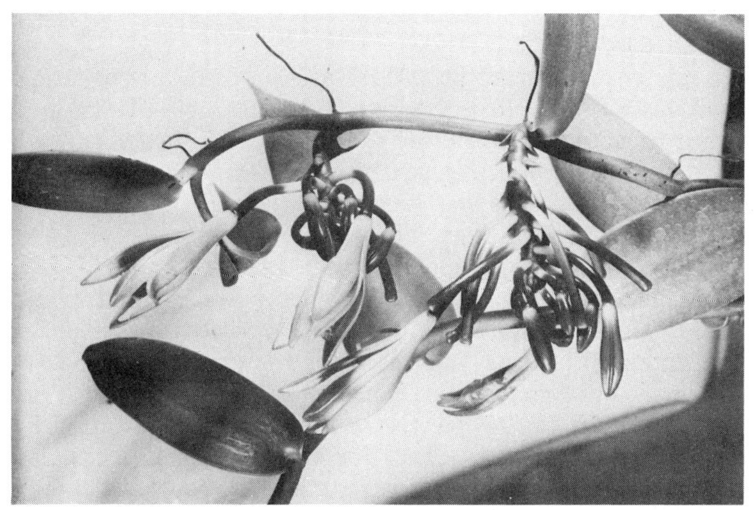

Courtesy of Information Service of Malagasy Republic

FIG. 6.1. VANILLA FLOWERS

carefully judged as the quality of the bean depends to a large extent upon the correct time of picking. The fresh pods are greenish-yellow and look not unlike runner-beans. When fully grown but as yet unripe they are picked by hand and sorted by size ready for the curing process. At this stage the beans are virtually odorless and have a decidedly unpleasant bitter flavor not in any way like that of the cured bean.

The Curing Process

The method of curing differs between the producing areas and this can have a significant influence on the quality and aromatic profile of the marketed bean. The methods employed have been reviewed by Sholto Douglas (1971) and Muralidharan and Balagopol (1973).

The Mexican method is the oldest still in use. The beans are first kept in sheds for a few days until they begin to shrink. They are then spread out on mats in the sun for a few hours, after which they are covered over. The beans are collected at night and placed in wooden containers lined with blankets in which they are allowed to sweat until the following day. This alternative heating and sweating is repeated daily for several weeks during which time the beans change to the familiar dark brown color. At this stage the beans are still too wet for use and are slowly dried, great care being taken not to disturb the fine aromatic balance nor to overdry to the point where the beans split.

The somewhat different procedure used in the Malagasy Republic has been described by Rosenbaum (1974). The sorted beans are first immersed in water at 65°C (150°F) for a few minutes to rupture the inner cell walls and possibly stimulate the enzyme system present. After draining, the still wet beans are packed into boxes and allowed to sweat for two days. Thereafter, the beans are laid out on trays, covered with a cloth and placed in the sun for about 6 hr a day over a period of about 1 week. The beans are covered over with blankets during the night. This sun-curing process is very uncertain and may take up to three months to complete, depending upon the prevailing weather conditions. Finally, a slow air-drying is carried out in sheds until the beans are in the correct condition for final sorting and tying into bundles. For the market, the bundles are usually packed in tin boxes lined with wax paper.

An alternative oven-drying method is gaining in favor as this enables a much closer control of temperature and relative humidity throughout the process.

Commercial vanilla beans are blackish-brown in color and vary in length from 6 to 10 in. and from ¼- to ½-in. in width. They are generally tied in bundles of 50 beans. On storage, good quality beans may become "frosted" with a mass of small needle-like crystals of vanillin (the "givre," as it is sometimes called). This is not necessarily a criterion of quality as is often assumed. Mexican beans seldom display this surface growth although they are the most esteemed from the point of view of flavor. The crystals do not appear in the producing areas where the climate is too hot and humid but develop only after shipment to cooler, dryer areas. The moisture content of most commercial samples is 30 to 50% but batches dried to a much lower moisture content are available and are preferred for the production of vanilla extracts.

The cured fruits of a wild vanilla, *Vanilla pompona*, Schiede, are sold under the name "vanillons." These are thicker and more fleshy than genuine vanilla beans. They are generally considered to be of poor quality for use in flavorings.

Profile of Vanilla Beans (Freshly-Minced)

Odor

Bourbon vanilla

Malagasy Republic	RICH, SMOOTH, FULL-BODIED.
La Réunion	SWEET, SPICY.

Comoro Islands	BALSAMIC with DRY back-note.
Seychelles	CHOCOLATE-like*; creamy. Slightly WOODY.
Mexican vanilla Mexico	SHARP, slightly PUNGENT. PENETRATING. SWEET, SPICY. DULL, TOBACCO-like.
Tahiti vanilla Tahiti Java	Strongly PERFUMED, FRAGRANT; flowery. HELIOTROPE-like. SWEET; to some, SICKLY.
Vanillon	Strongly PERFUMED; floral. HELIOTROPE-like. SWEET but lacking in body. SPICY, slightly ANISE-like.

*This response is one of association with a familiar product flavored with vanillin.

Flavor

The intrinsic flavor of vanilla beans is aromatic with a profile similar to that described above but unpleasantly SHARP and ACIDIC, slightly BITTER with a pronounced PUNGENCY. The after-taste is slightly ASTRINGENT. These characters are not normally observed in vanilla extracts owing to the modifying effects of added sugar, etc., but they are present in so-called vanilla oleoresins.

Components of Vanilla

Very little is known about the biochemistry of the flavor precursors in the developing bean nor of the reactions which take place during the curing process, although this is known to be enzymic and modified by the curing conditions. A preliminary study of the enzyme systems active during the growth period (Wild-Altamirano 1969) noted that glucosidase, peroxidase and polyphenoloxidase all reach a maximum at or near the ripening stage. It was concluded that these three systems may play a significant role in the formation of vanillin and nonvanillin aromatic components during the curing process.

The principal flavouring component is vanillin (3-methoxy-4-hydroxy benzaldehyde) which is present from 0.5 to 2.5% within the following ranges:

Bourbon vanilla	1.9–2.5
Mexican vanilla	1.3–2.0
Tahiti vanilla	1.0–2.0

Courtesy of Dr. Wm. Mitchell

FIG. 6.2. VANILLA BEANS READY FOR HARVESTING

Courtesy of Tropical Products Institute

FIG. 6.3. CURING VANILLA BEANS IN THE SEYCHELLE ISLANDS

Courtesy of Information Service of Malagasy Republic

FIG. 6.4. SORTING AND PACKING CURED VANILLA BEANS ACCORDING TO LENGTH OF THE BEAN

Courtesy of Bush Boake Allen, Ltd.

FIG. 6.5. A BUNDLE OF VANILLA BEANS WELL "FROSTED" WITH VANILLIN CRYSTALS

The precise figures quoted in the literature are difficult to interpret as they depend not only on the method of assay but also on the grade and moisture content of the beans tested. For comparative purposes it is necessary to quote the vanillin content either on a "dry" basis or on a standard moisture content of 25%. The quality of vanilla beans is not defined entirely by the vanillin content.

The following nonvanillin compounds also contribute to the aromatic profile:

vanillic acid (4-hydroxy-3-methoxy benzoic acid)
p-hydroxy benzoic acid
p-coumaric acid

In Tahitian vanilla the following additional compounds have been identified:

heliotropin (piperonal)
aubepin (*p*-methoxy benzaldehyde)
anisyl alcohol

Other constituents include reducing sugars (about 10%), fats and waxes (about 11%), gums and some resinous matter.

Vanilla Extracts

Before use, vanilla beans must be further processed as their form, color, texture and fatty nature makes direct incorporation into a food product of limited applicability. An homogenized purée of vanilla beans has been used in the flavoring of ice-cream but the presence of unsaturated fatty matter poses problems of stability and shelf-life. It is customary to use vanilla in the form of an alcoholic essence or extract and many such products are available commercially. By definition, the U.S.A. standard for vanilla extract requires the finished product to contain the extractable matter from 13.35 oz of vanilla beans (either *Vanilla planifolia* or *V. tahitensis*) having a maximum moisture content of 25% per U.S. gal. In metric terms this is stated as not less than 10 g of vanilla beans at 25% moisture, or 7.5 g on a moisture-free basis, per 100 ml. Such extracts are known as "single-fold." Higher strength extracts must contain a proportionate quantity of beans. The term "fold" is generally applied to vanilla extracts to indicate their relative strength, thus a double-strength extract is called "two-fold," etc. Such extracts normally contain about 35% of ethanol although nonalcoholic extracts, usually based on an acceptable solvent such as propylene glycol, are also available.

The much stronger ten-fold extracts are often made by dissolving an "oleoresin vanilla" in the appropriate strength alcohol. Oleoresin or resinoid vanilla is a dark brown semifluid extract obtained by alcoholic or other solvent extraction of the finely-chopped beans and involves the complete removal of the solvent under vacuum. Inevitably, the finer top-notes of the aromatic profile as well as some of the deeper undertones of the flavor are lost or modified; and diluted extracts made from these products do not have the full flavor character usually associated with a vanilla extract made directly from the bean.

The U.S.A. Standards of Identity also define the use of blended extracts containing both the natural essence and synthetic vanillin. In these, 1 oz of vanillin may be added per unit of vanilla extract. For example, a mixture claiming to be four-fold may consist of two-fold vanilla extract plus 2 oz per U.S. gal. of added vanilla. From the point of view of flavoring strength, 0.7 parts of vanillin is approximately equivalent to 10 parts of vanilla bean, but, as is to be expected, direct comparison is difficult as the two profiles are widely different. Although synthetic vanillin is distinctly "vanilla-like," it lacks body, tenacity and much of the fine flavor character associated with genuine vanilla bean extracts.

The problem of establishing the authenticity of vanilla extracts has, over the past decade, been the subject of considerable study by the Association of Official Analytical Chemists. A recent proposal by Martin *et al.* (1975) involves the determination of the concentration of potassium, inorganic phosphate and nitrogen and the calculation of the ratio of these factors against the vanillin content. The resulting "identification ratios" for a vanilla extract sample can be compared with those of a known authentic control sample to determine whether or not the product under test has been adulterated by the addition of vanillin or ethyl vanillin.

The creation of imitation vanilla flavors to replace the natural extract has now reached a very high level of quality and there are few products in which these flavorings cannot be used with complete success to give a product having a totally acceptable flavor profile at a considerable cost saving when compared with the natural extract. It is usual for such imitation flavors to be equivalent in usage to a ten-fold natural extract although other strengths are also available for specific purposes.

Vanillin

Any discussion on vanilla must take into account the very considerable tonnage of synthetic vanillin which is used as a source of this popular flavor in ice-cream, nonalcoholic beverages, table desserts, etc. It is now just over a century since vanillin was first precisely identified. The history and the many methods developed for the synthesis of this important synthetic flavoring material have been reviewed by Bedoukian (1967) and Rosenbaum (1974). The original synthetic route to vanillin started with eugenol derived from clove oil and until almost 40 years ago this method remained economically viable and resulted in a product of very acceptable quality. At that time, researchers in the paper-pulp industry discovered that vanillin could be made from raw sulfite liquors available in large quantities as a trade waste. Today this is the prime source of vanillin.

Ethyl Vanillin (Vanbeenol)

A closely-related compound, 3-ethoxy-4-hydroxy benzaldehyde, is not found in nature but is prepared synthetically from safrole. It has an intense vanilla-like odor and in flavoring strength is between 3–4 times more powerful than vanillin. Like vanillin, it is widely used in the blending of imitation vanilla flavors but too high a dosage level results in an unacceptable, harsh, "chemical" character. In practice, a maximum of 10% of vanillin may be replaced by ethyl vanillin without this objectionable note being obvious.

Profiles of Vanilla Extracts

At normal usage levels (0.0125%) the profile of vanilla is almost impossible to describe in precise terms as so much depends upon the medium in which it is tested. If sugar or any other sweetening agent is present in the extract, this, too, will have a modifying effect. In general, genuine vanilla extracts impart a smooth, rich, creamy character to products which may indeed be synergistic at subthreshold levels. Imitation vanilla flavors tend to be harsh by comparison and convey the "vanilla" note at a lower dosage.

BIBLIOGRAPHY

ANON. 1972. Vanilla bean production in the Malagasy Republic. Flavour Ind. *3*, No. 6, 307–309.

BEDOUKIAN, P. Z. 1967. Perfumery and Flavoring Synthetics. Elsevier Publishing Co., Amsterdam, London and New York.

BOHNSACK, H. 1967. Contribution to the knowledge of essential oils, fragrances and flavors. XVIII. The constituents of Bourbon vanilla pods. Part IV. Riechst. Aromen, Koerperpflegem. *17*, 133–134,136. (German)

BOHNSACK, H. 1971A. Contribution to the knowledge of essential oils, fragrances and flavors. XXV. On the constituents of Bourbon vanilla pods. Part V. Riechst. Aromen, Koerperpflegem. *21*, 125–126,128. (German)

BOHNSACK, H. 1971B. Contribution to the knowledge of essential oils, fragrances and flavors. XXVI. On the constituents of Bourbon vanilla pods. Part VI. Riechst. Aromen, Koerperpflegem. *21*, 163–164, 166. (German)

COWLEY, E. 1973. Vanilla and its uses. Tropical Prod. Inst. Conference Papers, London, 79–82.

FITELSON, J. and BOWDEN, G. L. 1968. Determination of organic acids in vanilla extracts by G.L.C. separation of trimethylsilylated derivatives. J. Assoc. Off. Anal. Chem. *51*, 1224–1231.

JACKSON, H. W. 1966. Gas chromatographic analysis of vanilla extracts. J. Gas Chromatogr. *4*, 196–197.

JAMINET, L. V. 1968. Vanilla, vanillin and vanilla flavours. Riechst. Aromen, Koerperpflegem. *18*, 232–233,236. (German)

MURALIDHARAN, A. and BALAGOPOL, C. 1973. Studies on curing of vanilla. Indian Spices. *10*, No. 3, 3–4.

MARTIN, G. E. *et al.* 1975. Determining the authenticity of vanilla extracts. Food Technol. *29*, No. 6, 54,56,58–59.

OLIVER, R. 1973. Methods for the study of the aromatic components of vanilla extracts. Tropical Prod. Inst. Conf. Papers, London, 49–54,55–60.

POTTER, R. H. 1971. Non-vanillin volatiles in vanilla extracts. J. Assoc. Off. Anal. Chem. *54*, 39–41.

ROSENBAUM, E. W. 1974. Vanilla extract and synthetic vanillin. *In* Encyclopedia of Food Technology. A. H. Johnson and M. S. Peterson (Editors). AVI Publishing Co., Westport, Conn.

STAHL, W. H. *et al.* 1960. Analysis of vanilla extracts. II. Preparation and reproduction of two dimensional fluorescence chromatograms. J. Assoc. Off. Anal. Chem. *43*, 606–610.

SULLIVAN, J. H. *et al.* 1960. Analysis of vanilla extracts. I. Organic acid determination. J. Assoc. Off. Anal. Chem. *43*, 601–605.

SCHOEN, K. L. 1972. Organic acids in vanilla by gas liquid chromatography: Interpretation of analytical data. J. Assoc. Off. Anal. Chem. *55*, 1150–1152.

SCHOFIELD, M. 1968. The genus *Vanilla* versus synthetic vanillin. Perfum. Essent. Oil Rec. *59*, 582–583.

SHOLTO, DOUGLAS, J. 1971. Producing vanilla beans. Flavour Ind. *2*, 405–407.

THEODOSE, R. 1973. Traditional methods of vanilla preparation and their improvement. Trop. Sci. *15*, No. 1, 47–57.

WILD-ALTAMIRANO, C. 1969. Enzymic activity during the growth of vanilla fruit. I. Proteinase, peroxidase and polyphenoloxidase. J. Food Sci. *34*, 235–238.

YANICK, N. S. 1963. Treatment of vanilla to increase flavor strength. (Assigned to National Dairy Products Corp.) U.S. Pat. 3,112,204, Nov. 26.

SECTION II
FLAVOR PRODUCTS

7

Flavoring Materials

Part 1: Introduction

The production of food has undergone considerable change over the past century, the emphasis of its preparation having moved progressively from the kitchen to the factory. This transition has been accompanied by significant changes in the types of food we eat, and the methods by which it is prepared, packaged and distributed. Much time and effort has been expended by manufacturers to recreate factory-processed food products having a traditional "home cooked" appeal. Such products must have many other attributes such as stability, convenience and safety which also are demanded by consumers. Even so, the apparent lack of flavor in many modern foodstuffs is regretted by those who claim to remember when everything tasted so much better than it does now. There is no lack of interest in matters affecting food and drink, particularly when it comes to flavor.

Many well-established food products and beverages could not be produced without recourse to the use of added flavorings. So much so, that it is estimated that by 1980 the sale of flavors and flavor enhancers in the United States will have reached some $476 million per annum representing about 40% of the total food additive market.

The choice between natural and synthetic flavorings is still a subject for debate; and this is likely to become more emotive as the demand for safe and wholesome food increases and a greater proportion of what we eat and drink becomes subject to mass production methods and its associated standardization. The problems of flavoring fabricated foods are numerous and complex (Downey and Eiserle 1970); they are not amenable to facile answers.

To cater for the needs of modern food technology, a very wide range of natural, synthetic and blended flavoring materials are available. In this chapter these will be discussed in convenient categories dependent upon physical form or some other well-established relationship.

BIBLIOGRAPHY

DOWNEY, W. J. and EISERLE, R. J. 1970. Problems in the flavoring of fabricated foods. Food Technol. *24*, No. 11, 38, 40–41.

Part 2: Materials Used to Make Beverages

The term "beverage" includes a very wide spectrum of liquids ranging from milk; through soft-drinks, which are being consumed in increasing quantity; to the hot drinks, such as chocolate, coffee and tea; to the alcoholic brews and distillations which are enjoyed universally. To review all of these would serve little purpose. For that reason, the discussion will be confined to those materials of most concern to the flavorist or the food technologist either as a source of flavoring, or as a target for imitation, or as a carrier for other flavoring effects.

CACAO (COCOA)

When Cortez and the conquestadors landed in Mexico in 1519, they found that the Aztecs had for centuries been using the ground roasted seeds of the cacao tree to make an extremely unusual and pleasant drink which they called "chocolatl." In due course, the tree was studied and classified as *Theobroma cacao*, L., which in Greek means "food for the gods." The beverage was first introduced into Spain from where its popularity spread throughout Europe. Today, it is universally appreciated and cacao is cultivated in most equatorial countries. The main commercial crops of cocoa beans come from Ghana, Brazil, Nigeria, Dominica, Ecuador and Venezuela with much smaller quantities from the West Indies.

To avoid confusion over nomenclature it is well to note the following descriptions:

Cacao	refers to the plant *Theobroma cacao*, L.
Cocoa	refers to the beans and the product ready for making the beverage; to its use as a flavoring material; or to its use for making the chocolate beverage.
Chocolate	refers to the product based on ground roasted cocoa beans, normally in a solid form; although powdered "drinking chocolate" is also available.

The cacao tree is a tall perennial evergreen bearing shiny leathery leaves and pod-like fruits on the main branches and on the trunk. The fruits, which are 2–4in. in diameter and 7–12in. long have a leathery skin which is initially dark green. Color changes from dark green through yellowish-orange to purple-red when fully ripe. Each contains up to 75 seeds embedded in a pinkish pulp. It is the seeds which, after processing, become the cocoa beans of commerce.

The ripe pods are cut from the trees, split open and the pulp and seeds removed. The mass is then covered and allowed to ferment for several days. This process is designed to kill the viability of the seed, to soften the pulp to facilitate subsequent separation and cleaning of the seeds, and to commence the complex enzymic reactions leading to the development of the characteristic flavor. The care with which this process is carried out has a significant effect upon the ultimate color and flavor of the roasted beans. The fermented beans are excessively wet and must be dried prior to storage or shipment. During this curing stage the beans change to the familiar brown color and are ready for further processing. As marketed, the cocoa beans consist of about 14% outer shell and 86% kernel or "nib."

According to Jacobs (1951) there are two main types of cacao, Crillo and Forastero, which produce beans having quite different flavor characteristics; but today the distinction is less pronounced due to considerable cross-breeding and hybridization. In commercial practice two categories are recognized. These are known as "bulk or basic beans" and "flavor beans." Within these groups there are many distinct varieties generally known by a recognized common name or by source (Lees and Jackson 1973; Allerton 1974). Selection of beans for roasting and ultimate processing requires a knowledge of the characteristics of these varieties; and blending them to achieve a particular flavor effect is a skilled operation. Basic or bulk beans have a strong, harsh flavor character whereas that of flavor beans is smoother and more aromatic.

The Cocoa Flavor

Raw cocoa nibs do not possess the distinctive aroma and flavor of cocoa, this being developed only as a result of the fermentation and roasting processes. Both of these have been shown to be necessary for the production of the characteristic profile (Rohan and Stewart 1967). The optimum conditions necessary to produce a high quality product are still a matter of expert judgement (Riedel 1974).

Considerable research has taken place into the nature of the flavor precursors in cocoa nibs and the involved reactions by which these are converted into the ultimate flavor complex. Three stages are recognized:

(a) *Fermentation*, during which the flavor precursors are formed. The proteins present in the nibs are degraded and and the level of free amino acids rises; sucrose is inverted to fructose and glucose which in turn are oxidized to alcohol and various acids; some theobromine and tannins are lost (Quesnel 1965; Rohan and Stewart 1967).
(b) *Drying*, during which the moisture content is reduced to about 8% with a loss of volatile acids and a consequent rise in pH.
(c) *Roasting*, during which the precursors are converted into aromatic compounds, the qualitative and quantitative nature of which determines the characteristic profile of the roasted cocoa beans. The nature of the reactions involved has still to be finalized, although many of the individual routes from precursor to aromatic compound have been reported (Maga and Sizer 1973).

The following groups of compounds have been reported as present in roasted cocoa (Van Praag *et al.* 1968; Vitzthun *et al.* 1975):

phenyl alkenals	pyrazines
oxazoles	quinoxalines
pyridines	quinoline
cycloalkapyrazines	

Cocoa beans are the source of three other important materials:
(a) Cocoa butter: a mixture of glycerides which are present from 52–56% in the nib.
(b) Theobromine: an alkaloid related to caffeine but having very little stimulant effect.
(c) Cacao purple: an astringent coloring matter formed in the fermenting bean.

Chocolate

Chocolate is the product made by grinding freshly-roasted and winnowed cocoa nibs. It contains 50–55% cocoa butter. When freshly made, it is liquid due to the heat generated by the milling process. This bitter chocolate liquor is usually cooled in molds to facilitate later handling. Milk chocolate is made from it by blending the liquor with sugar, milk solids and usually vanilla or some other flavoring such as cassia or cinnamon.

Cocoa Powder

Cocoa powder is a valuable flavoring material in baked goods and desserts as well as the basis of hot chocolate drinks. It is produced by finely milling the press-cake after partial removal of cocoa butter from the roasted nibs by hydraulic pressure. This removes most of the fats but still leaves about 22% in the press-cake. The flavor is little affected by this process and, in consequence, the cocoa notes are more concentrated in the powder than in the chocolate liquor.

The commercial powder varies in color and flavor, dependent upon the quality of the beans used, the degree of roasting and the precise method of processing.

A good quality cocoa powder has the following characteristics:

pH:	5.6 to 7.1 dependent upon whether or not the cocoa beans were procesed with alkali.
Fat:	about 24% and not less than 22%.
Moisture:	3–4%, but many be as high as 7%.

The standards imposed in the United States and the methods of analyses are discussed by Hart and Fisher (1971); and in the United Kingdom by Pearson (1970).

Cocoa Extract

A fine flavor extract can be prepared from ground, roasted cocoa nibs by maceration with 50% aqueous ethanol and subsequent pressing, concentration being achieved by a re-use of the liquor to extract further quantities of nibs. For an extract having the finest aroma, the use of heat is to be avoided. The high price of cocoa makes such a natural flavor very expensive, even for high quality liqueur work. Also, many good imitation cocoa and chocolate flavors are now available.

Imitation Flavorings

The reconstruction of a really close imitation of cocoa or chocolate flavor to replace the use of cocoa poses many problems (Broderick 1973). The majority of those currently available are designed for use with, and as, a partial replacement of cocoa powder in many types of products. The formulation of such flavorings depends upon the use of compounds which have been identified as being present in the natural flavor and which have been determined as safe for use in food flavorings. Many more have yet to be accepted and, in addition, there are numerous trace components, which probably have a significant influence upon the profile, and which are still to be identified and synthesized. Already a very impressive list of components has been reported as present in cocoa aroma (Marion *et al.* 1967; Flamant *et al.* 1967; van Straten and de Vrijer 1973); although the relative contribution which these make to the profile has yet to be established. Vanillin is frequently used in imitation cocoa flavors but this tends to soften the profile and makes it closer to that of chocolate.

COFFEE

The use of coffee as a refreshing beverage dates back at least 700 years although the first coffee houses did not appear in London and Paris until about 1650. Its use as a food and as a medicinal stimulant is lost in history. The roasting of coffee beans to give the aroma and flavor, which is universally recognized and appreciated, started during the 13th Century and the quality of the aromatic profile is still a matter of expertise in carrying out this process.

Coffee is a large evergreen shrub of which *Coffea arabica*, L. and *C. robusta*, L. are the most important commercial species existing in many varieties. Several other species are encountered but these are of less importance; the flavor of these beans is generally considered to be coarse, bitter and much less acceptable than that of *C. arabica*.

Coffee is a native of Ethiopia from where it spread initially to the other countries bordering the Red Sea, then to India and Sri Lanka. During the period of colonial expansion, it was successively introduced into Indonesia, Brazil, Colombia, El Salvador, the Phillipines and Mexico where the altitude and climate are most suitable for its growth. Presently, about 45% of the world's crop is produced in Brazil; 25% in Colombia; 6% in Mexico; and the remainder in numerous other countries of Central America, the northern part of South America and Africa.

The tree bears clusters of small, red, cherry-like fruits which contain two seeds. The fleshy berries are harvested almost continuously. Depending on the growing region, the whole fruits are either sun-dried or they are allowed a short period of fermentation after which the pulp is removed from the greenish-yellow seeds. These are cleaned and sun-dried to become the green coffee beans of commerce.

The Coffee Flavor

Green coffee beans are almost devoid of aroma and the flavor of a boiling water infusion is bitter and unattractive as a beverage. The pleasing, characteristic flavor is only developed when the beans are roasted. The source of the beans, the rate and method of roasting, and the final temperature achieved, all have a significant effect not only upon the aromatic profile but upon the solubility of the flavoring components present.

During the roasting process the beans change from the greenish-yellow color to dark brown. At about 204°C (400°F), just before completion of the roasting process, the beans suddenly expand to about twice their size. At this stage, the roasting is halted by rapid cooling under carefully-controlled conditions, a technique which demands considerable skill if the final product is to have a consistent flavor. The rate and depth of the color change is often used as a means of controlling the roasting process and, hence, of the resultant aromatic profile. Preferences range from a "light roast" which yields a soft reddish-brown powder and a light brown infusion to "high roast" which gives a very strong dark coffee with a blackish tint and a distinctly burnt odor note. The finest flavor is achieved in beverages made from freshly-ground beans. Unfortunately the fine aroma and flavor associated with freshly-ground coffee is subject to rapid deterioration. If the roasted beans are preground for sale, the product must be stored in sealed containers so as to minimize the volatile loss and other changes which result in coffee having a flat profile lacking the attractive characteristic top notes.

From the flavor point of view, coffee is very interesting. Of all the many varieties produced there are really only two main groups:

(a) Brazils: Coffee with a rich mellow aroma and flavor which is full-bodied and has a fine acidity. These are derived from Santos, Rio de Janero, Victoria and Bahia.

(b) Milds: All coffee grown elsewhere including those from Colombia, Venezuela, Guatemala, El Salvador, Nicaragua, Costa Rica, Cuba, Santo Domingo, Haiti, Jamaica, Kenya, Java, Sumatra and Hawaii, each of which produces coffee having a characteristic odor and flavor profile.

The chemistry of the components of both green and roasted coffee and the nature of the changes which take place during the roasting process, which is, in effect, a mild pyrolysis, have been the subjects of much research (Lockhart 1957; Gianturco 1967) and are, as yet, incompletely understood. At least 103 aromatic compounds have been reported in the literature (van Straten and de Vrijer 1973) but work is still in progress to establish their relative importance in the reconstruction of a coffee flavor. Most of the components which have been isolated and characterized are derived from the water-soluble components, mainly sugars, in the green beans. The fat, which is present to about 13% in the raw bean, is little changed by roasting; but it has the ability to absorb and so protect the aromatics as they are formed reducing their evaporation or degradation during the final stages of the roasting process.

Instant Coffee

A considerable amount of information about coffee volatiles has been obtained by detailed studies of the various stages of manufacture of instant, or soluble, coffee (Sivetz 1963). Soluble coffee, which is now universally accepted as a ready means of making a most acceptable beverage, is made by extracting the ground roasted beans with hot water and then removing this water either by spray drying or by the more sophisticated technique of freeze drying. Modern manufacturing methods include agglomeration and aromatization which result in a final product which is readily soluble and has an odor and flavor more nearly that of freshly ground coffee (Sivetz 1974).

Caffeine and Decaffeinated Coffee

As a beverage, coffee has no nutritive value but it is, nevertheless, an invaluable and refreshing stimulant which finds a definite place in helping to maintain our modern speed of life. This is due to the presence of 1.2–1.9% of caffeine, which exists in the raw beans and survives the roasting process with little loss. This crystalline alkaloid is tasteless and, being moderately soluble in water, is normally extracted in the production of a coffee beverage. In view of the demand for caffeine-free coffee, this alkaloid may be extracted from the dry bean, either before or after roasting, by nonpolar solvents without detriment to the flavor. Several patents exist for this process and about 1/6 of the instant coffee sold in the United States is now decaffeinated.

TEA

What is generally meant by "tea" is a hot-water infusion of the dried broken leaves of *Thea sinensis* (L) Sims, an evergreen shrub which grows throughout the tropics from sea level to about 6000 ft. The plants from the higher altitudes yield the most esteemed varieties.

As with most natural products, considerable differences exist between teas grown at different altitudes as well as teas collected at different seasons; but of more significance is the method by which the leaves are prepared for market. The trade distinguishes between the countries of origin and often between growing regions; but, in addition, teas are often classified by the method of postharvest handling and preparation for market. There are three main types:

(a) Fermented Teas (Black Tea)		Description
Derived from:	India, Sri Lanka, Java, Sumatra, Pakistan, Japan, Taiwan, Malawi, Kenya	Orange Pekoe Pekoe Souchung

(b) Unfermented Teas (Green Tea) — Description

Derived from: Japan, China, India, Indonesia.

Basket fired
Pan fired
Gunpowder
Imperials
Young Hysons

(c) Semifermented Teas (Ooling Tea)
Derived from: Taiwan, Hong Kong.

Black and Green Tea

Black and green teas result from different postharvest treatment. In the case of black teas, the freshly picked green leaves are spread out on mats and allowed to dry out slowly and wither after which they are mechanically rolled to break up the cells and liberate any enzymes which may be present. The rolled leaves are then fermented for several hours, either in baskets or spread out under damp cloths. During this stage the colorless tea tannins are partially oxidized changing to a reddish-brown color and some essential oil is developed. This process is stopped by subsequent hot-air drying (firing). The whole process determines the flavor, strength, smoothness and color of the resulting infusion and calls for skill and experience to achieve consistent results.

Green teas are not fermented. The freshly picked leaves are steamed which prevents fermentation and blackening during the final drying process. The tea tannins are, in consequence, largely in their original state and as no essential oil is developed unfermented teas lack the aromatic character of the fermented varieties.

Oolong teas have a half-way character; the leaves are allowed to wither and partially ferment before being rapidy dried at a high temperature. This results in a tea having a fine fragrant character.

The Flavor of Tea

The principal components which determine the aroma, flavor and physiological action of tea are:

(a) An essential oil: about 0.5%, which is probably formed during fermentation.

(b) Caffeine: 1.8 to 5.0%.

(c) Tannins: 13 to 18%.

Some 140 components have been reported as contributing to the aroma and flavor of tea (van Straten and de Vrijer 1973) and research in this field is continuing (Sanderson and Graham 1973; Nguyen and Yamanishi 1975).

The aim of making good tea is to obtain the maximum extraction of caffeine and the minimum amount of tannins. Most commercially available teas are blends, designed to satisfy the tastes of the customer. One would think that making a cup of tea would be quite a simple matter. This is far from so; there are many schools of thought as to how this should be done. These range from the very elaborate and formalized ritualistic proceedings of the Japanese tea cult, to the homely brew which is being increasingly appreciated around the world.

Instant Tea

The production of an "instant tea," to complement "instant coffee" in the ubiquitous vending machine, proved to be a very difficult technological problem. The currently available products do not occasion high praise but they are at least acceptable. They cannot really claim with any justification to satisfy the palate of those who enjoy a freshly brewed cup of tea. In spite of this, the sale of instant tea has shown an unprecedented growth over the past 20 years (Peterson 1974).

BIBLIOGRAPHY

Cocoa

ALLERTON, J. 1974. Chocolate and cocoa products. In Encyclopedia of Food Technology. A. H. Johnson and M. S. Peterson (Editors). AVI Publishing Co., Westport, Conn.

BRODERICK, J. J. 1973. Developing cocoa flavor by brainstorming. Food Prod. Dev. *7*, No. 9, 71,74.

FLAMANT, I. *et al.* 1967. Studies in aromas. Part 16. Cocoa aroma. Helv. Chim. Acta *50*, 2233–2243. (French).

HANSEN, A. P. and KEENEY, P. G. 1970. Comparison of carbonyl compounds in moldy and non-moldy cocoa beans. J. Food Sci. *35*, 37–40.

HART, F. L. and FISHER, H. J. 1971. Modern Food Analysis. Springer-Verlag, New York, Heidelberg and Berlin.

JACOBS, M. B. 1951. The Chemistry and Technology of Food and Food Products, 2nd. Edition. Interscience Publishers, New York.

LEES, R. and JACKSON, E. B. 1973. Sugar Confectionery and Chocolate Manufacture. Leonard Hill Books, London.

LOPEZ, A. and QUESNEL, V. S. 1973. Volatile fatty acid production in cacao fermentation and the effect on chocolate flavor. J. Sci. Food Agric. *24*, 319–326.

MAGA, J. A. and SIZER, C. E. 1973A. Cocoa flavor. CRC Critical Reviews in Food Technology *4*, 39.

MAGA, J. A. and SIZER, C. E. 1973B. The aroma and flavor of cocoa. J. Agric. Food Chem. *21*, 22.

MARION, J. P. *et al.* 1967. On the composition of cocoa aroma. Helv. Chim. Acta *50*, 1509–1516. (French).

PARSONS, J. G. *et al.* 1969. Identification and quantitative analysis of phospholipids in cocoa beans. J. Food Sci. *34*, 544–546.

PEARSON, D. 1970. The Chemical Analysis of Foods, 6th Edition. J. & A. Churchill, London.

PICKENHAGEN, W. *et al.* 1975. Identification of the bitter principle of cocoa. Helv. Chim. Acta *58*, 1078–1086.

REINECCIUS, G. A. *et al.* 1972A. Factors affecting the concentration of pyrazines in cocoa beans. J. Agric. Food Chem. *20*, 189–192.

REINECCIUS, G. A. *et al.* 1972B. Identification and quantification of the free sugars in cocoa beans. J. Agric. Food Chem. *20*, 199–202.

RIEDEL, H. R. 1974. Effects of roasting on cocoa beans. Confect. Prod. *40*, 193–194.

ROHAN, T. A. 1965. Application of gas chromatography to quantitative assessment of chocolate aroma potential of unroasted cocoa beans. Food Technol. *19*, No. 11, 122–124.

ROHAN, T. A. 1970. Food flavor volatiles and their precursors. Food Technol. *24*, No. 11, 20–37.

ROHAN, T. A. and STEWART, T. 1967A. The precursors of chocolate aroma: Production of free amino acids during fermentation of cocoa beans. J. Food Sci. *32*, 395–398.

ROHAN, T. A. and STEWART, T. 1967B. Production of reducing sugars during fermentation of cocoa beans. J. Food Sci. *32*, 399–402.

ROHAN, T. A. and STEWART, T. 1967C. Application of gas chromatography in following formation during fermentation of cocoa beans. J. Food Sci. *32*, 402–404.

VAN DER WAL, B. *et al.* 1971. New volatile components of roasted cocoa. J. Agric. Food chem. *19*, 276–280.

VAN PRAAG, M. *et al.* 1968. Steam volatile constituents of roasted cocoa beans. J. Agric. Food Chem. *16*, No. 6, 1005–1008.

VAN STRATEN, S. and de VRIJER, F. 1973. Lists of Volatile Compounds in Food, 3rd. Edition plus Supp. 1–6. Central Institute for Nutrition and Food Research T. N. O., Zeist, The Netherlands.

VITZTHUN, O. G. *et al.* 1975. Volatile components of roasted cocoa; basic fraction. J. Food Sci. *40*, 911–916.

WEISSBERGER, W. and KEENEY, P. G. 1971. Identification and quantification of several non-volatile organic acids of cocoa beans. J. Food Sci. *36*, 877–879.

Coffee

BUCHI, G. *et al.* 1971. Structure and synthesis of kahweofuran, a constituent of coffee aroma. J. Org. Chem. *36*, 199–200.

CORSE, J. *et al.* 1970. Isolation of chlorogenic acids from roasted coffee. J. Sci. Food Agric. *21*, 164–168.

FRIEDEL, P. *et al.* 1971. Some constituents of the aroma complex of coffee. J. Agric. Food Chem. *19*, 530–532.

GHOSH, J. J. and BHATTACHARYA, K. C. 1972. Ribonucleic acid from coffee beans. Phytochemistry *11*, 3349–3353.

GIANTURCO, M. A. 1967. Coffee flavor. *In* The Chemistry and Physiology of Flavors. H. W. Schultz, E. A. Day and L. M. Libby (Editors). AVI Publishing Co., Westport, Conn.

JO-FEN. T. KUNG. 1974. A new caramel compound from coffee. J. Agric. Food Chem. *22*, 494–496.

LOCKHART, E. E. 1957. Chemistry of coffee. *In* Chemistry of Natural Food Flavors. Quartermaster Food and Container Institute for the Armed Forces, Washington, D. C.

OBERMANN, H. and SPITELLER, G. 1975. 16,17-dihydroxy-9(11)-kauren-18-oic acid—a compound of roasted coffee. Chem. Ber. *108*, 1093–1100. (German)
PARLIMENT, T. H. *et al.* 1973. Trans-2-nonenal: Coffee compound with novel organoleptic properties. J. Agric. Food Sci. *21*, 485–487.
PETERSEN, E. E. *et al.* 1973. Influence of freeze-drying parameters on the retention of flavor compounds of coffee. J. Food Sci. *38*, 119–122.
PFLUGER, R. A. 1969. Aromatization of instant coffee. (Assigned to General Foods Corp.) Can. Pat. 823,142, Sept. 16.
SEGALL, S. *et al.* 1970. The effect of reheat upon the organoleptic and analytical properties of beverage coffee. Food Technol. *24*, No. 11, 54, 58.
SIVETZ, M. 1963. Coffee flavour identity and aroma/taste response. Coffee Tea Ind. *84*, Aug., 9–10.
SIVETZ, M. 1972. How acidity affects coffee flavor. Food Technol. *26*, No. 5, 70, 72, 74, 76–77.
SIVETZ, M. 1974. Coffee. *In* Encyclopedia of Food Technology. A. H. Johnson and M. S. Peterson (Editors). AVI Publishing Co., Westport, Conn.
TASSAN, C. G. and RUSSELL, G. F. 1974. Sensory and gas chromatographic profiles of coffee beverage headspace volatiles entrained on porous polymers. J. Food Sci. *39*, 64–68.
VITZTHUN, O. G. and WERKHOFF, P. 1974. Oxazoles and thiazoles in coffee aroma. J. Food Sci. *39*, 1210–1215.
VITZTHUN, O. G. and WERKHOFF, P. 1975. Cycloalkapyrazines in coffee aroma. J. Agric. Food Chem. *23*, 510–516.
WEIDEMANN, H. L. and MOHR, W. 1970. Specificity of roasted coffee aroma. Lebensm. Wiss. Technol. *3*, No. 2, 23–32. (German)
WINTEM, M. *et al.* 1975. Coffee flavoring agent. (Assigned to Firmenich & Cie.) U.S. Pat. 3,922,366, Nov. 25.

Tea

CO, H. and SANDERSON, G. W. 1970. Biochemistry of tea fermentation: Conversion of amino acids to black tea aroma constituents. J. Food Sci. *35*, 160–164.
COGGON, P. *et al.* 1973. The biochemistry of tea fermentation: Oxidative degallation and epimerization of the flavanol gallates. J. Agric. Food Chem. *21*, 684–692.
EYTON, W. B. 1972. The chemistry of tea. Flavour Ind. *3*, No. 1, 23–28, 36.
FUCKUSHIMA, S. *et al.* 1969. Studies on the essential oil of green tea. Isolation of dihydroactinidiolide and *p*-vinylphenol. J. Pharm. Soc., Japan *89*, 1727–1731. (Japanese)
GIANTURCO, M. A. *et al.* 1974. Seasonal variations in the composition of the volatile constituents of black tea. A numerical approach to the correlation between composition and quality of tea aroma. J. Agric. Food Chem. *22*, 758–764.
MARX, J. N. 1975. A new synthesis of theaspirone, an odiferous principle of tea. Tetrahedron *31*, 1251–1253.
NGUYEN, T.—T. and YAMANISHI, T. 1975. Flavor components in Vietmanese green tea and Lotus tea. Agric. Biol. Chem. (Japan) *39*, 1263–1267.
PANGBORN, R. M. *et al.* 1971. Analysis of coffee, tea and artificially flavored drinks prepared from mineralized waters. J. Food Sci. *36*, 355–362.
PETERSON, M. S. 1974. *Tea. In* Encyclopedia of Food Technology. A. H. Johnson and M. S. Peterson (Editors). AVI Publishing Co., Westport, Conn.
RENOLD, W. *et al.* 1974. An investigation of the tea aroma. 1. New volatile black tea constituents. Helv. Chim. Acta *57*, 1301–1308.

SANDERSON, G. W. *et al.* 1971. Biochemistry of tea fermentation: The role of carotenes in black tea aroma formation. J. Food Sci. *36*, 231–236.
SANDERSON, G. W. and GRAHAM, H. N. 1973. On the formation of black tea aroma. J. Agric. Food Chem. *21*, 576–585.
WICKREMASINGHE, R. L. *et al.* 1973. Gas chromatographic-mass spectrometric analysis of "flavory" and "non-flavory" Ceylon black tea aroma concentrates prepared by two different methods. J. Chromatogr. *79*, No. 5, 75–80.
WICKREMASINGHE, R. L. 1974. The mechanism of operation of climatic factors in the biogenesis of tea flavor. Phytochemistry *13*, 2057–2063.

Part 3: Herbs and Spices

Although the domestic culinary use of herbs and spices is well established and there are many specialist recipe books which cover this topic (Hayes 1961; McCormick & Company 1964), there still exists some uncertainty about the correct and most economical way to use them in the production of mass-processed foods, particularly those which may have to undergo long storage periods before ultimate use by the consumer. The housewife or chef, using traditional ground herbs and spices, can readily make adjustments in the quantity added to suit individual dishes with a knowledge of the degree of spicing which will give pleasure. The manufacturer, on the other hand, has a wider range of processed spices available to suit his technological needs and must ensure that the seasonings employed will result in a consistent product having a minimal consumer rejection level on this score (Heath 1972).

Because of their varied physical form and generally hard and somewhat woody nature, spices do not readily yield their flavoring components to the food in which they are incorporated without some prior treatment. There are few products in which whole spices can be used per se (e.g., mixed pickling spice, whole peppercorns in certain meat products, etc.). Spices are normally used in a finely ground condition and the leafy herbs in a well-broken state. These are purchased in the form most convenient for direct use in the end-product formulation.

Ground Spices

Spices may be milled to a wide range of particle size depending on their nature and ultimate use; the finer the powder the more readily available the flavor. In view of the many problems associated with purchasing, storing and handling spices, these are generally handled by specialist processors who have the necessary plant and skills to ensure that the raw materials are of the correct quality, that the grinding rate and the heat generated, as well as the methods of

FIG. 7.1. HARD WOODY STEMS OF DRIED HERBS ARE REMOVED AND THE BROKEN OR RUBBED HERB IS THEN SUITABLE FOR USE IN SEASONINGS

handling and packing the ground material, result in minimum volatile loss or degradation (Lissack 1975).

Spices, once ground, are liable to serious loss of volatile constituents with a consequent weakening of their flavoring power. These materials should always be stored in cool, dry conditions and turned over regularly. If these simple precautions are ignored, the resulting seasoning may produce an unacceptable flavor level in the end-product. In certain instances, a significant degradation of flavor quality may also occur during storage. This may be due to oxidation of the volatile oil constituents or of any fixed oil present, or, in certain instances, it may be the result of enzymic changes. One has only to smell a sample of ground pepper as purchased from a supermarket against that of a freshly-ground sample from a hand pepper mill to appreciate the great difference in aromatic quality between the two. The pungency of the pepper will not have changed but the odor most certainly has. With umbelliferous fruits (e.g., caraway, cumin, etc.) the onset of rancidity is the main cause of off-odor and off-flavor notes. In the case of ground paprika, this may also be accompanied by a fall-off in tinctorial power.

For this reason, many of the large food manufacturers prefer to purchase whole spices and store these in their unbroken condition; then only comminuting, blending or processing them into seasonings as required. Although there is still a significant use of traditional ground spices in food processing, the trend is toward alternative products which have marked technological advantages.

FIG. 7.2. SPICES ARE OFTEN PURCHASED ON PHYSICAL APPEARANCE ONLY
Here different lots of black pepper are under discussion.

Today, spice adulteration does not assume the same proportions as it once did but, even so, ground spices can more readily be admixed with components of lesser value. In the absence of specifications and precise methods of analysis, care should be taken in establishing an acceptable quality—bargain prices may be very deceptive.

In use, ground spices do have certain well-established advantages over other forms of seasoning. When used in baked goods (e.g., biscuits, cakes, etc.), the slow flavor release from the unbroken cellular tissues delays and reduces the loss of aromatics due to initial steam-distillation and ultimate high temperatures encountered during the baking process. Although losses do occur, the end-product is still adequately flavored without recourse to an excessive over-use of spice in the initial dough mix. The natural spices are of such a flavor level that they may be incorporated directly into food formulations and, although the quantities involved are generally small, they are readily manageable on the factory scale. From the food processor's point of view, spices are "natural" and pose no problems when it comes to label declarations.

However, in spite of their long usage, spices do present certain marked disadvantages to the food processor. In these days of close attention to consistent product quality, one cannot afford to ignore certain problems associated with their use in processed foods, as outlined below.

Variability of Odor and Flavor Strength and Quality.—Spices pos-

possess the variations characteristic of all natural products. With few exceptions, spices depend for much of their value upon the essential oil content. It is the quantity present which determines the flavoring power; and its composition determines the flavoring quality of the spice. The essential oil is a product of the plant's metabolic processes and is particularly subject to seasonal and climatic changes in the growing area; to the nature of the soil in each region; to the care taken in planting, husbandry and harvesting; and, finally, to the care taken during the preparation of the spice for marketing—which often may be after long periods of storage under adverse conditions in tropical warehouses. At all stages the essential oil content is at risk, giving rise to commercial samples of wide flavoring differences.

Whereas one would expect flavor differences between related plants of different species, perhaps it is not so widely realized that there can be even more significant differences in aromatic profile between plants of the same species but grown in different regions. This becomes obvious when one compares the flavor profiles of, say, ginger grown in Jamaica, Sierra Leone, Nigeria, Cochin, China, Australia or New Guinea. All are derived from *Zingiber officinale*, L.; but one certainly cannot transpose one source for another without endangering the flavor profile of the end-product.

The needs for a good working knowledge of the source of herbs and spices and the aromatic profiles of the commercial grades available are obvious.

Unfortunately, one cannot assume that any given lot of spice is uniform. No assessment of quality can ever be reliable unless the sample examined truly represents the bulk. Many spices are garden crops, or grown in small plantations, or even wild, so that shipments may represent very many individual sources. As these may not necessarily be bulked before shipment, sampling and quality assessment are very difficult. This, of course, does not apply in those areas where the spice production is under the control of the government or some other regulatory agency which determines grades and ensures conformity with the specifications. Meaningful specifications are hard to establish without being too restrictive to trade, but such are now being drafted by the International Standards Organization (I. S. O.). Until these are widely accepted by both the supplier and the user countries, the establishment of a realistic basis for quality assessment is likely to be difficult. So far as spices are concerned, they have always been highly speculative commodities in world trade. Too much emphasis on price can lead to the inevitable encouragement to treat the product as of secondary importance to commercial interests. This may result in spice bulks being retained in warehouses for far longer than is desirable.

They are Unhygenic.—In 1939, Fabian clearly demonstrated that spoilage in pickles was due to bacterial contamination derived from the spices used. Since then, successive investigations (Warmbrod and Fry 1966; Hadlok and Toure 1973) have proved conclusively that natural spices are heavily loaded with microorganisms and molds, many in spore forms. The precise figures obtained in various

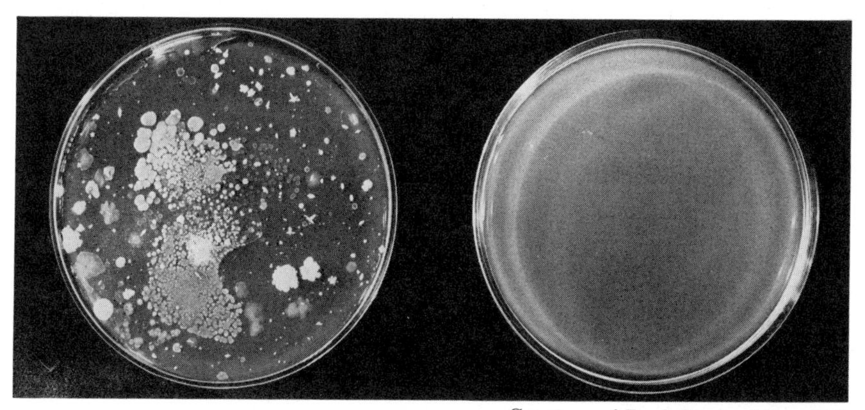

Courtesy of Bush Boake Allen, Ltd.

FIG. 7.3. NATURAL SPICES ARE UNHYGENIC AS IS SHOWN BY THE ABUNDANCE AND VARIETY OF BACTERIAL COLONIES ON THE LEFT-HAND PLATE WHEREAS SPICE EXTRACTS DISPERSED ON SALT ARE COMMERCIALLY STERILE AS IS SHOWN ON THE RIGHT-HAND PLATE

test methods are not of real significance as most of the bacteria present are nontoxic and are killed during any heat process to which the spice may be subjected. However, the presence of coliform organisms should be regarded as evidence of bad handling. Spores present may be thermophilic and these can lead to product spoilage and taints when the processing is inadequate or conditions for growth are right. At best, most spices carry a heavy load of microorganisms which can result in a reduced shelf-life for products in which they are used.

They Are Often Filthy.—Contamination is not confined to microorganisms. The generally negligent and unhygenic way in which many spices are handled from the time of harvesting until they are eventually ground for manufacturing use or retail sale results in the presence of such things as rodent hairs and droppings, insect parts, eggs and larvae, hessian sacking hairs, soil and stones, etc. The list of such filth is a long one and familiar to those who have control of the purchase and quality assessment of bulk spices.

They Are Easily Adulterated with Less Valuable Materials.—Adulteration is not a common practice; indeed every supplier would stoutly deny that it ever happens in the spice trade. But, whatever its extent, this is a subject of considerable interest to all concerned in the making and selling of wholesome food in which such materials may be used. This may take place in one of three different ways:

(a) Extension—the addition of exhausted material or diluents to extend the genuine material or the admixture of extractives or essential oils obtained from less valuable parts of the plant or even from related or unrelated plants.

(b) Weighting—the addition of foreign matter in whatever form.

(c) Substitution—the addition of a factitious article either in whole or in part. For example, the addition of ground chillies to enhance the pungency of a poor quality ginger, β-carotene to improve the color of ground turmeric, or dried *Schinus molle* berries to cheapen black pepper.

There are few limits to the degree of adulteration that can be achieved if the demand exists, particularly in those markets where there is an absence of regulatory legislation. Those who insist on purchasing at below accepted market values frequently lay themselves open to such practices. It is the responsibility of the user to ensure that his raw materials are those defined in the manufacturing specification or formulation sheet.

The Presence of Lipase Enzymes.—Many plant tissues contain enzymes which are of little or no consequence to the food processor as they are generally inactivated at some stage. Certain spices, however, contain lipase enzyme systems (Gross and Ellis 1969) which can result in the onset of rancidity in end-products containing fats, particularly in such products which are sold uncooked. The conversion of oil to free fatty acids in the case of ground black pepper was reported as being some ten times greater than for the other ground spices examined.

They Have Poor Stability on Storage.—Spices are normally ground to a fine powder before use in order to release the aromatic flavoring components. Ultra-fine grinding to less than 50μ is preferred to ensure uniform dispersion of the spice in food mixes. Such grinding generates considerable heat and, even when care is taken during milling by use of low-temperature techniques, a significant loss of volatiles can occur. Once ground, the essential oil is spread over the surface of the broken cellular tissue and is susceptible to loss or oxidation during ensuing storage. This necessitates their quick use or storage in well-sealed containers.

Undesirable Appearance Characteristics.—Unless micro-milled spices are used, "speckiness" almost invariably results from the use of traditional seasonings, particularly in smooth pale-colored meat products. This can lead to loss of eye appeal. On the whole, a product which looks smooth, uniform and free from dark specks is more likely to sell than one which looks dull, dirty and bitty. Ground black pepper has a fuller and more pleasing flavor profile than ground white pepper but the latter is often used in seasonings because it looks better in the end-product.

Discoloration May Result from the Presence of Tannins.—Clove and allspice (pimento) contain considerable quantities of tannins which, under some conditions of pH and in the presence of traces of iron and other metals, can result in the darkening of an end-product. The use of ground clove in the seasoning of a tomato ketchup certainly contributes to any black ringing which may occur in the neck of the bottle.

Finely-ground Spices May Be an Irritant.—Finely-ground pepper and capsicum, being sternutatory, are unpleasant to handle. On the factory scale, precautions must be taken to ensure that adequate protective masks are worn to reduce the irritation and discomfort of the process worker during the preparation or handling of finely powdered seasonings.

The culinary value of ground herbs and spices is beyond dispute but each of the above disadvantages presents problems for the food manufacturer. More satisfactory alternative processed spice products have been developed (Fig. 7.4); these are increasingly replacing ground spices in a wide spectrum of food products. In consequence, spice usage has moved from the traditional approach to the highly technological from which much of the guesswork has been removed.

Sterilized Spices

One of the major objections to traditional spices is that of their poor hygiene. This objection may be overcome by the use of ground spices which have been exposed to mixtures of ethylene oxide and carbon dioxide (or other bactericidal gases) in order to eliminate, or at least reduce, the bacterial load to acceptable levels (Rauscher *et al.* 1957; Mayr and Suhr 1973). Such spices show a marked improvement over untreated spices but the technique does not get rid of any extraneous filth which, however sterile, is still objectionable in a finished food product.

The sterilization technique involves precise operating conditions

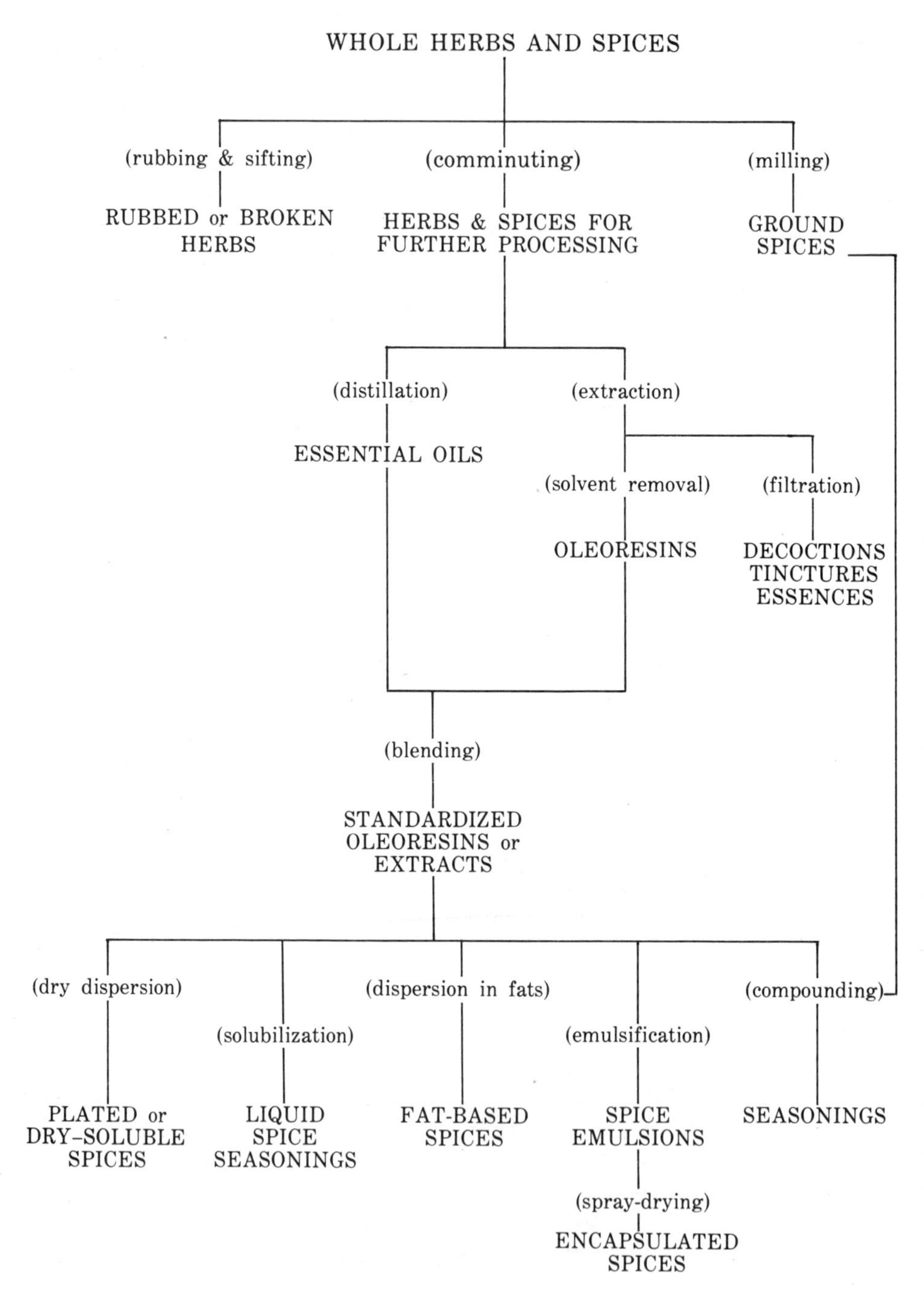

FIG. 7.4. SPICE PRODUCTS USED IN FOOD PROCESSING

(Peat 1963), particularly with regard to temperature and humidity control. The concentration of gases used rarely penetrates sufficiently to kill all of the bacteria and bacterial spores present. If the technique is correctly handled, then ethylene oxide residues are minimal and no change in the flavor profile of the spice can be observed. Some loss of essential oils may occur as the ground spices must be held under vacuum prior to admission of the sterilizing gases and finally spurged with air to remove residual gases.

During the sterilization process, ethylene and propylene oxide may react with chlorine present as organic chlorides in the ground spice in the presence of moisture to form toxic chlorohydrins (Wesley *et al.* 1965; Coretti and Inal 1969). The amounts formed are variable and very small, but their presence in a material to be used in foods gives rise to some concern. The safety of ethylene oxide as a sterilant has received some attention by F.D.A. and B.I.B.R.A. (Charlesworth 1975). It is generally felt that the problem may perhaps have been

Courtesy of Griffith Laboratories, Ltd.

FIG. 7.5. A RETORT USED FOR THE GAS-STERILIZATION OF SPICES

Courtesy of Griffith Laboratories, Ltd.

FIG. 7.6. BAGGED SPICES BEING REMOVED FROM THE RETORT AFTER GAS-STERILIZATION

overstated as the chlorohydrins are not persistent in subsequent food processing conditions and are unlikely to present any hazard to the consumer.

Alternative methods of sterilization such as γ-irradiation are, as yet, not accepted; and, at effective levels, the spice flavor is adversely modified. Reactions may be induced in the flavor complex which could lead to the development of objectionable taints and off-notes during later processing.

Scharf (1967) patented a process of sterilization involving treatment of the ground spice with acids followed by neutralization. It was claimed that the resulting product was sterile but no claims were made concerning the effects upon the flavor of the spice.

Spice Essential Oils

The bulk of all spices is cellular and contributes little or nothing to the aroma or flavor of the spice. With a few exceptions (e.g., capsicum), spices are, in part and in some cases, entirely dependent upon their essential oil content for their characteristic aromatic profile. In consequence, the spice essential oils have for long been used in seasoning formulations for their obvious advantages over the ground spices. Essential oils are generally uniform in flavor quality irrespective of the quantity present in the original spice. They are sterile and free from all extraneous contaminants. They can readily be incorporated into liquid concentrates and emulsions, the flavor character and strength of which can be standardized to within acceptable limits. They can be admixed with fats and oils which are frequently present in food products; and aid in uniform dispersion throughout the mix. They are normally stable and show only slight deterioration over long periods if correctly stored. Being concentrated, they show a saving in both storage space and handling costs. With all these advantages one may wonder why they have not completely replaced traditional spices in food processing.

Spice essential oils are, however, not the complete answer to seasoning problems for they too have disadvantages and limitations as described below.

The Flavor is Good but Often Incomplete.—Spice oils do not necessarily fully represent the flavor complex of the spices from which they were distilled. In the case of black pepper and ginger, the essential oil has the full aroma and flavor associated with the freshly-ground spice but is bland and quite devoid of the characteristic pungency of these spices.

The Flavor is Often Unbalanced.—The composition and, hence, the profile of an essential oil is determined to a large extent by the method of distillation. Steam-distilled oils are often richer in the lighter, low boiling terpenoid constituents so that the profile is fresher and less full-bodied than that of the ground spice. The oxygenated components, being high boiling, involve long and often uneconomic distillation times to recover so that the proportion of these tends to be greater in the ground spice (Eiserle and Rogers 1972).

Some Are Readily Oxidized.—Many essential oils contain high percentages of terpenes which tend to oxidize and resinify under adverse storage conditions, resulting in a marked turpentine-like odor and an unacceptable flavor.

Natural Spice Antioxidants Are Absent.—Some herbs and spices contain significant quantities of antioxidant components which improve the color of meats (Berner *et al.* 1973). These compounds are nonvolatile and do not occur in the essential oil.

They May Readily Be Adulterated.—No other group of products has been subjected to such a level of adulteration, substitution and extension as the essential oils. The increasing knowledge of their chemical composition and the very cheap and readily available synthetic versions of the components has lead, over the years, to quality and authenticity being sacrificied to low costs. Current legislation has considerably reduced this practice.

They Are Very Concentrated.—Essential oils are present in plant material to only a small percentage so that they are very much more concentrated in their odor and flavor strength than the original spice. This poses problems of weighing and incorporation into food mixes as a minor error may result in a major change in the flavor profile of the end-product.

They Are Difficult To Disperse in Product Mixes.—Being concentrated and, hence, used in small quantity, the essential oils are difficult to disperse uniformly in a bulk mix particularly where this is a dry powder. Special mixing techniques may be necessary to ensure homogeneity.

Subject to Loss During Processing.—Essential oils are steam-volatile and may be lost from products exposed to high temperatures during processing.

The spice essential oils possess many desirable advantages over the equivalent ground spices making them eminently suitable for use in liquid seasonings for products such as mayonnaise, ketchup and sauces; but they are generally less satisfactory in overall quality than the corresponding ground spice in processed meats, canned goods, etc., where a full-bodied seasoning is required.

Spice Oleoresins

Since spice essential oils fall short on many counts, the next alternative to be considered is the use of extracts or oleoresins which consist of a blend of the essential oil and other flavoring components of the spice which are soluble in the solvent used in their manufacture. Such products come nearer to the profile displayed by the freshly-ground spice which make them an acceptable form of seasoning in a wide range of food products without loss of quality (Staniforth 1972).

Production of Oleoresins.—Oleoresins are prepared from the ground herb or spice by extraction with a selected solvent or solvents which must be totally removed from the final product by distillation under vacuum. The unit operations involved include (Sabel and Warren 1972):

(a) Preparation of the raw material. This involves grinding or comminuting to a suitable particle size depending on the physical nature of the herb or spice.

(b) Exposure to solvent. This is a three-stage process which involves the addition of the solvent and its penetration into the mass,

Courtesy of Bush Boake Allen, Ltd.

FIG. 7.7. A BATTERY OF EXTRACTORS USED FOR THE PRODUCTION OF SPICE OLEORESINS AND OTHER VEGETABLE EXTRACTS

the achievement of equilibrium and the replacement of the solute with fresh solvent. This is carried out in extractors designed to handle specific products.

(c) Separation of the solute or miscella from the extracted mateial, or marc, which is carried out batch-wise or as a continuous recycling process depending upon the plant available.

(d) Solvent Stripping. This is a stage of immense importance to the quality of the end-product. The solvent must be evaporated from other low-boiling constituents as rapidly as possible at the lowest practicable temperature so as to avoid loss or damage to the essential oil. The limits of residual solvent permitted in the United States as published in the Federal Register (1968) under Food Additive Regulations (121.1040–121.1045) are:

	Maximum Permitted in Parts per Million
Acetone	30
Methanol	50
Isopropanol	50
Hexane	25
All chlorinated solvents	30

These residual limits have been widely adoped in other countries.

(e) Recovery of residual solvent from the marc.

(f) Disposal of inert waste material.

The detailed technique employed may differ somewhat from the above depending on the spice and the type of oleoresin required, but in all cases considerable expertise is necessary to achieve satisfactory products having a consistently high quality (Cripps 1972).

Solvents

Since the nature of the oleoresin is largely determined by the solvent used in its extraction, a brief review of those solvents most widely used for this purpose is warranted. These may be divided into three categories:

(a) Polar solvents containing hydroxy- or carboxy- groups. These are relatively reactive chemicals with high dielectic constants and are miscible with water (e.g., alcohols, acetone).

(b) Nonpolar solvents which are chemically inert, having low dielectic constants and being generally immiscible with water (e.g., petroleum hydrocarbons, benzene).

(c) Chlorinated hydrocarbons. These are solvents with very low boiling points, heavier than (a) or (b) and immiscible with water.

In handling large volumes of solvents, which is inherent to the commercial production of oleoresins, flammability and toxicity are of major importance, particularly if there is any risk of an accumulation of vapors in the processing areas. The nonflammable chlorinated solvents have many advantages in this respect but are generally more toxic. Manufacturers are well aware of this health hazard and take all precautions to avoid vapor loss from the extraction system.

In herbs and spices, flavor is due to both volatile and nonvolatile components. Most constituents are hydrophobic and are best extracted with nonpolar solvents such as the hydrocarbons; but others are hydrophilic and, hence, likely to be more fully recovered with a polar solvent such as acetone. This being so, it is necessary to determine for each herb and spice the solvent which produces the optimum balance of extractable flavoring components. The polar solvents are powerful and dissolve a wider variety of substances than do the nonpolar solvents. Acetone and ethanol are obvious solvents of choice as they have a wide spectrum; but the former is highly flammable and requires a specially-designed plant operated in a flashproof area; whereas the latter is too expensive due to high customs duty in most countries. Hexane has a significantly good solvent action on essential oils and fats but yields oleoresins which lack "body" owing to the absence of hydrophilic components; again, it is highly flammable and poses a fire hazard.

Oleoresins are now extensively produced by the use of the chlorinated hydrocarbon solvents such as methylene dichloride. Here, one has the marked advantage of nonflammability; and since they are low boiling point materials, they can readily be removed from the end-product. Care must be exercised in their handling.

Recent work has been directed to the use of liquified gases such as carbon dioxide, nitrous oxide, or ethane as solvents. The dry gas is used to extract the aromatic fraction and humid gas for the hydrophilic flavoring components. Then the partially extracted spice is finally washed with water to recover any residual flavorings (Vitzthun and Hubert 1972). To date, no commercial samples of the products have been available for evaluation.

Water has been used as a solvent to produce stable spice extracts having a flavor of the fresh herb or spice. The extraction is carried out on the frozen comminuted material which is treated with acetic acid before hydraulic expression. Salt may be added to the resulting liquor which may also be mixed with starch and dried (Dragoco, G.m.b.H., 345 Holzminden, West Germany).

Oleoresins now form an important source of flavoring in a wide spectrum of food products as they are free from many of the inherent disadvantages of ground spices; in particular, they are hygenic and can be standardized for flavor quality and strength to within acceptable limits by blending of the essential oil and resinous fractions. Unlike the essential oils, these extracted products contain any natural antioxidants which may be present in the original spice—a valuable attribute when they are used in the seasoning of meat products where the red color of the end-product is of importance (Eiserle 1971). With certain exceptions (e.g., oleoresins of dill, celery, cumin and paprika), they have a very good shelf-life and retain their color and flavoring power for long periods if stored in a cool place in well-closed containers. The exceptions are those spices which contain high percentages of fixed oils which tend to oxidize and go rancid.

There are disadvantages to the use of oleoresins, namely:

(a) Although the flavor is full-bodied and generally representative of the ground spice, the oleoresins are as variable as the spice from which they were extracted. The selection of the material for extraction is of considerable importance to the quality of the resulting oleoresin (Lewis 1972). If there is an imbalance between the volatile and nonvolatile components in the spice, as may occur with old or poor quality material, this will be reflected in the final oleoresin.

(b) Oleoresins are concentrated and are from 10 to 50 times stronger than the spice itself. This necessitates the weighing of small quantities. The amount remaining in a weighing vessel may significantly affect the level of flavor in the end-product.

(c) Depending on the spice and the solvent used, the resulting oleoresin may vary in physical form from a light, mobile oil (e.g., oleoresin coriander) through a viscous paste (e.g., oleoresin Dalmatian sage) to a friable solid (e.g., oleoresin turmeric). Such products cannot readily be incorporated directly into food mixes without the danger of local concentrations or "hot spots."

(d) In certain oleoresins (e.g., oleoresin clove and allspice) tannins are present in the original spice and are also present in the extract, unless specially processed to remove them.

(e) The flavor profile of the oleoresin is closely determined by the nature of the solvent used and the care with which it is removed. Industrial solvents frequently contain traces of higher boiling fractions which may remain in the oleoresin and give rise to off-odors and off-flavors. Excessive vacuum distillation or heat processing to remove the last traces of solvent can easily result in damage to the thermolabile aromatic components or, in extreme cases, to an almost complete loss of the top-notes of the essential oil present resulting in an oleoresin which has a decidely "flat" profile (Eiserle and Rogers 1972).

As a source of flavoring, therefore, the spice oleoresins are valuable in modern food processing; but, because of their powerful effects, it is necessary for them to be diluted prior to incorporation into a product mix. This is achieved by the use of one of the following liquid or dry powder products: essences, emulsions, solubilized spices, plated dispersions, or encapsulated spices.

Each of these products has marked advantages in usage over the equivalent ground spices, the essential oils or the oleoresins; although none is entirely free from disadvantages in meeting the demands of processing and storage.

Liquid Spice Flavorings

Diluted products, whether they be liquid or solid, overcome the danger of weighing and incorporating small quantities of powerful flavoring materials; a small error in which can have a profound effect upon the profile of the end-product. Liquid spice flavorings are prepared by diluting blends of essential oils and/or oleoresins in a suitable solvent which is acceptable in the product formulation. Such solvents are proscribed in the food additives legislation and include glycerol, isopropyl alcohol, propylene glycol, etc., all of which are GRAS although there may be problems of acceptability in some countries and in certain classes of foods (e.g., meat products) where the

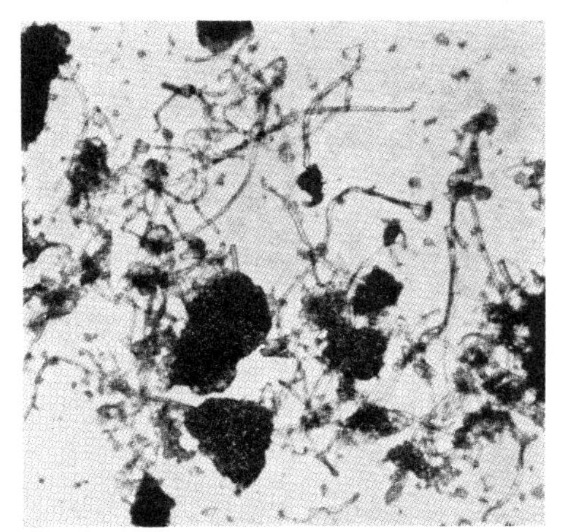

1

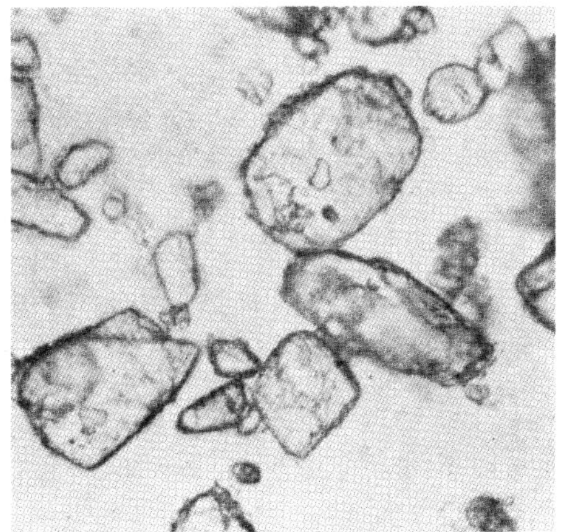

2

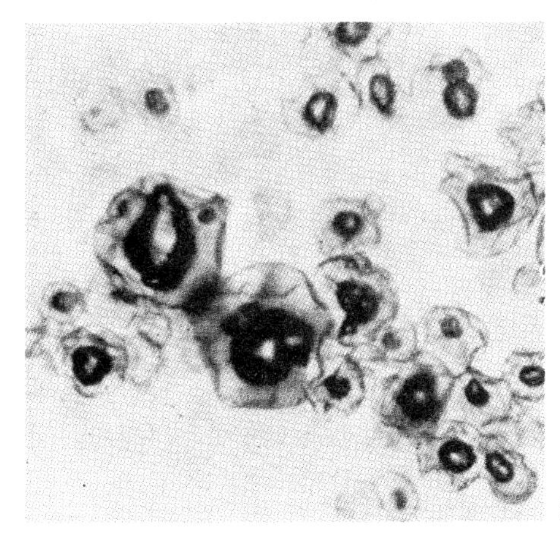

3

Courtesy of Bush Boake Allen. Ltd.

FIG. 7.8. PHOTOMICROGRAPHS OF DALMATIAN SAGE IN (1) GROUND FORM, (2) DISPERSED EXTRACT ON SALT AND (3) MICROENCAPSULATED BY SPRAY DRYING IN A MODIFIED STARCH BASE

legislation is particularly precise. Liquid spice flavors or blends as seasonings have enjoyed long popularity for use in canned goods and pickle liquors where automatic dosage is employed to facilitate uniform and continuous addition of the seasoning. However, the presence of the solvent may be regarded with disfavor by some manufacturers.

Liquid flavors cannot, of course, be readily incorporated into dry mixes (e.g., soup powders, gravy mixes) as the presence of the solvent would most likely result in balling and unacceptably lumpy product. In such products a dry powder seasoning blend is necessary.

An alternative to the use of solvents is the presentation of the blended essential oils and/or oleoresins in a water-dispersable form as an emulsion. Gum acacia (Arabic) or one of the modified starches is used as the emulsifying agent and a stabilizer may be added to prevent creaming. Such products have a real practical value in the manufacture of soups and in other canning operations, although their use is strictly limited in most meat processing techniques. Most of the essential oils are composed of highly reactive chemicals which are susceptible to hydrolysis in the presence of excess water and may undergo radical changes in profile during storage. Many of the commercially available seasoning emulsions are prepared from essential oils since the oleoresins are much more difficult to emulsify and are more costly to process with less chance of achieving a stable product. Spice emulsions are very prone to microbiological contamination and require appropriate techniques to avoid this. Once containers are opened the contents must be used. Seasoning emulsions are best prepared freshly as required or used within a few days of production, otherwise it becomes necessary to add a preservative.

Another liquid seasoning can be prepared from the essential oils and some of the oleoresins by admixture with solubilizing agents such as one of the polyoxyethylene esters (e.g., polysorbate 80), although these are not universally acceptable, particularly for use in meat products. Depending on the relative concentration of oils to polysorbate, such products will give a clear solution when mixed with water, brine or dilute solutions of acetic acid; a marked advantage in certain products in which a cloud or haze would detract from the visual appearance. Solubilized spices are generally quite stable on storage but may give to the product a slight soapy note and can, under certain circumstances, result in foaming during processing.

In the pickle and sauce industry, it was formerly a common practice for spiced brines and vinegars to be prepared by boiling whole or coarsely ground herbs and spices in the brine or vinegar. Such decoctions result in a clear pickling liquor but, as water is not a good solvent for most of the flavoring components of the spices used, only a small percentage of them is extracted by this method. The boiling of herbs and spices with water inevitably results in volatile losses by steam distillation and, in consequence, little of the true aroma of the herb or spice survives in the end-product. This method of incorporating seasonings into pickles and sauces is strongly to be deplored on purely economic grounds. It has largely been replaced by the use of solubilized seasonings, the formulation of which must be established experimentally depending on the level of spicing required.

Dry Processed Spices

Most of the dry processed spices commercially available are designed to replace the ground spices in existing seasoning formulations either on an equivalent weight-for-weight basis or at a stated level of concentration. It should be appreciated that the appearance and physical character which such products give in the final formulation may be entirely different from that given by the equivalent ground spices so that their adoption may result in a significant visible change in the end-product.

There are three main types of such products currently being marketed either as single-named spices or as blended seasonings: (a) dispersed, plated or so-called "dry soluble" spices, (b) encapsulated spices, and (c) fat-based spices.

Dispersed Spices.—In this range of products, the total flavor components of the herb or spice, usually in the form of the oleoresin or a blend of essential oil and oleoresin to achieve a specific aromatic profile, are uniformly dispersed onto the surface of an edible carrier such as salt, dextrose, milk whey, flour, etc., so that the flavor strength of the dispersion is equal to that of the freshly-ground equivalent spice. Such plated spice products have the advantage that they can generally be used to replace the natural ground spice on a one-for-one basis in existing formulas although the nature of the carrier, particularly if this be salt, must be taken into account in the total formulation. The claims for equivalency made by the several manufacturers of these products should be checked out experimentally under processing conditions to ensure that the resulting profile is satisfactory and that the actual usage level is economically viable.

Dispersed spices have become well established in a wide spectrum of food processing operations; they display considerable advantages over the equivalent ground spices. Although comparatively straightforward, their production requires considerable expertise in the opperation of suitable stainless-steel blenders the design of which is determined by the quantity and physical nature of the oleoresins and/or essential oils to be dispersed and the absorbancy of the base carrier. Some oleoresins (e.g., celery, coriander and paprika) are fluid and are readily dispersed in cold; others (e.g., black pepper, Dalmatian sage, etc.) are either viscous fluids or pastes and require the application of gentle heat to facilitate uniform dispersion. Certain oleoresins (e.g., capsicum) require special safety precautions because of their intensely irritant nature. In all cases the resulting product must be sieved to ensure freedom from local concentrations of inadequately mixed oleoresin. The blending technique employed is similarly dictated by the nature of the constituents. For instance, the incorporation of small quantities of highly concentrated essential oils (e.g., oil of garlic) requires the preparation of a premix; the dispersion of a very sticky oleoresin (e.g., nutmeg or mace) requires careful attention to mixing times, otherwise a marzipan-like mass can result. The manufacturers of such products have so refined their techniques that plated spices of a consistently high quality are now readily available to satisfy almost all seasoning needs.

When deciding whether or not to adopt this form of seasoning the following disadvantages should first be considered:

(a) In dispersed spice products, the aromatic components are spread over a very considerable internal surface area, hence are liable to volatile losses on storage. If full advantage is to be taken of these products, short storage times and a quick and systematic turnover of seasonings is essential. This disadvantage can, to some extent, be reduced by the use of seasoning blends in impermeable unit packs which remain sealed until required. Storage in a cool dry place away from direct heat and sunlight is necessary to minimize losses.

(b) They do not resist the high temperatures encountered during baking so that volatile losses are significant due to steam-distillation during the initial stages of the process. Ground spices have a built-in natural encapsulation in the form of unbroken cells which reduces this loss of flavor.

A complete range of herbs and spices is now available in this form, either singly or as seasoning blends, and reference should be made to the manufacturers' literature and specifications. At the time of their introduction such products represented a marked technological advance in the technique of spicing and, although there is still some resistence to their use in certain specialist products, their use is now widely accepted throughout the food industry.

Dispersed, plated or dry-soluble spices offer many advantages to the food processor. The method of formulation gives to the manufacturer a means of standardizing his added seasonings, resulting in a far better continuity of flavor strength and profile than can be achieved by using ground spices. Probably of equal importance to most food manufacturers is that such products are virtually free of bacteria and mold spores or, at worst, have only low total plate counts, depending upon the carrier used (e.g., flour); and they are free from any enzymes and extraneous filth. Hygenically, therefore, they are infinitely more acceptable than the equivalent ground spices, even where these have been gas-sterilized.

The flavor imparted by processed spice products is not exactly the same as that of the ground spices; in many cases, they are fresher and have a marked initial impact but lack some of the full-bodied character. But, unlike the ground spices, the flavor is instantly available in the product. In products which involve an open cooking or heating stage during processing, it is an advantage to add any plated seasoning after this is completed as this reduces the chances of volatile loss. In a sealed system, such as canning, the problem does not arise and any change in profile due to heat degradation can be established and adjusted experimentally.

Encapsulated Spices

One way in which the loss of volatiles can be obviated is to encapsulate the flavoring components by spray drying or some other method of obtaining an impermeable shell which locks in the aromatics and thereby increases considerably their shelf-life.

Microencapsulation is the technique of enclosing a minute quantity of a liquid, solid or gas in a continuous shell. The ultimate particle size of such products may range from 5 to 2000 μ but generally is about 5 to 400 μ so that they are very fine free-flowing powders. For inclusion in food products, the carrier or encapsulant must itself be acceptable as a food additive. The techniques used in their manufacture may be either simple or complex, depending to a large extent upon the nature of the flavoring material to be encapsulated (McKernan 1972–1973). By far the most widely used process is that of spray drying an emulsion of the flavoring materials using either gum acacia (Arabic) or one of the modified starches as the emulsifying agent and, ultimately, the encapsulant. The techniques employed were reviewed by Masters (1968) and Todd (1970).

There are many advantages to the use of encapsulated flavoring materials, among them are the following:

(a) The flavor strength and quality is fully protected from evaporation and oxidation and is retained on storage, even at elevated temperatures, over periods in excess of 12 months.

(b) They can be incorporated directly into dry mixes, using normal

TABLE 7.1
ADVANTAGES AND DISADVANTAGES OF SPICE PRODUCTS

Traditional Ground Spices

Advantages

Slow flavor release in high temperature processing.
Easy to handle and weigh accurately.
No problems of labeling declaration.

Disadvantages

Variable flavor quality.
Variable flavor strength.
Unhygienic.
Often contaminated by filth in one form or another.
Ready adulteration with less valuable materials.
Presence of lipase enzymes.
Flavor loss and degradation on storage.
Undesirable appearance characteristics in end-products.
Flavor distribution poor, particularly in thin liquid products (sauces).
Discoloration due to tannins.
Herbs usually have hay-like aroma.
Dusty and unpleasant to handle in bulk.

Essential Oils

Advantages

Hygienic, being free from all bacteria, etc.
Reasonable standard flavoring strength.
Flavor quality consistent with the source of the raw material.
Do not impart any color to the product.
Free from enzymes.
Free from tannins.
Are stable on storage under good conditions.

Disadvantages

Flavor good but incomplete.
Flavor often unbalanced.
Some readily oxidize.
No natural antioxidants present.
Readily adulterated.
Very concentrated—hence, difficult to handle and weigh accurately.
Not readily dispersable, particularly in dry products.

Oleoresins

Advantages

Hygienic, being free from bacteria, etc.
Can be standardized for flavor strength.
Contain natural antioxidants.
Free from enzymes.
Have long shelf-life under ideal conditions.

Disadvantages

Flavor good but as variable as the raw material.
Very concentrated—hence, difficult to handle and weigh accurately.
Range from liquids to viscous solids which are difficult to incorporate into food mixes without "hot spots."
Tannins present unless specially treated.
Flavor quality depends upon the solvent used.

Dispersed Spices

Advantages

Standardized flavor effect.
Standardized flavor quality by specifying source of the raw material.
Hygienically excellent being free from bacteria, filth and other impurities.
Free from enzymes.
Readily handled and weighed with accuracy.
Readily dispersed in food mixes.
Usually free from dust during handling.
Contain natural antioxidants.
Have low water content.
Do not contribute unwanted specks or color to the end-product.

Disadvantages

Allowances necessary for the base used.
Lose volatiles on long storage, particularly if ambient temperature high.
Do not resist high temperature processing under open conditions.

Encapsulated Spices

Advantages

Aromatics fully protected from volatile loss and degradation.
Long shelf-life under all conditions.
Readily incorporated into food mixes.
Free from objectionable odors (particularly applies to garlic and onion).
Hygienically excellent, being free from bacteria, filth and other impurities.
Free from enzymes.
Low water content.
Do not contribute unwanted specks or color to the end-product.

Disadvantages

Concentrated usually 10-fold, so that weighing is difficult.

commercial blending methods, to give a uniform and homogeneous dispersion. The only limitation is that such mixes should have a narrow range of particle size so as to reduce separation during handling. The average particle size of most spray-dried products is 50 μ.

(c) Being encapsulated, they are free from the strong and often objectionable odors associated with certain spices and essential oils (e.g., oil of garlic).

(d) They are free from salt, enzymes and filth and, although they do have a low total plate count, they are generally hygenically excellent.

(e) They have a very low water activity (0.2–0.3) and are thus compatible with dehydrated food materials in dry food mixes.

(f) They are nonhygroscopic and under normal storage conditions remain free-flowing and lump-free throughout their shelf-life.

There are certain disadvantages to be considered. In the case of the spices, most of the commercially available spray-dried products are formulated to be ten-fold the flavoring strength of the equivalent herb or spice. In consequence, weighing may be difficult and the direct addition of such strong flavoring materials may pose problems on the factory floor. They can, of course, be diluted with any finely powdered ingredient such as flour or starch without loss of their flavoring characteristics but, as indicated above, care has to be taken to ensure uniformity of particle size.

As the encapsulant is water-soluble, the shell of microencapsulated flavorings breaks down on admixture with water and liberates the flavor load uniformly throughout the product mix. Apart from stability on storage, there is, therefore, little advantage to be gained from employing encapsulated products in wet formulations (e.g., canned goods, sauces, etc.) or in baking where the volatile constituents are unprotected at the stage when losses are greatest.

Fat-based Spice Products

In this type of product, the spice extractives or essential oils are blended with a fatty carrier which may be either fluid (e.g., vegetable oil) or solid (e.g., hydrogenated oil). In the latter case, the product may be in the form of a solid block which must be melted before use or it may have been spray-cooled to give a free-flowing powder. Such products are designed for incorporation directly into any other fatty or oily component of the end-product (e.g., pastry for pie crusts or the oil phase of mayonnaise). Generally, they are formulated to be either 3- or 4-fold the flavoring strength of the original spice. Those made with a liquid vegetable oil base are particularly suitable for spraying onto the surface of freshly baked biscuits or cookies conveying both a flavor and glaze to the product.

Seasonings

The blending of herbs and spices tends to be traditional and imprecise. At the domestic culinary level it is all too easy to adjust the nature and quantity of any added seasoning to give just the flavor required; but when it comes to factory processing and the development of a seasoning formulation which will give consistent results, it is necessary to approach the problem more scientifically. So often, food products are either grossly under-seasoned and hence not so attractively flavorful as they could be or else they are over-seasoned and the true flavor destroyed. In either case, the acceptance level is reduced. It is not uncommonly assumed that if a little of a particular seasoning is good then more will be better; frequently, this is not the case. A seasoning can become the main flavor of the product over-riding the more delicate primary flavors of the meat, fish or vegetables present. A good seasoning should be subtle and not obvious in its effect and should be designed specifically to enhance and bring out those pleasurable flavor characteristics of the food items present. Only in very few specialist dishes should the spices used impose an entirely different or new flavor effect (e.g., curry). So far as herbs and spices are concerned, these require skillful blending, taking into account the flavor profile of all of the ingredients in the product, the aim being to fortify those characters which improve the flavor profile of the end-product or suppress those which detract from a high level of acceptability. This is both an art and a science demanding a good knowledge of the raw materials available, an understanding of how flavor is affected during processing and an ability to assess consumer needs and preferences.

In creating seasonings it is necessary to appreciate that all the herbs and spices and the products made from them do not have the same flavoring power. Some are excessively strong and over-ride other flavors readily (e.g., capsicum, thyme, etc.); others are weak and just as easily swamped or suppressed (e.g., parsley herb). In Fig. 7.9 the herbs and spices have been listed in order of their aromatic strengths with the weakest at the top and the strongest at the bottom. Alongside these are listed the main protein sources also in descending flavor strength. This illustration may be used to establish those herbs and spices most appropriate for use in a seasoning for each protein source. In general, the weaker the flavor of the main staple item the lower the level of added seasoning required to achieve a satisfactory balance of flavor in the end-product. For example, with fish it is usual to add only a lightly balanced herbal seasoning with only a very low level of a pungent spice such as pepper; on the other hand, with ham the intrinsic flavor is so strong

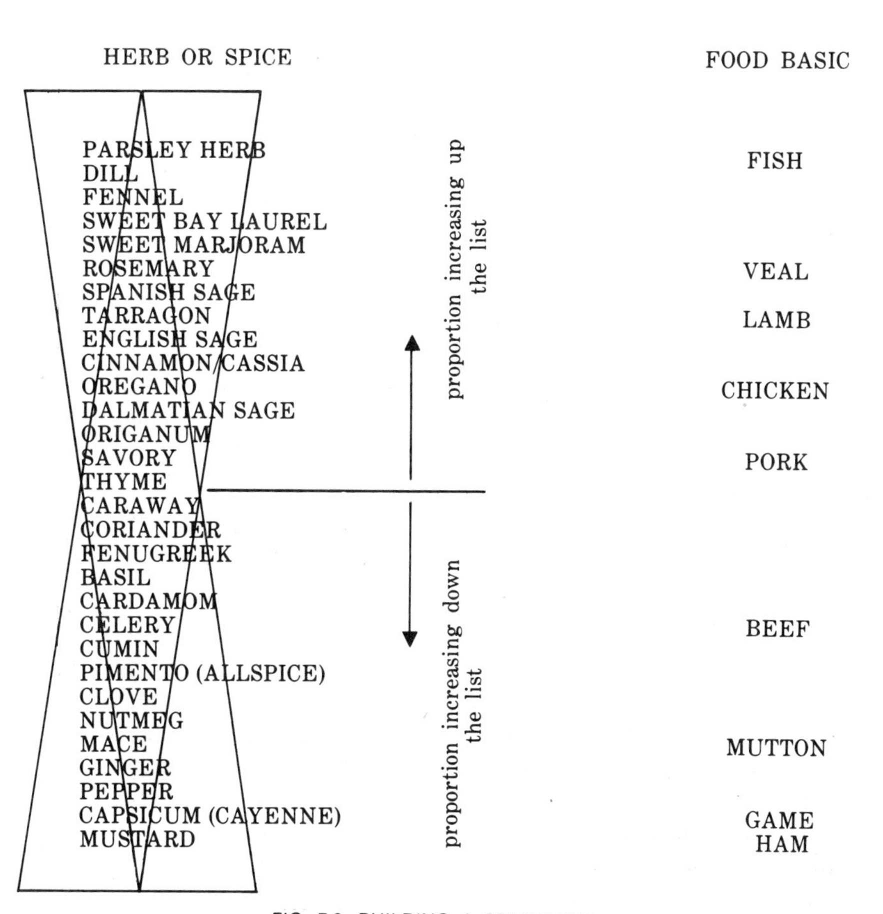

FIG. 7.9. BUILDING A SEASONING

that only a strongly pungent spice such as mustard is necessary to bring out the full flavor. In the medium range, a wide spectrum of spices may be used to produce the desired flavor effect. In all cases, salt is an essential component of the seasoning and its level of usage can significantly alter the overall profile.

Not all of the aromatic components present in a broken herb or ground spice are necessarily wanted in a well-balanced seasoning. In this respect, the use of the essential oils or oleoresins, assuming that these have been prepared from genuine material, can result in a cleaner and better-defined flavor profile in the end-product. Cloves, for instance, have, in addition to their high percentage of essential oil, some 12–13% tannin (quercitannic acid) which accounts for their marked astringency. A clove extract from which the tannin has been eliminated overcomes this objection without imparing the clean phenolic odor and flavor of the clove oil. On aging, ground pepper develops a strong smell of ammonia with a skatole-like back-note which detracts from the overall aromatic character of any seasoning in which it is incorporated. This unpleasant off-odor does not develop when dispersed oleoresin pepper is used to replace the ground spice. Nutmegs and mace of poor quality are intensely bitter and leave an unpleasant lingering after-taste; again, the use of nutmeg oil or oleoresin is indicated to overcome this disadvantage in a blended seasoning. Regarding the herbs, with few exceptions these develop strong hay-like profiles during drying and the unpleasant dried-grass notes can easily carry through into the end-product. This character tends to dull and may even completely swamp the distinctive fine fresh notes associated with the green herb.

One of the greatest problems of achieving a satisfactory flavor balance is brought about by changes that take place in the flavor profiles of the individual constituents during processing. All too often a seasoning is created in the laboratory or test kitchen and is evaluated in a simple neutral soup base or white sauce. At this stage, the seasoning may appear to be well balanced and pleasing but when it is incorporated into, say, a canned meat with vegetables or the gravy for a frozen dinner, it is not unusual to find that some item of the seasoning is preferentially absorbed or suppressed so that the resulting flavor profile becomes unbalanced even possibly to the point of making the product unacceptable. One cannot judge blended seasonings or other added flavors other than in the product for which they were designed, with the product processed exactly as it will be in the plant and evaluated under the same conditions as those employed by the ultimate consumer. Only then can one be sure that the correct seasoning effect has been achieved. This particularly applies if the end-product is to be sold under chilled or deep-frozen conditions. Unless the test conditions reproduce the processing, storage and reheating conditions, totally unrepresentative results will be obtained.

In an earlier section, each of the herbs and spices was discussed on the basis of its aromatic character each of which is indicative of the several attributes which collectively make up the distinctive profile. Although such profiles are useful, it should be remembered that herbs and spices can only be used effectively when one has a first-hand knowledge of their organoleptic effects; mere descriptions can be misleading. Each food processor has his own ideas on the nature of the products to be offered to the public and generally develops a product which uniquely satisfies the established needs of potential consumers. However there is a big difference between product concept and ultimate production and marketing.

Product development can be very time-consuming; but it is essential to ensure continuity of product quality. This can only be achieved by the use of good quality raw materials and flavorings having a high degree of reproducibility. There are no set rules for the use of herbs and spices in seasonings; it is just a matter of experience. However, there is an increasing use of alternatives rather than traditional ground spices as these products are proving invaluable in meeting the demands of modern food processing technology.

The food industry covers an enormous range of products, the processing of which involves, in many cases, special techniques and complicated in-line production plants. For this reason, it is not feasible to generalize on the creation of seasoning formulations nor to give specific examples; what may be quite satisfactory for one product under one set of processing and marketing conditions may be totally valueless under different conditions.

One should not too readily assume that handling herbs and spices and the many products made from them in food processing is elementary. In fact, there are many pitfalls for the unwary. Spices have always been highly speculative in world trade and it is not surprising that there is often ambiguity in names, descriptions, grades and quality; a shortage of official specifications does not help in this respect. In the past, adulteration and sophistication of genuine material were widely practiced; but today, this is much less common. Commerce in whole spices is largely in the hands of specialist companies which are well aware of the need for good quality control, and the substitution of inferior material is difficult to achieve. The problems are accentuated when it comes to ground spices and seasonings and, in particular, in spice essential oils and oleoresins where materials of a different geographical source or even different botanical entity can be used to replace the genuine spice to satisfy commercial ends. This is particularly so when price is considered to be of more importance than quality or flavor value. For instance, the aroma and flavor of cinnamon leaf oil bears little resemblance to that of genuine cinnamon bark oil and, yet, there are many blends available at a far lower price than the genuine bark oil. These may be made from both bark and leaf oils plus such added synthetic materials as cinnamic aldehyde. Strictly, these are imitation flavorings but the blends are often very sophisticated and not readily detected by normally available quality procedures. Another example is the cheaper pimento leaf oil and clove stem and leaf oils which are closely similar in profile but are cheaper than genuine berry and bud oils respectively; yet all are widely used in the making of blends to suit a price constraint. For those concerned with establishing the quality of traditional and processed spices, it is necessary to correlate sensory and instrumental test results to ensure authenticity (Heath 1968; Stahl 1972).

BIBLIOGRAPHY

ANON. 1975. Spice purchasing specifications. Usage applications determine spice form and strength. Food Process. *36*, No. 5, 41–44.

ASHURST, P. R. *et al.* 1972. A new approach to spice processing. Tropical Prod. Inst. Conf. Proc., London, 209–214.

BERNER, D. L. *et al.* 1973. Spice anti-oxidant principle. (Assigned to Campbell Soup Co.) U.S. Pat. 3,732,111, Oct. 1.

BHALLA, K. and PUNEKAR, B. D. 1975. Incidence and state of adulteration of commonly consumed spices in Bombay city. II. Mustard, black pepper and asafoetida. Indian J. Nutr. Diet. *12*, 216–222.

CHARLESWORTH, F. A. 1975. Ethylene oxide residues in sterilized medical devices. Br. Ind. Biol. Res. Assoc. Info. Bull. *14*, 227–228.

CORETTI, K. and INAL, T. 1969. Residue problems with the cold treatment of spices with T-gas (ethylene oxide). Fleischwirtschaft *49*, 599–604.

CRIPPS, H. M. 1972. Spice oleoresins, the process, the market, the future. Tropical Prod. Inst. Conf. Proc., London. 237–242.

EISERLE, R. J. 1971. Gemini Rising—a natural flavoring and stabilizing system for food. Food Prod. Dev. *5*, No. 6, 70–71.

EISERLE, R. J. and ROGERS, J. A. 1972. The composition of volatile oils derived from oleoresins. J. Am. Oil Chem. Soc. *49*, 573–577.

GROSS, A. F. and ELLIS, P. E. 1969. Lipase acivity in spices and seasonings. Cereal Sci. Today *14*, 332–335.

HAYES, E. S. 1961. Herbs, Flavours and Spices. Faber and Faber, London.

HADLOK, R. and TOURE, B. 1973. Mycological and bacteriological studies of "sterlized" spices. Arch. Lebensmittelhyg. *24*, No. 1, 20–25. (German)

HEATH, H. B. 1968. The evaluation of herbs and spices. Can. Inst. Food Sci. Technol. J. *1*, No. 1, 29–36.

HEATH, H. B. 1972. Herbs and spices for food manufacture. Tropical Sci. *14*, 245–259.

HEATH, H. B. 1973. Herbs and spices—a bibliography. Part I. Flavor Ind. *4*, 24–26.

INAL, T. *et al.* 1975. Sterilization of spices by γ-rays. Fleischwirtschaft *55*, 675–677 (23 refs). (German)

JULSETH, R. M. and DEIBEL, R. H. 1974A. Microbial profile of selected spices and herbs at import. J. Milk Food Technol. *37*, 414–419.

JULSETH, R. M. and DEIBEL, R. H. 1974B. Indian Spices *11*, No. 3/4, 6–11.

KOLLER, W. D. 1976. Temperature—an important factor in the storage of ground natural spices. Z. Lebensm. Unters. Forsch. *160*, 143–147. (German)

KRISHNASWAMY, M. A. *et al.* 1974. Microbiological quality of certain spices. Indian Spices *11*, No. 1/2, 6–11.

LAW, D. 1973. The Concise Herbal Encyclopedia. John Bartholomew and Son, Edinburgh, Scotland.

LEWIS, Y. S. 1973. The importance of selecting the proper variety of a spice for oil and oleoresin extraction. Tropical Prod. Inst. Conf. Proc., London, 183–188.

LISSACK, W. 1975. Increased yield—cold milling of spices. Ernaehrungswirtschaft *7*, 438, 440–443. (German)

MacLEOD. A. J. 1973. Spices. Chem. Ind. No. 19, 778–780.

MASTERS, K. 1968. Spray-drying: The unit operation today. Ind. Eng. Chem. *60*, Oct., 53–63.

MAYR, G. E. and SUHR, H. 1973. Preservation and sterilization of pure and mixed spices. Tropical Prod. Inst. Conf. Proc., London, 201–207.

McCORMICK & COMPANY 1964. Spices of the World Cookbook. Penguin Books, Baltimore, Md.

McKERNAN, W. M. 1972–1973A. Microencapsulation in the flavour industry. Part I. Flavour Ind. *3*, 596–598, 600.

McKERNAN, W. M. 1972–1973B. Microencapsulation in the flavour industry. Part II Flavour Ind. *4*, 70, 72–74.

NEARLE, M. W. 1963. Spices in food product manufacture. Canner/Packer *132*, No. 3, 41–47.

PEAT, M. R. 1963. Method of stabilizing spice material and the resulting product. (Assigned to Wm. J. Stange & Co.) U.S. Pat. 3,095,306, June 25.

POWERS, E. M. *et al.* 1975. Microbiology of processed spices. J. Milk Food Technol. *38*, 683–687.

PROVATOROFF, N. 1973. Some details of the distillation of spice oils. Tropical Prod. Inst. Conf. Proc., London, 173–181.

RAUSCHER, H. *et al.* 1957. Ethylene oxide for cold sterilization. Food Manuf. *32*, 169–172.

SABEL, W. and WARREN, J. D. F. 1973. Theory and practice of oleoresin extraction. Tropical Prod. Inst. Conf. Proc., London, 109–192.

SALZER, U.-J. 1975. Analytical evaluation of seasoning extracts (oleoresins) and essential oils from seasonings. I. International Flavours and Food Additives, *6*, 151–157; II. *Ibid.* 206–210; III. *Ibid.* 253–268.

SCHARF, M. M. 1967. Sterilization of spices by acid treatment. U.S. Pat. 3,316,100. Apr. 25.

STAHL, W. H. 1972. Oleoresin quality analysis—fact or fancy? Tropical Sci. *14*, 335–345.

STANIFORTH, V. 1973. Spices or oleoresins: A choice? Tropical Prod. Inst. Conf. Proc., London, 193–197.

STOBART, T. 1970. Herbs, Spices and Flavourings. International Wine and Food Publishing Co., London.

THANGAMANI, H. *et al.* 1975. Microbial contamination of spices. Indian Food Packer *29*, No. 2, 11–13.

TODD, R. D. 1970. Microencapsulation and the flavour industry. Flavour Ind. *1*, 768–771.

VITZTHUN, O. and HUBERT, P. 1972. Process for producing spice extracts of natural composition. (Assigned to HAG AG.) Process for producing spice extracts of natural composition. Ger. Fed. Rep. Pat. Appl. 2, 127, 611.

WALFORD, J. 1976. Solubilizers for essential oils in flavour formulations. Food Manuf. *51*, No. 2, 35–37.

WARMBROD, F. and FRY, L. 1966. Coliform and total bacterial counts in spices, seasonings and condiments. J. Assoc. Off. Anal. Chem. *49*, 678–680.

WESLEY, F. *et al.* 1965. The formation of persistant toxic chlorohydrins in foodstuffs by fumigation with ethylene oxide and propylene oxide. J. Food Sci. *30*, 1037–1042.

Part 4: Fruit Juices, Concentrates and Pastes

Reference has already been made to the citrus fruits as a source of strongly aromatic essential oils, but citrus and many soft fruits are primarily processed commercially for their juice. This may be either consumed directly as a drink or further concentrated to produce a flavorful base of value in beverages, yogurt, preserves and several other categories of food products. For those concerned with the processing and handling of fruit juice products, several authoritative texts are available (Berk 1969; Hulme 1970; Nagy *et al.* 1977; Tressler and Joslyn 1971). This section deals only with those aspects of fruit juices, concentrates and pastes as flavoring materials; the products involved are shown in Fig. 7.10.

A fruit juice is defined as the clear or uniformly cloudy, unfermented liquid recovered from sound ripe fruits by pressing or other mechanical means. Not all fruits are of equal value as a source of juice. The selection of fruit on the basis of variety, ripeness and quality demands considerable expertise as so many variables determine the initial quality, potability, flavor character and keeping properties of the resulting juice.

The method of recovering fruit juices depends upon the structure of the fruit, the position and character of the tissues in which the juice is located and the nature of the juice itself, particularly its pH.

Extraction of Juice from Soft Fruits

The mechanics employed for the extraction of juice from soft fruits is the vertical basket hydraulic press, usually fabricated in stainless steel. As pressure is applied, the juice is slowly expressed, passes through a coarse mesh screen and is collected. This single basket press is still very widely used but has the disadvantage of

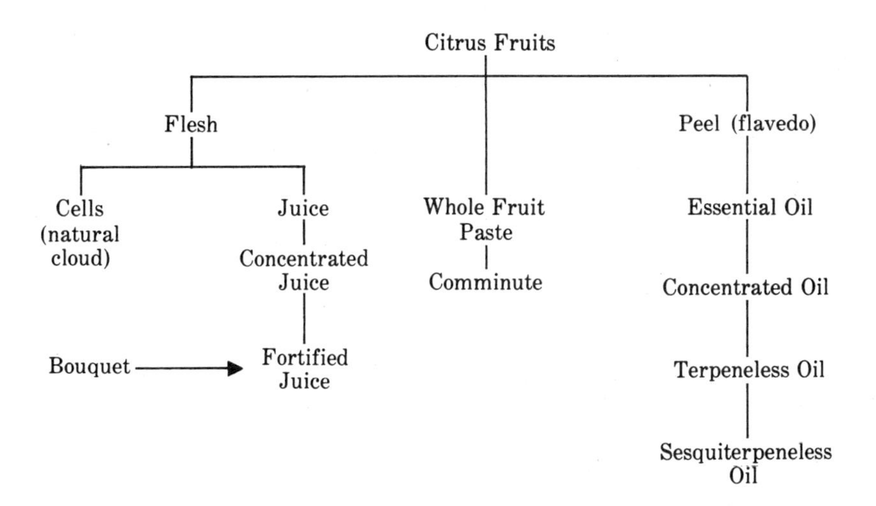

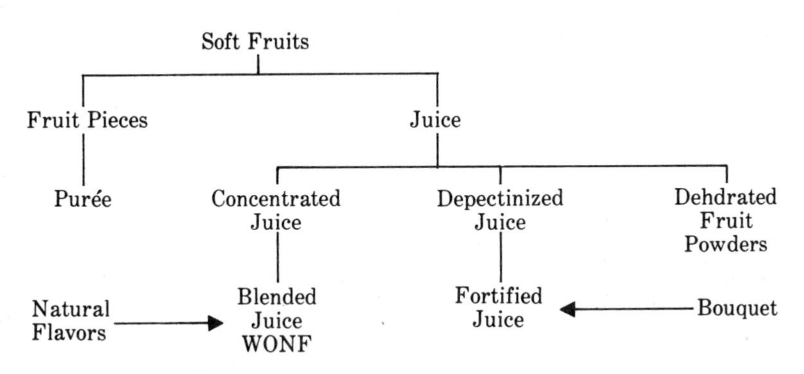

FIG. 7.10. FRUIT PRODUCTS USED AS FLAVORINGS

batch operation and slow through-put. Continuous presses are now more favored. In these, the fruit is fed into a screw conveyor within a tapering perforated cage. The juice escapes as the mass passes along the screw bed with the mass becoming increasingly compressed as it does so. Such presses enable the processor to handle large tonnages of fruit rapidly and minimizes any deterioration during handling. However, screw presses tend to grind the pulp and this may result in a juice having a high cellular content. An alternative process consists of comminuting the fruit and passing the resulting fine pulp through a screen press; but, again, the juice tends to be high in suspended solids.

After expression, the dry press-cake is usually discarded; but in order to increase the yield of juice, it may be disintegrated, mixed with water and re-pressed. Such procedure certainly increases the yield but does little for the quality of the end-product. A second hot pressing may give a product of value for blending into flavorings; but such secondary products are normally kept quite separate from the main bulk of cold pressings.

Soft fruits differ considerably in the ease with which they can be processed. The presence of pectins and proteins may give rise to a viscous juice which is only slowly released under pressure and is difficult to pass through a fine screen. In certain instances, slight prefermentation may be allowed prior to pressing. This has a dual effect of improving the flavor and allowing some cellular breakdown which facilitates the release of juice. The addition of pectinase is also a means of achieving a better yield of juice, but care must be taken to ensure that any added enzymes are effectively removed from the final product.

Extraction of Citrus Juices

The processing of citrus fruit for juice requires special equipment designed to reduce the amount of peel oil included in the recovered juice. Although citrus fruits display a wide range of size, the equipment used is remarkably similar in design; although there are several varieties of press used throughout the industry, they differ only in detail. They are mainly based on the reaming principle: the fruits are halved and the halves are held in a cup against the increasing pressure of a rotating cone which removes the juice and a proportion of the cellular matter. In another process, the whole fruit is positioned in a holding cup, a stainless steel tube is inserted into the fruit and the juice expelled by pressure. In the case of limes, the small size of the fruit makes reaming difficult so many processors crush the whole fruit in roller presses or screw presses, after which

Courtesy of Citrus Central, Inc.

FIG. 7.11. JUICE EXTRACTED FROM DIFFERENT VARIETIES OF ORANGES IS BLENDED FOR FLAVOR, COLOR AND SWEETNESS

juice is screened to remove fragments of cellular tissue and then centrifuged to separate the peel oil.

After extraction, the juice is strained and screened as necessary to remove pips and other unwanted solids. In the case of orange juice, the product desirably contains a uniform cloud of suspended solids; whereas with lemon and lime juices, a clear product is preferred and the juice is either centrifuged or filtered to achieve this. Freshly extracted juices almost always contain extraneous matter in the form of a fine suspension of pectins, gums and proteinaceous substances, many of which are almost colloidal in character. All irregular particles such as pieces of peel or skin, pips or broken seeds, etc., are readily removed in the initial clarification and screening, the amount and nature of any residual suspended matter being determined by the required nature of the end-product. Finely divided fruit solids, in some cases derived from the press-cake, may be added to improve or modify the natural cloud but care is necessary in using such added materials as the stability of the cloud may be affected.

Preservation of Juices

Fruit juices not intended for immediate use must be preserved to prevent deterioration and ultimate spoilage due to yeast growth in the product. Yeasts are the chief source of spoilage as they cause fermentation. The pH, moisture and total solids present are a determinant of the type of organisms most likely to develop.

Rigid standards of cleanliness and hygiene in manufacturing and storage of fruit juices must be maintained. The juice must be treated in a manner that will keep it from fermenting or undergoing any other form of deterioration. The principal method employed is pasteurization at 60°–88°C (140°–190°F) in hermatically sealed containers for periods of time ranging from 5 to 30 min. Some change in flavor profile is inevitable; but fortunately, most juices have a relatively low pH in which bacteria do not readily multiply and, in consequence, heat processing can be limited to that necessary to destroy yeasts and molds and also to inactivate any enzyme systems which may lead to flavor deterioration and loss of cloud stability on storage. Preservation by freezing is also popular and most fruit juices are available in cold-pack units.

The retention of flavor in fruit juices during storage is of considerable importance, particularly where the juice is sold for direct consumption (e.g., orange juice). Canned citrus juices, in particular, develop off-flavors after short periods of storage at room temperature (20°–25°C) and may become unpalatable after a few months. Most canned juices are best stored at below 4°C (40°F) under which conditions the flavor is retained almost indefinitely. Although the fresh quality is preserved in frozen juices until the product is thawed prior to use, the defrosted juices must be used quickly as they are relatively unstable.

Chemical preservation is permitted in most countries and sodium benzoate at 0.1–0.3% or sulfur dioxide at 0.03–0.08% are both effective in preventing spoilage. In the latter case, there is a gradual loss of SO_2 during storage and in bulk containers it is often necessary to add further quantities of sulfite to maintain an adequate level of preservative. The use of these additives is controlled by legislation which should be checked, particularly with regard to labeling declaration.

Concentrated Juices

The further processing of fruit juices depends upon their ultimate usage, e.g., bottled or canned for drinking as such or manufactured into fruit juice syrups for beverages, etc. Most juices consist of 85–90% water and have a flavor strength which is acceptable for drinking per se; however, as a source of flavor in beverages and other processed foods, the natural juices are much too weak and are generally replaced by concentrated products.

Ideally, the concentration process should result in the removal of water without in any way affecting the delicate aroma and flavor profile or other characteristics of the natural juice either by the loss of volatile components or by the development of a "cooked" note in the flavor profile. The most convenient method of concentration is by vacuum evaporation and several types of evaporator, embodying the latest techniques of heat exchange, are currently in use (Shacklady 1969). In such plants, most fruit juices can be processed without undue change in their basic flavor character, although inevitably there will be a loss of some low-boiling components by steam-distillation. Various techniques are now available for the recovery of these aromatics from the distillate by the use of special traps and the recovered material can be added back to the concentrated juice. This considerably improves the final product and gives to it much of the fresh character of the original juice. The overall effect of concentration is to produce a more rounded profile due to changes in the sugars and acids during heat processing.

When dilute fruit juices are frozen the resulting ice crystals are pure water until the eutectic point is reached, when the mass solidi-

Courtesy of DEC International, Inc.

FIG. 7.12. AN INDUSTRIAL EVAPORATOR USED IN THE PRODUCTION OF CONCENTRATED CITRUS JUICES

fies. Making use of this, freeze concentration has been used as a means of achieving a concentrated product without the use of heat which retains the top-notes which would otherwise be lost. The process is costly and involves the formation of a frozen slush from which the ice crystals are removed by centrifuging or rotary vacuum filtration. The freezing-separation cycle is continued until the right degree of concentration is attained. This is usually limited to about 50° Brix. At this level of concentration, the solids are not sufficient for satisfactory shelf-life and it is usual to add sugar to raise the solids to the necessary degree.

Brix Value

The specifications for most fruit juice products include a value for soluble solids in terms of "degrees Brix." This figure is based on the relationship between the specific gravity of the product and the percentage of dissolved solids it contains. The Brix Tables quoted in the literature are based on an accurate determination of these relationships using a solution of sucrose and it is assumed that all the dissolved solids will have a similar effect upon the specific gravity of the juice. This is not strictly so; but is sufficiently close for most practical purposes. The specific gravity of the product may be determined by using a hydrometer calibrated directly in degrees Brix or a refractometer, the determined refractive index then being converted into degrees Brix from the appropriate Tables. The two methods are not strictly comparable but, again, are sufficiently so for control purposes. The Brix value is strictly a measure of sucrose concentration, but since fruit juices contain more sugar than other soluble solids the result acts as a ready guide.

Blended and Fortified Juices

Many natural and concentrated juices are either inadequate as a source of flavor or are too expensive for use in low-priced products. To cater to the needs of beverage and food processors, two types of product, based on concentrated juices, have been developed and are widely used:

(a) With other natural flavors (WONF). These comprise at least 50% of the flavor strength from extractives of the named fruit source and not more than 50% from other natural flavors.

(b) Fortified concentrates. These are products in which the flavor strength is fortified or the bouquet intensified by the addition of, in the main, natural constituents. Any synthetic chemicals added are generally, but not necessarily, identical to those found in the natural fruit.

These products find wide acceptance but reference should be made to the manufacturer to establish the relative strength of such concentrates and their compliance with current legislation.

Depectinized Juices

All fruit juices contain pectin which is generally removed during processing as an aid to clarification and stability. Depectinization has a further advantage; it results in a concentrate being miscible with diluted alcohol of the strength normally employed in the manufacture of liqueurs, whereas the natural juice would precipitate under the same conditions. Such products are normally preserved with 0.05–0.06% benzoic acid.

Fruit Pastes and Comminutes

For use in ice cream, frozen confectionery and certain types of beverages, citrus fruits may be passed through a colloid mill to form an homogenous paste. These finely ground products are made from the whole fruit to which may be added further juice and natural peel oil emulsion in order to achieve some measure of flavor standardization. The product is generally flash pasteurized to inactivate any pectolytic enzymes and is then filled aseptically into sterile cans which are sealed and kept refrigerated until required. Citrus pastes and comminutes do not contain any preservative, added sugar or coloring matter. If correctly stored, the full fresh flavor is very stable.

Similar pulpy products made from soft fruits (e.g., apple, apricot, peach, etc.) are called nectars.

Dehydrated Fruit Juices

Over the years, many attempts have been made to present fruit juices in a dehydrated form as such products are attractive to the food processor for inclusion in jellies, table desserts and dry mixes as well as for use in beverages and "instant" soft drinks. The processes involved are not technically difficult but the sugars present in the juices tend to dehydrate as a plastic mass rather than as discrete crystals or granules and the resulting products are very deliquescent. The following systems have been used with varying degrees of success (Shacklady 1969):

(a) Roller drying—the standard method of drying a film of the juice on heated rollers. It has limited application, as the product tends to have an unacceptable cooked note and an abnormally dark color.

(b) Vacuum drying—an improved version of roller drying. Lower evaporation temperatures achieved yield a product which is friable, easily reconstituted and of good flavor quality.

(c) Spray drying—this involves the drying of the juice, with or without the addition of a carrier such as dextrose, at a relatively high temperature for a very short time. The equipment used may be a conventional spray drier or a Birs tower drier in which the temperatures employed are much lower.

(d) Foam-mat drying—this involves the formation of a foam stabilized with proteinaceous matter or glyceryl monosterate. The foam is dried on a conveyor belt either in a stream of hot air or by blowing air through the foam. The product has a good fresh character which is somewhat prone to oxidation owing to the considerably increased surface area produced. The changes occurring during this method of processing have been studied by Bolin and Salunkhe (1971).

(e) Freeze drying—this process involves the sublimation of ice directly from the frozen juice which is maintained at a very low temperature under high vacuum with just sufficient heat input to ensure the removal of the water. The products are exceptionally good but the operating costs are relatively high.

Dried fruit juice products are difficult to handle in view of the speed with which they absorb atmospheric water. The advice of the manufacturer should be sought as to the best method of storage and incorporation into end-products.

BIBLIOGRAPHY

BAKER, R. A. 1976. Clarification of citrus juices with polygalacturonic acid. J. Food Sci. *41*, 1178–1200.

BERK, Z. 1969. Industrial processing of citrus fruits. *In* Food Ind. Studies *ID/SER 1/2*. U.N. Ind. Dev. Organ., Rome.

BERRY, R. E. *et al.* 1972. Preparation of instant orange juice by foam-mat drying. J. Food Sci. *37*, 967–968.

BERTRAND, J. 1970. Fruit flavors. Present processes for industrial recovery. Fr. Ses Parfums *13*, No. 69, 234–236. (French)

BOLIN, H. R. and SALUNKE, D. K. 1971. Physiocochemical and volatile flavor changes occurring in fruit juices during concentration and foam-mat drying. J. Food Sci. *36*, 665–668.

BROCKMAN, M. C. 1974. Freeze drying. *In* Encyclopedia of Food Technology. A. H. Johnson and M. S. Peterson (Editors). AVI Publishing Co., Westport, Conn. (31 refs.)

CHARON, M. A. 1969. Freeze concentration and lyophilization. Two complimentary techniques. Fr. Ses Parfums *12*, No. 66, 401–406. (French)

CRUSE, R. R. and LIME, B. J. 1970. How to make citrus whole fruit purée. Food Eng. *42*, 109, 111–112.

FLINT, J. and KAREL, M. 1970. Effect of process variables on retention of volatiles in freeze-drying. J. Food Sci. *35*, 444–447.

FODA, Y. H., HAMED, M. G. E. and ABD-ALLAH, M. A. 1971. Preservation of orange and guava juices by freeze-drying. Food Technol. *24*, No. 12, 74–80.
GIERSCHNER, K. and REINERT, C. 1970. Pectin and pectin enzymes in the preparation of orange and apple juices with steady cloudiness. Fluess. Obst. *37*, No. 1, 6–14. (German)
HOLLAND, R. R. *et al.* 1976. Cloud stability test for pasteurized citrus juice. J. Food Sci. *41*, 812–815.
HULME, A. C. (Editor). 1970. The Biochemistry of Fruits and Their Products, Vol. 1 and 2. Academic Press, London, New York.
ISHII, S. and YOKOTSUKA, T. 1972. Clarification of fruit juices by pectin *trans* eliminase. J. Agric. Food Chem. *20*, 787–791.
ISHII, S. and YOKOTSUKA, T. 1973. Susceptibility of fruit juices to enzymatic clarification by pectin lyase and its relation to pectin in fruit juice. J. Agric. Food Chem. *21*, 269–272.
KRAMER, A. *et al.* 1971. Freeze concentration by directional cooling. J. Food Sci. *36*, 320–322.
MANNHEIM, C. H. and PASSY, N. 1975. Aroma recovery and retention in fruit juices. Int. Flavours Food Additives *6*, 323–328.
MATSUURA, T., BAXTER, A. G. and SOURIRAJAN, S. 1974. Studies in reverse osmosis for concentration of fruit juices. J. Food Sci. *39*, 704–711.
MERORY, J. 1968. Food Flavorings. Composition, Manufacture and Use. AVI Publishing Co. Westport, Conn.
MIZRAHI, S. and BERK, Z. 1970. Physico-chemical characteristics of orange juice cloud. J. Sci. Food Agric. *21*, 250–253.
MORGAN, A. I., Jr. 1974. Foam-mat drying. *In* Encyclopedia of Food Technology. A. H. Johnson and M. S. Peterson (Editors). AVI Publishing Co., Westport, Conn.
MOY, J. H. 1971. Vacuum-puff freeze-drying of tropical fruit juices. J. Food Sci. *36*, 906–910.
NAGY, S., SHAW, P. E. and VELDHUIS, M. K. 1977. Citrus Science and Technology, Vol. 2—Fruit Production, Processing Practices, Derived Products and Personnel Management. AVI Publishing Co., Westport, Conn.
ROTHSCHILD, G. and KARSENTY, A. 1974. Cloud loss during storage of pasteurized citrus juices. J. Food Sci. *39*, 647–652.
SAINTE-HILAIRE, P. and SOLMS, J. 1973A. Freeze-drying of orange juice. I. Influence of chemical composition on the temperature of sublimation. Lebensm. Wiss. Technol. *6*, 170–173. (German)
SAINTE-HILAIRE, P. and SOLMS, J. 1973B. Freeze-drying of orange juice. II. Influence of the freezing technique. Lebensm. Wiss. Technol. *6*, 174–178. (German)
SARAVACOS, G. D. 1970. Effect of temperature on viscosity of fruit juices and purée. J. Food Sci. *35*, 122–125.
SARAVACOS, G. D. and MOYER, J. C. 1968. Volatility of some aroma compounds during vacuum drying of fruit juices. Food Technol. *22*, No. 5, 89–93.
SHACKLADY, J. 1969. Fruit juices and fruit juice beverages. *In* Food Industries Manual, 20th Edition. A. H. Woolen, (Editor). Leonard Hill & Co., London.
TRESSLER, D. K. and JOSLYN, M. A. 1971. Fruit and Vegetable Juice Processing Technology, 2nd Edition. AVI Publishing Co., Westport, Conn.

Part 5: Essential Oils

In an earlier section the flavor profiles and characteristics of essential oils derived from herbs and spices were discussed, particularly in relation to their usage as alternatives to ground natural spices. Many other aromatic plants are also a source of essential oils which find wide application throughout the food, beverage and confectionery industries; it is with these oils that this section is concerned. A list of the essential oils used in flavorings, as opposed to fragrances, is given in Table 7.2, and their botanical affiliation is given in Table 7.3. The botanical source of the essential oils is given in Table 7.4. These oils are individually described in standard texts such as Arctander (1960), Fenaroli (1975) and Guenther (1952).

Description of Essential Oils

An essential oil, or a volatile oil, as the two expressions are used synonymously, is a volatile mixture of organic compounds derived from odorous plant material by physical means. A specific oil is derived from one botanical source with which it agrees in name and odor. Most essential oils exist per se in the source material but certain oils are formed only as a result of an enzymatic reaction after the plant tissues have been crushed or macerated with water (e.g., onion, mustard). The oil formed by these reactions can then be recovered by traditional methods.

Essential oils may be found throughout the plant cellular tissue or in special cells, glands or ducts located in several parts of the plant, i.e., in the leaves, barks, roots, flowers, fruits or seeds, sometimes confined to special structures and in others not localized. The type of structure of the secretory tissues is one of the characteristics of a botanical family (Hardman 1972) and may be grouped as follows:

(a) Oil cells	*Graminae*	Citronella
	Zingiberaceae	Ginger
	Piperaceae	Black pepper
	Magnoliaceae	Star anise
	Myristaceae	Nutmeg
	Lauraceae	Cinnamon/cassia
(b) Schizolysigenous oil glands	*Rutaceae*	Citrus fruits
	Myrtaceae	Allspice, clove
(c) Oil canals	*Bursuraceae*	Myrrh
(d) Vittae or secretory oil ducts	*Umbelliferae*	Anise, dill, coriander
(e) Glandular trichomes or hairs	*Labiatae*	Peppermint, basil

With few exceptions (e.g., clove), the essential oil is present at a low percentage and constitutes only a small fraction of the total plant weight. The quantity and composition of the essential oil var-

TABLE 7.2
PRINCIPAL ESSENTIAL OILS USED IN FLAVORINGS

Common Name	Botanical Source	Principal Producing Countries	Main Crop Period	Average Yield of Essential Oil (%)	Chief Aromatic Constituents	Regulatory Status
Allspice (pimento)	Fruits of *Pimenta dioica*, L.	West Indies (Jamaica)	July-Aug.	3–4.5	Eugenol	GRAS
Ambrette	Seeds of *Hibiscus abelmoschus*, L.	Java Martinique		0.2–0.6	Farnesol Ambrettolide	GRAS
Angelica	Root and seeds of *Angelica archangelica*, L.	Europe (Germany, France)	July	Root: 0.3–1 seeds: 0.6–1.5	β-Phellandrene	GRAS
Angostura	Bark of *Galipea officinalis*, Hancock. (*G. cuspania*, DC)	South America (Venezuela)		1–2		GRAS
Anise	Fruits of *Pimpinella anisum*, L.	Europe U.S.S.R.	Aug.-Sept.	1.5–4	Anethol	GRAS
Artemisia	Herbaceous tops of *Artemisia absinthium*, L.	Central Europe Asia North America	Aug.	0.3–0.4	Thujone	GRAS, thujone free
Asafetida	Oleo-gum-resin from *Ferula asafoetida*, L. and related species	Iran Afganistan	July	7–9	Disulfides	GRAS
Balm (lemon balm)	Herbaceous tops of *Melissa officinalis*, L.	Southern and Central Europe North Africa	Aug.	0.1	Geraniol Citronellol Citral	GRAS

TABLE 7.2 (Continued)

TABLE 7.2 (Continued)

Common Name	Botanical Source	Principal Producing Countries	Main Crop Period	Average Yield of Essential Oil (%)	Chief Aromatic Constituents	Regulatory Status
Balsam Peru	Oleoresin from *Myroxylon pereirae*, Klotzsch.	El Salvador	Nov.–June	45–55 (from the extracted balsam)	Benzyl cinnamate Benzyl benzoate	GRAS
Basil, sweet	Herbaceous tops of *Ocimum basilicum*, L. Several varieties are distilled:	Mediterranean U.S.A.	Sept.	0.1	Methyl chavicol Linalool Cineol	GRAS
	Réunion-type	Malagasy Republic				
	Methyl cinnamate-type	Bulgaria Haiti				
	Eugenol-type	Java U.S.S.R.				
Bay laurel (sweet bay)	Leaves of *Laurus nobilis*, L.	Mediterranean	Sept.–Mar.	0.5–1	Cineol	GRAS
Bay, West Indian	Leaves of *Pimenta racemosa* (Mill) J. W. Moore (formerly *Myrcia acris*, Wight	West Indies (Dominica) Venezuela	Jan.–June	1–2	Eugenol	GRAS
Bergamot (bergamot orange)	Peels of *Citrus bergamia*, Risso or *C. aurantium*, L. sub-species *bergamia*, Wright and Arn.	Calabria (Italy)	Dec.–Mar.	0.5 (approx)	Linalyl esters	GRAS

Birch bark (sweet birch)	Bark of *Betula lenta*, L.	Canada U.S.A. Eastern Europe, U.S.S.R.	Oct.–Apr.	0.5	Methyl salicylate	
Bitter almond	Partially de-oiled press-cake from the kernels of *Prunus amygdalus*, Batsch, var. *amara*.	U.S.A. Middle East Southern Europe North Africa		0.5–0.7	Benzaldehyde (oil is treated to remove HCN)	GRAS, free from HCN
Bois de Rose	Wood of *Aniba rosaeodora*, Ducke.	South America (Brazil, Peru, Guiana)	Jan.–June	0.7–1.2	Linalool	GRAS
Buchu	Leaves of *Barosma betulina*, Bartl and Wendl, *B. crenulata* (L) Hook., or *B. serratifolia*, Willd.	South Africa		0.8–3.5 (depending on species)	Diosphenol	Good mfg practice
Cacao	Fermented and roasted seeds of *Theobroma cacao*, L.	Central and South America Africa		0.001 (approx)	Linalool	GRAS
Camomile (German chamomile)	Flowers of *Matricaria chamomilla*, L.	Europe Asia U.S.A. Australia	May–July	0.3–1	Chamazulene	GRAS
Camomile (Roman or English chamomile)	Flowers of *Anthemis nobilis*, L.	Europe	July–Oct.	0.3–1		
Camphor, tree	Wood of *Cinnamomum camphora* (L) Eberm.	Far East (China, Taiwan, Japan)			Camphor and linalool depending on variety	

TABLE 7.2 (Continued)

TABLE 7.2 (Continued)

Common Name	Botanical Source	Principal Producing Countries	Main Crop Period	Average Yield of Essential Oil (%)	Chief Aromatic Constituents	Regulatory Status
Cananga	Flowers of *Cananga odorata*, Hook and Thoms.	Islands in Indian Ocean Java	Oct.–Dec.	0.5–1	Sesquiterpene alcohols	GRAS
Caraway	Fruits of *Carum carvi*, L.	Europe (Holland, Hungary) North America	Aug.–Sept.	3–6	Carvone	GRAS
Cardamom	Seeds of *Elettaria cardamomum*, Maton.	India Guatamala	India: Aug.–Oct. Guatamala: Jan.	5 (approx) 3–6	Cineole d-α-Terpineol Terpinyl acetate	GRAS
Carrot	Seeds of *Daucus carota*, L.	France	Sept.	0.4–0.8	Carotol	GRAS
Cascarilla	Bark of *Croton eluteria*, Bennett	Bahamas		1.5–3		GRAS
Cassia, Chinese	Bark of *Cinnamomum cassia*, Blume.	Southeast China	June–July	0.3 (approx)	Cinnamic aldehyde	GRAS
Cedar, white (thuja)	Leaves and twigs of *Thuja occidentalis*, L.	U.S.A. Canada Japan	Apr.–Nov.	0.6–1	α-Pinene Thujone	GRAS (thujone free)
Celery seed	Fruits of *Apium graveolens*, L.	France Holland Hungary India	Sept.	1.5–2.5	d-Limonene Sedanolide	GRAS
Cherry, wild bark	Bark of *Prunus serotina*, Ehrh. or *P. canadensis*, Poiret.	Canada U.S.A. Europe				GRAS (mainly as extract)
Chervil	Herbaceous tops of *Anthiscus cerefolium* (L) Hoffm.	Southern Europe U.S.A.				GRAS

Cinnamon bark, Ceylon	Bark of *Cinnamomum zeylanicum*, Nees.	Sri Lanka	July–Aug. and Dec.–Jan.	0.5–0.8	Cinnamic aldehyde	GRAS
Cinnamon bark, (cassia)	Bark of *C. cassia*, Blume *C. loureirii*, Nees. *C. burmanni*, Blume	 China Saigon Batavia		0.5–3 (depending on species)	Cinnamic aldehyde	GRAS
Cinnamon leaf, Ceylon	Leaves of *Cinnamomum zeylanicum*, Nees.	Sri Lanka	July–Sept.	0.5–0.7	Eugenol	GRAS
Citronella, Ceylon	Fresh or dried herb *Cymbopogon nardus*, Rendle	Sri Lanka	Three crops per year	0.5 (approx)	Geraniol Citronellal	GRAS
Citronella, Java	Fresh herb *Cymbopogon winterianus*, Jowitt.	Java	3–4 crops per year	0.7 (approx)	Geraniol Citronellal	GRAS
Clove bud	Unopened buds of *Eugenia caryophyllata* Thunb.	Tanzania Malagasy Republic East Indies	Throughout the year	14–21	Eugenol	GRAS
Clove leaf	Leaves of *Eugenia caryophyllata*, Thunb.	East Indies		1.5–2	Eugenol	GRAS
Clove stem	Stems of *Eugenia caryophyllata*, Thunb.	East Indies		5 (approx)	Eugenol	GRAS
Clover	Flowers of *Trifolium* species	Europe Central Asia		0.005–0.02		GRAS
Cognac oil	Wine lees and dregs or pressed grape skins	France		0.04–0.1 (depending on source)	So-called ethyl aenanthate	GRAS
Coriander	Fruits of *Coriandrum sativum*, L.	North Africa Russia India	Sept.	0.1–1	Linalool	GRAS
Cubeb	Unripe berries of *Piper cubeba*, L.	Indonesia Africa	Apr.–May	13 (approx)	Cineole	Good mfg practice

TABLE 7.2 (Continued)

TABLE 7.2 (Continued)

Common Name	Botanical Source	Principal Producing Countries	Main Crop Period	Average Yield of Essential Oil (%)	Chief Aromatic Constituents	Regulatory Status
Cumin (cummin)	Ripe fruits of *Cuminum cyminum*, L.	Mediterranean countries India U.S.S.R.	Sept.	2.5 (approx)	Cuminic aldehyde	GRAS
Davana	Flowers of *Artemisia pallens*, Wall.	Southern India	Apr.–May	0.1–0.6 (approx)		Good mfg practice
Dill (seed)	Fruits of *Anethum graveolens*, L.	Europe Pakistan U.S.A.	Sept.	2.5–4	Carvone	GRAS
Dill (weed)	Green fruiting tops of *Anethum graveolens*, L.	Eastern Europe U.S.A.	July	0.3–1.5	Carvone	GRAS
Dill, Indian	Fruits of *Anethum sowa*, Roxb. or *Peucedanum graveolens*, Benth & Hook.	India		1.2–3.5	Carvone Limonene Dillapiol	Good mfg practice
Elder flowers	Flowers of *Sambucus canadensis*, L. and *S. nigra*, L.	Hungary Germany France	June			GRAS
Elemi	Resin from *Canarium commune*, L. or *C. luzonicum*, Mig.	Phillipines	June–Oct.	20–30 (from the crude resin)		GRAS
Erigeron	Herbaceous tops of *Erigeron canadensis*, L.	Canada U.S.A. Southern Europe	July–Nov.	0.3–0.6	Terpinyl acetate	
Estragon (tarragon)	Herbaceous tops of *Artemisia dracunculus*, L.	Central and eastern Europe	July–Aug.	0.3–1.5	Methyl chavicol (estragole) Anethol	GRAS

TABLE 7.2 (Continued)

Eucalyptus	Leaves of *Eucalyptus globulus*, Labill. and many other *Eucalyptus* sps. (Guenther 1949)	Spain Portugal Brazil Zaria		0.75–1.25	Cineole	GRAS
Fennel, sweet	Fruits of *Foeniculum vulgare*, Mill. var. dulce	Germany Romania U.S.S.R. North Africa India	Sept.	4–6	Anethol	GRAS
Fir (pine)	Needles, twigs and cones of: Siberian Fir (*Abies sibirica*, Lebed); Silver Fir (*A. alba*, Mill.); Japanese Fir (*A. Sachilenensis*, Masters or *A. mayriana*, Miyabe & Kudo)	U.S.S.R. North America Europe Japan		0.3 (approx)		Good mfg practice
Galangal	Rhizomes of *Alpinia officinarum*, Hance.	China		0.5–1		GRAS
Galbanum	Dried gum-resin from *Ferula galbaniflua*, Boiss & Buhse and other *Ferula* sps.	Iran Turkestan		10–22 (from the crude resin)		Good mfg practice
Garlic	Bulbs of *Allium sativum*, L.	U.K. Italy Egypt U.S.A.		0.1–0.25	Allyl sulfides	GRAS

TABLE 7.2 (Continued)

Common Name	Botanical Source	Principal Producing Countries	Main Crop Period	Average Yield of Essential Oil (%)	Chief Aromatic Constituents	Regulatory Status
Geranium	Leaves and stems of *Pelargonium graveolens*, L'Her. and other *Pelargonium* sps.	Réunion North Africa Central and western Europe	Mar.–Oct. (depending on region)	0.15–0.2 (from fresh material)	Geraniol Citronellol	GRAS
Ginger	Rhizomes of *Zingiber officinale*, Roscoe	Jamaica West Africa India China Australia	Jamaica: Febr.–Mar.	0.25–1.25	Zingiberene Citral	GRAS
Grapefruit	Peel of *Citrus paradisi*, Macf. fruit, by cold expression.	U.S.A. East Africa West Indies Brazil				GRAS
Guaiac	Wood of *Guaicum officinale*, L. *G. sanctum*, L. or *Bulnesia sarmienti*, Lor.	South America Jamaica Cuba		approx 5 (much lower in local field stills)	Guaiol	Good mfg practice
Hemlock (spruce)	Needles and twigs of *Tsuga canadensis* (L) Carr. T. *heterophylla*, (Raf) Sarg. *Picea glauca* (Moench) Voss. or *P. mariana* (Mill) B.S.P.	North America China Japan			α- and β-pinenes (terpenes approx 45%) Bornyl acetate	Good mfg practice
Hops	Flower catkins of *Humulus lupulus*, L.	Europe North America	Sept.	0.3–0.5	Myrcene Humulene	GRAS

Horsemint (Monarda)	Herbaceous tops of *Monarda punctata*, L.	U.S.A.		1–3	Thymol (approx 60%)	GRAS
Hyssop	Herbaceous tops of *Hyssopus officinalis*, L.	Southern Europe	Aug.	0.1–0.3	α-Pinene Sesquiterpene alcohols	GRAS
Immortelle	Flowering tops of *Helichrysum angustifolium*, DC.	Mediterranean countries (Yugoslavia)		less than 0.1	Nerol Neryl acetate	GRAS
Jasmine	Flowers of *Jasminum officinale*, L. or *J. grandiflorum*, L.	Mediterranean countries Asia	June–Sept.	3 of concrete containing 40–50% essential oil	Nerol Terpineol Jasmone	GRAS
Juniper berry	Berries of *Juniperus communis*, L.	Europe (Italy) Asia North America	Aug.–Sept.	0.5–0.6 (up to 2.5 from dried berries)	α-Pinene	GRAS
Laurel (sweet bay)	Leaves of *Laurus nobilis*, L.	Mediterranean countries	Sept.–Mar.	0.5–1	Cineole	GRAS
Labdanum	Gum resin from the leaves and twigs of *Cistus ladaniferus*, L. and other *Cistus* sps.	Spain Cyprus	July–Aug.	1–2 (from the gum resin)		Good mfg practice
Lavandin	Flowering tops of *Lavandula hybrida*, Rev. being a hybrid of *L. latifolia*, Vill. and *L. officinalis*, Chaix.	Southern France	Aug.	0.5–1	Linalyl acetate (20–22%)	GRAS
Lavandin (Abrialis type)	Flowering tops of a selected strain of *L. hybrida*, Rev.	Southern France	Aug.	0.5–1	Linalyl acetate (30–32%)	

TABLE 7.2 (Continued)

TABLE 7.2 (Continued)

Common Name	Botanical Source	Principal Producing Countries	Main Crop Period	Average Yield of Essential Oil (%)	Chief Aromatic Constituents	Regulatory Status
Lavender	Flowering tops of *Lavandula, officinalis,* Chaix.	Mediterranean countries U.K. North Africa	July–Aug.	0.5–1	Linalool Linalyl acetate	GRAS
Lavender, spike	Flowering tops of *Lavandula latifolia,* Vill. (*L. spica,* DC)	Spain France	Aug.	0.75–1	Linalool Linalyl acetate	GRAS
Lemon	Peel of *Citrus limon,* (L) Burm. fruit by cold expression.	Italy California Spain Israel	Dec.–Apr. through to Oct.	4 (approx)	Limonene (90%)	GRAS
Lemongrass	Herbaceous tops of *Cymbopogon citratus,* DC or *Andropogon nardus* var. *ceriferus*	West Indies (Malagasy Guatemala)	Aug.	0.3–0.4	Citral (approx 70%)	GRAS
	C. flexuosus, (Nees.) Stapf. or *Andropogon nardus* var. *flexuosus*	East Indies (India Indonesia Sri Lanka)	May–June	0.2–0.3	Citral (80–85%)	
Lemon verbena	Leaves and twigs of *Lippia citriodora,* H.B.K.	Europe	June–Oct.	0.1–0.2	Citral	Good mfg practice in alcoholic beverages only
Lime	Peels of *Citrus aurantifolia,* Swingle fruit by cold expression or by distillation as a by-product of lime juice production	West Indies Mexico Florida	Throughout the year	0.1 (approx)	Citral	GRAS

TABLE 7.2 (Continued)

Linaloe	Wood of *Bursera delpechiana*, Poiss. and other *Bursera* sps.	Mexico	Jan.–May	8 (approx)	Linalool (60–75%)	Good mfg practice
Lovage	Rhizomes of *Levisticum officinale*, Koch.	France Germany Czechoslovakia	Sept.	fresh: 0.1–0.2 dried: 0.5–1	α-Terpineol	Good mfg practice
Mace	Arilus of the fruits of *Myristicum fragrans*, Houtt.	West Indies East Indies	Throughout the year	4–15		GRAS
Mandarin (tangerine)	Peels of *Citrus reticulata*, Blanco. fruit by cold expression or by steam-distillation.	Italy U.S.A.		0.5 (approx)	Decylaldehyde	GRAS
Marjoram, sweet	Herbaceous tops of *Majorana hortensis*, Moench. (*Origanum majorana*, L.)	Europe	Aug.–Sept.	0.2–0.3	Terpen-4-ol α-Terpineol	GRAS
Mentha Arvensis (corn mint)	Leaves of *Mentha arvensis*, L. var. *piperascens*, Holmes or *Mentha arvensis*, L. var. *glabrata*, Holmes.	Brazil Japan China	Brazil: Jan.; Mar.–Apr. Japan: Aug. and Oct.	0.5–1 (depending on source)	Commercial oil is partially de-mentholated to a residual menthol level depending on source	Not permitted in U.S.A. but widely used elsewhere
Mustard	De-oiled press-cake of the seeds of *Brassica nigra* (L) Koch.	U.K. Europe	Sept.–Oct.	0.6–1.2	Allyl isothio-cyanate	GRAS
Myrrh	Oleo-gum-resin from *Commiphora molmol*, Engl., *C. abyssinica*, Engl. and other *Commiphora* sps.	Somalia Ethiopia		3–8 (from the gum)		Good mfg practice

TABLE 7.2 (Continued)

Common Name	Botanical Source	Principal Producing Countries	Main Crop Period	Average Yield of Essential Oil (%)	Chief Aromatic Constituents	Regulatory Status
Myrtle	Leaves of *Myrtus communis*, L.	Mediterranean countries	July–Aug.	0.2–0.3	α-Pinene Camphene Cineole	Good mfg practice
Nutmeg	Kernels of the fruits of *Myristica fragrans*, Houtt.	East Indies West Indies Sri Lanka		6.5–13 (depending on source)	α- and β-Pinenes Limonene 4-Terpineol Safrole Myristicin	GRAS
Ocimum	Herbaceous tops of *Ocimum* sps. Thymol-type: *O. gratissimum*, Eugenol-type: *O. gratissimum*,	 West Africa Seychelles Indonesia Brazil			 Thymol Eugenol	
Onion	Bulbs of *Allium cepa*, L.	U.K. Europe U.S.A.		0.005–0.01	n-Propyl disulfide	GRAS
Orange, bitter	Leaves and twigs (petitgrain); flowers (neroli bigarde) or peel of *Citrus aurantium*, L. subspecies *amara*, L.	Sicily Spain Guinea Brazil		Petitgrain: approx 0.2 Neroli approx 0.1 Peel: 0.4–0.5	Linalyl acetate	GRAS

TABLE 7.2 (Continued)

Orange, sweet	Leaves and twigs (petitgrain); flowers (neroli) or peel of *Citrus sinensis*, L. Osbeck. or *C. aurantium*, L. var. *dulcis*, L.	Italy Spain North Africa California Florida		Peel: 0.4–0.5	Limonene (90%) Decylaldehyde	GRAS
Origanum (wild marjoram)	Herbaceous tops of *Origanum vulgare*, L. and other *Origanum* sps. (Arctander 1960)	U.S.S.R. Central Europe Italy	Aug.	1 (approx)	Thymol or carvacrol depending on sps. or variety	GRAS
Orris	Peeled rhizomes of *Iris germanica*, L. or *I. pallida*, Lam.	Italy	Peeled rhizomes are aged for 2 yr	0.2–0.3	Irone, fatty acid esters	Good mfg practice
Palmarosa	Herbaceous tops of *Cymbopogon martini*, Stapf. var *motia*.	Indonesia India Seychelles	Java: every 4 months India: Oct.–Sept.		Geraniol (approx 90%)	GRAS
Parsley Seed	Ripe seeds of *Petroselinum sativum*, Hoffm.	Europe U.S.A.		1.5–3.5	Apiol	GRAS
Parsley herb	Young green tops of *Petroselinum sativum*, Hoffm.	U.K. Hungary Germany U.S.A.		0.05 (approx)		GRAS
Patchouly	Leaves of *Pogostemon cablin*, Benth. or *P. heyneanus*, Benth.	Indonesia Malaysia Brazil	Dec.–Jan.	1.5–3		Good mfg practice
Pennyroyal	Herbaceous tops of *Hedeoma pulegioides* (L) Pers. (American) or *Mentha pulegium*. L. (European)	U.S.A. Mediterranean countries	May–Sept.	0.5–1	Pulegone	Good mfg practice

TABLE 7.2 (Continued)

Common Name	Botanical Source	Principal Producing Countries	Main Crop Period	Average Yield of Essential Oil (%)	Chief Aromatic Constituents	Regulatory Status
Pepper, black	Dried immature berries of *Piper nigrum*, L.	India Indonesia Brazil	Aug. and Nov. (depending on region)	1–2.7	Terpenes	GRAS
Peppermint	Flowering tops of *Mentha piperita*, L.	U.S.A. Italy Japan	Sept.	0.3–0.7 (depending on source and variety)	Menthol Menthone Menthyl acetate	GRAS
Petitgrain	Leaves and twigs of Citrus sps. (*q.v.*)					
Pimento (allspice)	Fruits of *Pimenta dioica*, L. (*Pimenta officinalis*, Lindl.) An oil is also distilled from the leaves	West Indies	July–Aug.	3–4.5	Eugenol (70–80%)	GRAS
Pine, dwarf	Needles and twigs of *Pinus mugo*, Turra. var. *pumileo* (Haenke) Zenari	Central Europe		0.3–0.4	α-Pinene	Good mfg practice
Pine, Scotch	Needles and twigs of *Pinus sylvestris*, L.	Europe Asia North America			Borneol Bornyl acetate	Good mfg practice
Pine, white	Needles and wood of *Pinus strobus*, L.	North America (but of no commercial value)			α- and β-pinene	Good mfg practice in alcoholic beverages only

TABLE 7.2 (Continued)

Rose	Flowers, buds and leaves of *Rosa damascena*, Mill., *R. alba*, L., *R. centifolia*, L., *R. gallica*, L. and vars. of these sps.	Bulgaria Turkey Morocco France Italy	June	Depends on distillation method (Guenther 1949)	Rhodinol (30–60%)	GRAS
Rosemary	Whole plant or leaves of *Rosmarinus officinalis*, L.	Spain and other Mediterranean countries	May–Sept.	0.5–1.2	Cineole Borneol	GRAS
Sage, Dalmatian	Leaves of *Salvia officinalis*, L.	Yugoslavia	Sept.	0.5–1	Cineole	GRAS
Sage, Greek	Leaves of *Salvia triloba*, L.	Greece Turkey	Sept.		Cineole	GRAS
Sage, Spanish	Leaves of *Salvia lavandulaefolia*, Vahl.	Spain	Sept.		Cineole	GRAS
Sandalwood, white (East Indian)	Wood from *Santalum album*, L. or *S. spicatum* (R. Br) and *Eucarya spicata*, (R. Br) Sprag & Summ.	India (Mysore) Australia		Mysore: 4.5–6 Australia: 1.4–2.6	Santalol (90%)	Good mfg practice
Sandalwood, West Indian (Amyris)	Wood from *Amyris balsifera*, L.	South America Haiti Jamaica		2.5–3	β-Caryophyllene	Good mfg practice
Sassafras	Root bark and leaves of *Sassafras albidum*, (Nutt) Nees.	U.S.A. Canada	Aug.–Sept.	2–3	Safrole	Not permitted in food in many countries
Sassafras, Brazilian	Wood from *Ocotea cyrubarum*, H.B.K. (*O. pretiosa*)	Brazil	July–Oct.	1.2–1.5	Safrole (90%)	
Savory, summer	Herbaceous tops of *Satureia hortensis*, L.	Southern Europe U.S.A.	July–Aug.	0.1 (approx)	Carvacrol	GRAS

TABLE 7.2 (Continued)

Common Name	Botanical Source	Principal Producing Countries	Main Crop Period	Average Yield of Essential Oil (%)	Chief Aromatic Constituents	Regulatory Status
Savory, winter	Herbaceous tops of *Satureia montana*, L. freed from woody portions	Spain Morocco Yugoslavia	Aug.	0.2 (approx)	Carvacrol	GRAS
Schinus Molle (Californian or Peruvian pepper)	Dried berries of *Schinus molle*, L.	Mediterranean countries	Aug.	5–7	β-Phellandrene	GRAS
Spearmint	Leaves of *Mentha spicata*, Houds or L.	North America Europe	Sept.	0.6 (approx)	1-Carvone	GRAS
Spike lavender	Flowering tops of *Lavandula latifolia*, Vill (*L. Spica*, DC)	Spain France	Aug.	0.75–1	Linalool Linalyl acetate	GRAS
Star anise	Fruits (seeds) of *Illicium verum*, Hook.	China Vietnam	Throughout the year; max. Aug.–Sept.	8–9	Anethole (85–90%)	GRAS
Storax (styrax)	Balsamic exudate from *Liquidambar orientalis*, Mill. (Levant or Asian) or *L. styraciflua*, L. (American)	Asia Minor Southern U.S.A. Mexico Honduras Guatemala		0.5–1 (from the crude balsam)		Good mfg practice
Tagetes (marigold)	Flowering tops of *Tagetes patula*, L., *T. erecta*, L. or *T. minuta*, L (*T. gladulifera*, Schrank)	India Southern Mexico South Africa Australia		0.3–0.4	Ocimene Tagetone (50–60%)	Good mfg practice

Tarragon (estragon)	Herbaceous tops of *Artemisia dracunculus*, L.	Central and eastern Europe	July–Aug.	0.3–1.5	Methyl chavicol (estragole) Anethol	GRAS
Thyme	Herbaceous tops of *Thymus vulgaris*, L., *T. zygis*, L. var. *gracilis*, Boiss., (white thyme) or *T. serpylllum*, L (wild thyme) or *T. capitatus*, Hoff & Link. (Spanish 'origanum') (Arctander 1960)	Mediterranean countries Central Europe North America	May–Sept.	0.5–1.5 (depending on source)	Thymol Carvacrol	GRAS except Spanish "origanum" which is good mfg practice
Tolu balsam	Balsamic exudate of *Myroxylon balsamum* (L) Harms.	Venezuela		2–7 (from the crude balsam)	Benzyl benzoate Benzyl cinnamate	
Tuberose	Flowers of *Polyanthes tuberosa*, L.	France Morocco	Aug.–Sept.	3–6 (from the extracted concrete)		GRAS
Turmeric (curcuma)	Dried rhizomes of *Curcuma longa*, L.	India Malaysia Sri Lanka Japan		1.5–5	Turmerone	GRAS
Wintergreen	Leaves of *Gaultheria procumbens*, L.	Canada U.S.A.	June–Sept.	0.5–0.7	Methyl salicylate	
Ylang-ylang	Flowers of *Cananga odorata*, Hook. and Thoms.	Phillipines Malagasy Réunion		1 (approx)	Alcohols and esters (50–60%) Sesquiterpenes (30–35%)	GRAS
Zedoary	Rhizomes of *Curcuma zedoaria*, Rosc.	India		1 (approx)	Cineole	GRAS

SOURCE: In addition to Arctander (1960) and Guenther (1949), data was procured from numerous articles and papers which have appeared in *Flavor Industry* and *International Food Additives and Flavours*.

TABLE 7.3
BOTANICAL CLASSIFICATION OF ESSENTIAL OILS

Family *Rutaceae*		
Genus *Citrus*		
Lemon	Petitgrain bigarade	Citron
Petitgrain (lemon leaf)	(bitter orange leaf)	Lime
Sweet orange	Neroli bigarade	Mandarin (or tangerine)
Neroli (orange blossom)	(bitter orange blossom)	Grapefruit
Bitter orange	Bergamot	
Other Genera		
Buchu	Rue	Sandalwood, West Indian
Family *Labiatae*		
Balm	Marjoram, wild	Spearmint
Basil	Monarda	Rosemary
Hyssop	Origanum	Sage
Lavender	Patchouly	Clary Sage
Lavandin	Pennyroyal	Spanish Sage
Spike lavender	Peppermint	Savory
Majoram, sweet	Mint	Thyme
Family *Gramineae*		
Palmarosa	Lemongrass	Vetivert
Gingergrass	Citronella	
Family *Lauraceae*		
Bois de Rose (rosewood)	Massoi	Cassia
Sassafras	Cinnamon	Camphor
Laurel (sweet bay)	Cinnamon leaf	

TABLE 7.4
BOTANICAL SOURCES OF ESSENTIAL OILS

Derived from	Name of Oil		
Flowers			
	Carnation	Mimosa	Violet
	Cassie	Narcissus	Ylang-Ylang
	Cananga	Neroli	Geranium
	Hyacinth	Reseda	Lavender
	Jasmin	Rose	Lavandin
	Jonquil	Tuberose	Spike Lavender
Herbs			
	Basil	Origanum	Sage
	Camomile	Parsley	Clary Sage
	Dill	Pennyroyal	Savory
	Fennel	Peppermint	Spearmint
	Estragon (tarragon)	Mint	Tansy
	Lovage	Rosemary	Thyme
	Marjoram	Rue	Verbena
Leaves			
	Bay, Sweet	Cedar	Wintergreen
	Bay, West Indian	Patchouly	Eucalyptus
	Cinnamon	Petitgrain	
Woods			
	Amyris	Laurel	Camphor
	Cade	Linaloe	Santal
	Cedar	Birch	Sassafras
	Guaiac	Rosewood	

TABLE 7.4 (Continued)

Derived from	Name of Oil		
Leaves, needles, twigs			
	Pine (silver, Sylvestris, Siberian, Abies)	Cassia	Cypress
		Cedar	
	Cajuput		
Barks			
	Birch	Cascarilla	Cinnamon
	Cassia		
Fresh fruits (cold-expressed)			
	Bergamot	Lemon	Orange (sweet and bitter)
	Citron	Mandarin	
	Grapefruit		Tangerine
Fresh fruits (distilled)			
	Limes		
Grasses			
	Citronella	Lemongrass	Palmarosa
	Gingergrass		
Seeds			
	Cardamon		
Dired fruits			
	Anise	Coriander	Parsley
	Angelica	Cumin	Nutmeg
	Celery	Dill	Macc
	Carrot	Fennel	
Dried buds			
	Clove		
Dried berries			
	Cubeb	Pepper	Pimento
	Juniper		
Roots and rhizomes			
	Angelica	Ginger	Vetivert
	Calamus	Orris	Turmeric
	Costus	Valerain	
Balsams			
	Copaiba	Peru	Gilead
	Labdanum	Tolu	
Gums			
	Galbanum	Myrrh	Opopanax
	Mastic	Olibanum	Storax

ies not only with the type of plant but also, in particular, with the conditions prevailing during its growth, i.e., climate, soil, altitude, etc.

The function of essential oils in the living plant tissue is far from completely understood. Odors of flowers, for instance, may be directly associated with insect attraction or repulsion and so influences pollination and to some extent natural selection. Some are thought to act as a form of protection against parasites; and others have such a repulsive odor as to afford the plant protection from animal depredation. Whatever their botanical usefulness, the essential oils are a wonderful source of blended aromatics of inestimable value to the flavorist and the perfumer.

Distillation Methods

The recovery of an essential oil from plant material is basically one of steam-distillation, the only exception being the citrus peel oils which are obtained by cold-expression. The choice of the method of distillation depends to a large extent on the nature of the source material. Three methods are available:

(a) Steam Distillation.—The whole or comminuted material is packed into a tall cylindrical still, frequently on spaced supporting grids; and live steam, often at pressures higher than atmospheric, is introduced through a perforated coil below the charge. The oil-laden steam passes through a suitable condenser and the oil separated from the condensate in a specially designed vessel. In some stills a re-boiler unit can be incorporated so that the separated condensate can be re-used, not only as a further source of steam, but also to ensure an efficient stripping of any residual aromatic fractions. The method is suitable for most plant materials provided they are not too finely ground to cause channelling.

(b) Water and Steam Distillation.—In this case the plant material is supported on a grid immediately above the boiling water. This method is used for green plant material such as freshly cut herbs.

Courtesy of Bush Boake Allen, Ltd.

FIG. 7.13. A BATTERY OF STEAM STILLS

(c) Water Distillation.—The comminuted plant material is completely immersed in the water which is maintained at boiling while the still contents are mechanically stirred. Condensation conditions are similar to those used in steam distillation but the condensate water may be returned to the still by way of a special trap.

Standardization of distillation methods is impossible as every material requires individual consideration and treatment to achieve the optimum yield of essential oil. Distillation times vary widely, ranging from as little as 1 hr for some of the seeds and fresh herbs to as much as 100 hr or more for hard woody materials such as sandalwood. There are few guides to the best method to be adopted; it is a question of long experience both with the raw materials and the quality of the resulting essential oil.

The techniques of distillation have changed little over the centuries. Even today, one may find extremes: primitive cultivation and inefficient field-stills, to intensive farming and ultra-modern, purpose-designed distillation plants. Most aromatic plants grow wild or, at best, are cultivated by out-dated methods in small quantities as garden or patch crops by natives with a few acres. Those associated with the use of essential oils realize only too well that very wide variance in odor and flavor quality can exist in commercially available essential oils from the same prime source. Those involved in

Courtesy of Bush Boake Allen, Ltd.

FIG. 7.14. A BATTERY OF WATER STILLS USED FOR THE DISTILLATION OF A WIDE VARIETY OF AROMATIC PLANT MATERIALS

distillation fully appreciate that skill and long experience are important in the cultivation, harvesting and predistillation handling of plant material to yield an oil of consistent quality.

Only a few aromatic plants, used as a source of essential oil, are cultivated by methods based on modern agronomic practice involving selective plant breeding, mechanized methods of harvesting and a skilled labor force. The reason for this is the high capital investment involved, the uncertainty of cropping and the generally low price structure which governs trade in these materials. Such an outlay can only be justified when the demand for the oil is large and universal. The floral oil industry centered in Grasse in southern France, the citrus oil industry in Florida, California and Italy, and the immense peppermint and spearmint growing operation in Western United States are all examples of such concentrated expertise resulting in the production of uniformly high quality oils.

Constitution of Essential Oils

Before considering the further processing of the prime essential oils as flavoring materials for specific applications, it is necessary to establish something of their composition. The essential oils are very complex mixtures of several compounds although, in certain instances, one main constituent may predominate to the extent of 85% or more (e.g., oil of clove bud).

The constituents may be classified into three groups:

(a) Hydrocarbons of the general formula $(C_5H_8)_n$; where
  $n = 2$ they are called terpenes
  $n = 3$ they are called sesquiterpenes
  $n = 4$ they are called diterpenes

(b) Oxygenated compounds derived from these hydrocarbons, including alcohols, aldehydes, esters, ethers, ketones, phenols, oxides etc.

(c) Other specific compounds containing either sulfur or nitrogen

An understanding of these compounds helps to explain some of the well-recognized characteristics of the essential oils.

Terpenoid Hydrocarbons

The terpenoid hydrocarbons are characterized by their:

(a) Poor solubility in dilute alcohol. An essential oil having a high percentage of terpenes is relatively insoluble in dilute alcohol whereas those rich in oxygenated compounds are more readily soluble. This character is often used as a guide to quality.

(b) Tendency to oxidize with consequent deterioration in odor and flavor quality. Oxidation may be accompanied by polymerization and even resinification which can be seen as a thickening of certain oils on storage. Use is made of this, in the case of turpentine, in the paint and varnish industry.

(c) Low contribution to the flavor profile. It is probably not quite true to say that they have no flavor value but, in comparison with the flavor strength of the associated oxygenated compounds, their contribution is insignificant. They do, however, influence the odor of the oil by acting as a bridge or carrier for less volatile components and deterpenated oils are in consequence always flat and lack the freshness of the original oil.

Processing of Essential Oils

The reasons for the further processing of the prime essential oils are: (a) redistillation—the removal of undesirable components from the oil (e.g., metallic contamination, excessive color, etc.); (b) rectification—the removal of unpleasant and/or unwanted odor or flavor components or fractions; and (c) de-terpenation—the partial or complete removal of terpene hydrocarbons in order to achieve a concentration of flavoring effect, improvement in the solubility of the oil in dilute alcohol and increase its stability and shelf-life.

Courtesy of Bush Boake Allen, Ltd.

FIG. 7.15. A VARIETY OF HIGH-VACUUM STILLS USED FOR THE RECTIFICATION OF ESSENTIAL OILS

Redistilled Oils

Apart from drying and a period of maturation, many essential oils are sold and used as originally distilled; but certain oils require redistillation to remove undesirable characters or components present in the single distilled product. Redistillation is carried out in specially designed stills under carefully controlled conditions which provide for the separation and collection of specific fractions.

The following examples are of importance in flavoring:

Cassia.—Cassia oil is distilled in the producing areas in southeastern China. It is often very dark colored and may be contaminated with lead derived from the seams of containers, as well as adulterated with such things as kerosene (Guenther 1949). In this state, as imported, it is quite unacceptable for use in foods and the color and contaminants must be removed by redistillation. The percentage recovery is determined by the quality of the start material.

Peppermint.—Single-distilled peppermint oils as received from the grower/distillers frequently have an objectionable odor and flavor due to the presence of weedy and unpleasant off-notes (mostly dimethyl and other polysulfides). These unwanted characters, which detract from the sweetness and clean freshness of a good quality oil, are best removed by redistillation. This reprocessing is required for a peppermint oil to comply with U.S.P. standards.

Redistillation is sometimes referred to as though it were the complete answer to turning a bad oil into a good one—this is just not the case. Redistillation, in effect, partially concentrates an oil and, in so doing, may accentuate any abnormalities or in-balance in the main constituents rather then eliminate them.

Rectified Oils

The rectification of an essential oil is also redistillation but with a more specific purpose aimed at the separation of high and low volatile components to provide an end-product having a defined character. The process may be carried out in one of two ways: either by dry distillation in vacuo or by steam injection under reduced pressure. The plant and techniques used are fully described in the literature.

Most rectification is carried out by the dry method in which the oil is redistilled under a high vacuum (5–10 mm Hg), the oil boiling at a sufficiently low temperature to ensure the minimum of damage to thermolabile constituents. The vapors rise through a suitable fractionating column which enables fractions of a defined boiling range to be separated and collected. Depending on the nature of the starting material and the desired character of the end-product, the various fractions may be re-blended as necessary and any unwanted fractions rejected.

Rectification of Peppermint Oil.—Rectification of peppermint oil can be carried out to varying degrees dependent on the quality of the single distilled oil and the purpose for which the redistilled oil is intended. Commercially, there are two principal products:

Oil of Peppermint, Single-Rectified (Sometimes Called Double-Distilled).—A simple redistillation in which the last 2–5% of the oil remaining in the still is rejected. The product is almost colorless and is free of the harsh resinous notes of the natural oil.

Oil of Peppermint, Double-Rectified (Sometimes Called Triple-Distilled).—A rectification in which the first fractions of distillate, up to whatever percentage is deemed necessary but usually 5–10%, are rejected as well as the last 2–4% of still residues. The product is water-white and has a much improved odor and stability.

Single-rectification removes most of the highest boiling components whereas the so-called double-rectification removes, in addition, the low boiling terpenes. The figures quoted for the percentage rejections are averages and typical of current practice; they are by no means fixed, as many oils are rectified to satisfy a user's particular requirements.

The single-rectified oils have a variety of uses in flavorings such as chocolate cream centers and in toothpastes where the clean fresh cooling flavor of peppermint is all-important. The double-rectified oil is much more satisfactory where solubility is of prime importance (e.g., production of crème de menthe).

The effect of these different methods of rectification are best exemplified in the odor profiles.

Odor Profile of Single-Rectified Oil of Peppermint

Origin	Commercial sample of Oil of Peppermint, U.S.P.	
Odor	Initial:	SWEET, CLEAN; freshly aromatic. COOLING, FRESH; light. MENTHOLIC, MINTY. SMOOTH; creamy, toffee-like. STRONG, PUNGENT; irritating.
	Persistence:	Very lingering, the background notes becoming progressively prominent.

	Dry-out:	BUTTERY. Old oils may show a marked COCONUT-like character.

Odor Profile of Double-Rectified Oil of Peppermint

Origin	Commercial sample of Oil of Peppermint, American double-rectified.	
Odor	Initial:	STRONGLY MENTHOLIC. CLEAN, FRESH but not sharp. WARM; sweetly rounded.
	Persistence:	Lingering, the initial fresh top-notes give way to a more rounded profile which persists unchanged for a long time.
	Dry-out:	DULL, FLAT but distinctly MINTY.

Odor Comparisons

Origin: Commercial samples of oils from Oregon, U.S.A.

Single Distilled	Single Rectified	Double Rectified
MINTY	SHARPLY MINTY	STRONGLY MINTY
TOFFEE-like	CLEAN, COOLING	SMOOTH, COOLING but WARM
FULL-BODIED	SMOOTH	SWEETLY ROUNDED
GREEN, HERBY	CREAMY	

The dementholized oil derived from *Mentha arvensis*, L., is certainly improved by rectification. The pungent, harsh flavor notes appear to reside in the terpene fraction and the characteristic dull, oily notes in the high boiling resinous residue. Double rectification smooths the aromatic profile very considerably and brings the odor much nearer that of *piperita* oils with which it blends well. However, an expert or a trained panel can readily detect as little as 10% of rectified *arvensis* oil in a blend.

Rectification of Spearmint Oil. —Because of its almost universal popularity, natural or single-distilled spearmint oil is widely used as a flavoring. However, spearmint oils are much improved by rectification. The terpenes have a strongly muddy character which seem to repress the true flavor of this oil. It may well be that this explains the strange paradox of spearmint which is so popular as a flavor but primarily only in two types of product—chewing gum and toothpaste. It could well be that the unpleasant features of the natural oil are absorbed or even neutralized by the nature of the chicle gum, thus effectively suppressing them; similarly, this could happen in an alkaline toothpaste base.

Rectified spearmint oil certainly has a more attractive odor and flavor than the natural oil and is even better when deterpenated. It is this quality of spearmint oil which has achieved some popularity in chewy Dutch caramels.

Deterpenated Oils

The chemistry of the terpenes is complex and beyond the scope or purpose of this book. In the living plant, they have a particular affinity for oxygen giving rise to a wide range of aromatic compounds which provide the main odor and flavor characters of any essential oil. Once the plant is harvested, the whole biochemical balance is upset. This applies equally to any essential oil present in the cells and ultimately recovered by distillation. Under these changed conditions, oxidative reactions in the terpenoid components give rise, not to desirable flavor components, but to unwanted off-notes. For this reason, the removal of terpenes from an essential oil is advantageous and improves its stability, enhances its flavor quality and increases its shelf-life. The deterpenation of an oil is necessary to get the full benefit of the oxygenated aromatic components.

The term "terpeneless" or "terpene-free" is usually applied to essential oils from which the low-boiling monoterpenes have been removed leaving the sesquiterpenes and the oxygenated components. Further processing is required to remove the high-boiling sesquiterpenes to give a "sesquiterpeneless" oil. Many essential oils contain negligible quantitites of sesquiterpenes and, in these cases, the two terms are synonymous.

Methods of Deterpenation.—Three methods are widely used to produce deterpenated products: (a) fractional distillation under high vacuum; (b) selective extraction of the soluble oxygenated compounds in dilute alcohol; and (c) chromatography using silica-gel.

Fractional Distillation.—The bulk of terpeneless oils are produced commercially by fractional distillation under high vacuum. The technique has already been described and differs from rectification only in the degree of care necessary to ensure minimal damage to sensitive flavoring constituents during the long distillation period required to remove relatively large proportions of terpenes.

The vacuum employed must be as good as possible and preferably in the order of 1–2 mm Hg. At this low pressure, the boiling points of the constituents are much lowered but are also brought closer together, making efficient separation more difficult. However, these are problems of still design and reflux balance which can be solved for each particular essential oil.

An alternative method employing steam injection has been used; but this has the disadvantage of increasing the processing time and, thereby, causing even more damage to the flavoring constituents. In addition, the presence of water leads to hydrolysis of certain constituents inducing deleterious changes in the odor and flavor profile of the end-product.

In Sicily, a third alternative has been developed in which the bulk of the terpenes are removed by normal vacuum distillation and the residual oil is steam-distilled under vacuum to give the terpeneless oil. This is claimed to have a much fresher top-note than oils produced by fractional distillation alone.

Selective Solvent Extraction.—The alcoholic-solvent extraction technique for producing terpeneless oils depends on the large differences in the solubilities of the terpenes and the oxygenated flavoring constituents in dilute alcohol (30–35% ethanol by volume). The method involves the distillation of the natural essential oil with 3 or 4 times its volume of dilute alcohol in a still fitted with a separator and a return trap to allow the condensate to be continuously re-cycled. During the process, the terpenes distill over with the alcohol vapor and ultimately separate on the surface of the distillate, the oxygenated fraction being progressively dissolved in the weak alcohol in the still. When no further separation of terpenes takes place, the re-cycling is discontinued and the higher boiling oxygenated constituents are fractionated and recovered as the terpeneless oil.

Another technique, which makes use of differential solubilities in immiscible polar and nonpolar solvents as a means of separating terpenes was patented by Van Dyck and Ruys (1937), but has not been developed commercially.

The cold-solvent extraction of oxygenated compounds from citrus oils is attractive, as these oils are particularly sensitive to heat degradation. The most widely used method consists of mixing the natural oil with alcohol (95% ethanol by volume) and adding, with vigorous agitation, the calculated volume of water to reduce the alcohol to about 35%. The terpenes are insoluble in this strength of alcohol and, after a period of standing, separate as an upper layer. The lower layer may be run off and processed directly as a soluble flavoring essence; or it may be further diluted with water, saturated with sodium chloride to break any emulsion and the terpeneless oil separated and dried. This method, although very time-consuming produces good quality terpeneless oil having the full bouquet of the original oil, if not its freshness.

Chromatographic Deterpenation.—Based on a much earlier concept, a chromatographic method of deterpenation, using silicic acid as an absorbant for the oxygenated fraction, was developed by Braverman and Solomiansky (1957) and has since found commercial acceptance for the production of high quality terpeneless orange oil. The technique involves the passage of natural citrus oil through a column of absorbant followed by elution of the column with a non-polar solvent (e.g., hexane). This removes residual terpenes and the terpeneless fraction can be recovered from the silica gel by extraction with ethyl acetate or some other low-boiling polar solvent. The process is exothermic and precautions are necessary to ensure that the temperature of the column is controlled so as to minimize heat damage to the flavoring constituents.

Concentrated and Terpeneless Citrus Oils

Concentrated citrus oils, from which only part of the terpenes have been removed, are widely used as flavoring materials. They have an improved stability, a lower usage rate and retain much of the freshness associated with the natural oil. Their concentration or "fold," in terms of the flavoring effect of the natural oil, is usually quoted as part of the descriptive title of these products. The subject of the level of concentration achieved in partial or complete deterpenation of citrus oils is one which can lead to a great deal of controversy. Published claims by the manufacturers of these products show widely different figures for total aldehyde content and equivalent flavoring strength. The following figures are illustrative of a range of lemon oils:

Product	Total Aldehydes (as Citral) (%)	Claimed Concentration
Lemon oil, Sicilian	3–5	—
Lemon oil, concentrated 5–fold	20	x 5
Terpeneless lemon oil, Sicilian	60	x 25

The technical facts and commercial claims are apparantly at variance. If one takes, for example, the terpeneless oil with a citral content of 60% derived from a natural Sicilian oil having a citral content of say 5%, then, based on the citral figures, the concentration is 12-fold. Mathematically, this is correct but the method of basing the claim on just one component, in this case the total aldehydes calculated as citral, completely ignores other flavoring components of equal or possibly greater importance in the overall effect of the oil. This is substantiated in practice and it is found that the actual concentration is often well above the theoretical 12-fold, although not necessarily as high as that claimed in the manufacturer's literature.

In marketing these products, the concentration, in terms of flavoring effect, is best determined from the actual yield of concentrated or terpeneless oil. Analytical specifications are important as a means of quality control, but the results can be totally misleading when applied to a determination of the usage level of the product. The most satisfactory way of establishing this is for the user to make comparative trials, with both the natural oil and the concentrated oil side by side, in a finished product formulation under normal processing conditions. Only then can the actual equivalence be established.

There is little value in attempting to publish meaningful odor and flavor profiles for concentrated, terpeneless and sesquiterpeneless citrus oils. The range of available products is much too wide and, within each fruit category, displays a broad spectrum of flavoring effects. The aromatic quality and flavoring strength of each product is dependent on that of the start material, the method used for deterpenation and the degree of concentration achieved. The manufacture of these products calls for considerable expertise at all stages of processing to ensure consistency; for reliable evaluation, therefore, they should be used in accordance with the manufacturers' recommendations.

Essential Oil Blends

The designation of commercial essential oils is not always an accurate indication of their source or quality. "Clove oil," for instance, may be genuine clove bud oil or one obtained from clove stems or leaves and rectified to bring its profile and analytical characters nearer to that of the genuine bud oil; or, it may be a blend from all these sources. One has only to examine critically a range of samples of what is supposedly the same oil to realize the wide spectrum of profile that exists. In this very speculative area of trading, it is not unusual to find essential oils that have been sophisticated or extended by the addition of synthetic versions of the major components. This practice in North America at least, is now less prevalent since current labeling legislation makes it an offense to mislabel as "natural" an essential oil which has been modified in this manner—such products must now be clearly marked "imitation."

When dealing with essential oil blends, it is a matter of taste and the effect required in the end-product. There is no need to use a very expensive essential oil if a cheaper grade will do the job just as well. In many cases, it is a matter of what one expects to get for one's money. Whatever the blend may be, it is pretty certain that any pretensions to quality or authenticity will depend entirely on the amount of genuine oil which is present. The quality control of essential oils calls for a determination of physical characters, an assessment of the odor profile over a period of 24 hr supported by an examination by gas-liquid chromatography. This data should enable samples to be screened so that any showing abnormal characters can be questioned. The physical characters of the major, commercially available essential oils are published by the Essential Oil Association of the U.S.A. (1976) in their *E.O.A. Specifications and Standards* which are regularly up-dated. The many sources and varieties of these products are fully discussed by Guenther (1949) and Arctander (1960).

Uses of Essential Oils

Essential oils have amazingly wide and varied industrial applications for the flavoring of foods, beverages, candy and sugar confectionery, cookies and flour confectionery, and in tobacco. In addition, they are used in the creation of fragrances for cosmetics, toiletries and industrial re-odorants. Many of them have an acknowledged physiological action and are powerful external and internal antiseptics, carminatives, etc., of proved medicinal value. There is barely an industry that does not make some use of these versatile natural products.

Isolates

The oxygenated compounds present in essential oils can be considerably concentrated by rectification and deterpenation to the point where the product is predominantly the major component (e.g., citral in terpeneless lemon oil). However, there are many other constituents present which influence the ultimate flavor profile of such an oil so that a solution of citral at the same concentration is, by comparison, lacking in flavor character. By further processing using both physical and chemical techniques, many of the prime components of an essential oil may be separated and purified to give chemical entities which are called "isolates." In many cases, these isolates can be produced more economically than an identical compound prepared synthetically, although this will depend on the current market value of the start material and its availability.

The following are typical natural isolates:

Menthol—isolated from Japanese mint oil by freezing.

Eucalyptol (cineole)—isolated from eucalyptus oil by freezing.

Linalool—isolated from oil of Bois de Rose by fractional distillation.

Eugenol—isolated from clove oil by a simple chemical process and re-distilled.

There are many more examples (e.g., safrole, santalol, camphor, thymol, geraniol, citronellol, etc.) some of which are available as both natural isolates and synthetic chemicals. In all cases, the isolates exist in the natural material in the same chemical identity as they are recovered. These materials can then be used for the production of further derivatives (e.g., iso-eugenol from eugenol).

BIBLIOGRAPHY

Standard Tests on Essential Oils

ARCTANDER, S. 1960. Perfume and Flavor Materials of Natural Origin. Steffen Arctander, Elizabeth, N. J.

FENAROLI, G. 1963. Natural Aromatic Materials, Vol. 1. Hoepli, Milan, Italy.

FENAROLI, G. 1968. Aromatic Synthetics and Isolates, Vol II. Hoepli, Milan, Italy.

FURIA, T. E. and BELLANCA, N. (Editors). 1975. Fenaroli's Handbook of Flavor Ingredients, 2nd Edition, Vols. I and II. CRC Press, Cleveland.

GILDERMEISTER, E. and HOFFMAN, F. R. 1960–1963. Die Aetherischen Oele. Vols. IIIa–IIIc, Akademie-Verlag, Leipzig, Germany. (German)

GUENTHER, E. 1949. The Essential Oils. Vols. I–VI. D. Van Nostrand & Co., New York.

Specifications for Essential Oils

ESSENTIAL OIL ASSOC. OF U.S.A. 1976. E.O.A. Specifications and Standards. Essential Oil Assoc. of U.S.A., 60 E. 42nd St., New York.

Specific Topics

BRAVERMAN, J. B. S. and SOLOMIANSKY, L. 1957. Separation of terpeneless essential oils by chromatographic methods. Perfum. Essent. Oil Rec. *48*, 284.

van DIJCK, W. J. D. and RUYS, A. H. 1937. Terpeneless oils: A new method of production. Perfum. Essent. Oil Rec., *28*, 91.

GREEN, R. J. 1975. Peppermint and spearmint production in the United States—progress and problems. Int. Flavour Food Additives *6*, 246–247.

GUENTHER, E., GILBERTSON, G. and KOENIG, R. T. 1975. Essential oils and related products (annual review). Anal. Chem., *47*, No. 5, 139R–157R.

HARDMAN, R. 1973. Spices and herbs; their families, secretory tissues and pharmaceutical aspects. Tropical Prod. Inst. Conf. Papers, London.
HERRICK, A. B. and TROWBRIDGE, J. R. 1961. Process for the manufacture of terpeneless essential oils. (Assigned to Colgate-Palmolive.) U.S. Pat. 2,975,170, Mar. 14.
HODGE, W. H. 1975. Survey of flavor producing plants. Int. Flavour Food Additives *6*, 244–245.
KOHLER, H. 1971. Formation of essential oil in plants. Reichst. Aromen, Koerperpflegem. *21*, No. 2, 60. (German)
MOSS, G. P. 1971. The biogenesis of terpenoid essential oils. J. Cosmet. Chem. *22*, 231–248.
PICKETT, J. A., COATES, J. and SHARPE, F. R. 1975. Distortion of essential oil composition during isolation by distillation. Chem. Ind. *14*, 612–614.
PINTAURO, N. 1975. Peppermint and citrus oil processing. *In* Flavor Technology. Noyes Data Corp., Park Ridge, N.J.
QUARRE, J. 1975. Cultivation of mint in the U.S.A. Parfums Cosmet. Aromes *2*, Mar./Apr., 27–30, 33–35. (French)
RUCKER, G. 1973. Sesquiterpenes. Angew. Chem. (International Edition in English) *12*, 793–806.
SHOLTO, D. J. 1971. Cultivating condiment and seasoning plants for flavours and essential oils. Flavour Ind. *2*, 222–225.
SHOLTO, D. J. 1971. Growing medicinal plants for essential oils and flavours. Flavour Ind. *2*, 697–698.
TEISSEIRE, P. 1969. Biogenesis of natural substances. Fr. Ses Parfums *12*, No. 65, 299–311. (French)
VASSILIEV, G. 1972. Terpenes. Their theoretical and practical importance. Parfums, Cosmet. Savons Fr. *2*, 195–204. (French)
VIRMANI, O.P. and DATTA, S. C. 1970. Oil of *Mentha piperita* (Oil of peppermint). Flavour Ind. *1*, 59–63, 111–133.

Part 6: Imitation Flavorings

NATURE OF FLAVORINGS

The value of natural flavors has already been well established and it is accepted that nature provides a very wide spectrum of flavorful raw materials which may safely be incorporated into food products and beverages, thereby enhancing their acceptability. There has been rapid development of more sophisticated food products, involving complex processing techniques together with built-in customer convenience in preparation. Many of the natural flavorings have proved inadequate to meet these needs. As a result, processors have turned increasingly to imitations of the natural flavors, the composition and nature of which can be specifically designed to satisfy the conditions involved. Such flavorings are composed of aromatic chemicals which may be identical with those found in the natural flavor or they may have an aromatic character which is similar to, or associated with, the desired flavor profile.

The naturally-occurring flavor components embrace the whole

range of organic groupings such as alcohols, acids, esters, aldehydes, ketones, ethers, oxides, phenols, hydrocarbons, etc. Individual flavors are, themselves, as diverse in their complexity. If one takes, for example, the flavor of an extract of vanilla beans, the predominant component is vanillin but this alone does not give the full flavor effect of the naturally derived product. Many other chemicals, often present in a low level of concentration, modify and give a distinctive character to the effect of the major component. A typical analysis of a natural vanilla bean extract is given in Table 7.5. In order to

TABLE 7.5
VOLATILE COMPOUNDS IDENTIFIED IN VANILLA FLAVOURING CONCENTRATES

Hydrocarbons	
Aliphatic	
Nonane	Heptadecane (branched)
Decane	Octadecane (branched)
Undecane (branched)	Eicosane (branched)
Duodecane	Docosane (branched)
Dodecane (branched)	χ-Decene (branched)
Tetradecane	χ-Dodecene
Tetradecane (branched)	χ-Tetradecene
Pentadecane	χ-Eicosene
Hexadecane (branched, 2 isomers)	
Aromatic	
Benzene	Ethylbenzene
Toluene	Propylbenzene
Dimethyl benzene	*p*-Ethyltoluene
(3 isomers)	Styrene
Trimethyl benzene	Napthalene[1]
(3 isomers)	
Terpenoid	
α-Pinene	α-Terpinene
β-Pinene	*p*-Cymene
Limonene	α-Curcumene[2]
Myrcene	δ-Cadinene[2]
β-Phellandrene	α-Muurolene[2]
Alcohols	
Aliphatic	
Butane-2,3-diol	Heptan-1-ol
Pentan-1-ol	Octan-1-ol
2-Methylbutan-1-ol	Nonan-2-ol
3-Methylbutan-1-ol	Dedecan-1-ol
Pentan-2-ol	Prenol
Hexan-1-ol	1-Octen-3-ol
3-Methylpentan-1-ol[2]	
Aromatic	
Benzyl alcohol	Phenylethyl alcohol
p-Methoxybenzyl alcohol[1]	
Terpenoid	
Myrtenol	Citronellol
α-Terpineol	Nerol
Terpinen-4-ol	Geraniol
Linalool	β-Bisabolol[2]

(TABLE 7.5 continued)

Carbonyl Compounds	
Aliphatic	
Pentan-1-al	3-Penten-2-one
Hexane-2-one	3-Octen-2-one[3]
Heptan-2-one	Octa-4,6-dien-3-one[3]
Octan-2-one	3-Hydroxybutan-2-one
Nonan-2-one	1-Hydroxypentan-2-one[2,3]
Decan-2-one	1-Hydroxyhexan-2-one[2,3]
	5-Hydroxyheptan-2-one[2,3]
Aromatic	
Benzaldehyde[1]	*p*-Methoxybenzaldehyde[1]
Salicylic aldehyde[1]	Vanillin[1]
p-Hydroxybenzaldehyde[1]	Heliotropin[1,2]
Terpenoid	
β-Cyclocitral	6,10,14-Trimethylpentadecan-2-one[2]
Esters, Lactones	
Aliphatic	
n-Amyl acetate	Methyl nonanoate
n-Hexyl acetate	Methyl laurate
Methyl valerate	Methyl myristate[2]
Propyl valerate	Methyl pentadecanoate[2]
Isopropyl valerate	Methyl palmitate[2]
Butyl valerate	Methyl heptadecanoate[2]
Isobutyl valerate	Methyl glycolate[2]
Ethyl 2-methylbutyrate[2]	Ethyl methoxyacetate[2]
Methyl caproate	Methyl lactate[2]
Ethyl caproate	Ethyl lactate[2]
Methyl haptanoate	Ethyl levulinate[2]
Aromatic	
Methyl benzoate	Methyl *cis*-cinnamate
Ethyl benzoate	Methyl *trans*-cinnamate
Methyl salicylate	Monoethyl phthalate
Benzyl formate	Diethyl phthalate
Benzyl acetate[1]	Di-n-propyl phthalate
Phenethyl acetate	Di-n-butyl phthalate
Methyl phenylacetate[2]	Methyl vanillate[2]
Various	
γ-Butyrolactone	Methyl cyclohexane-carboxylate
Dihydroactinidiolid[2]	
Acids	
Aliphatic	
Formic acid	Octanoic acid[1,2]
Acetic acid[1]	Decanoic acid[1,2]
Propionic acid	Lauric acid[2]
Butyric acid	Myristic acid[2]
Valeric acid	Glycolic acid[2]
Isovaleric acid[1]	Methoxyacetic acid[2,3]
Caproic acid	Lactic acid[2]
Aromatic	
Salicylic acid[2]	
Phenols, Phenol Ethers	
Phenol	1,2-Dimethoxybenzene
p-Cresol[1]	Anisole[2]
Guaiacol[1]	Diphenyl ether[2]

(Continued)

TABLE 7.5 (continued)

Creosol	*p*-Hydroxybenzyl methyl ether[2,3]
p-Cresyl isopropyl ether[2]	*p*-Hydroxybenzyl ethyl ether[2]
p-Ethylguaiacol[2]	4-Hydroxy-3-methoxybenzyl-methyl ether[2]
p-Vinylphenol[2]	4-Hydroxy-3-methoxybenzyl-ethyl ether[2]
p-Vinylguaiacol[2]	
Vanillin 2,3-butyleneglycol acetal (2 isomers)[2]	
Heterocyclics	
Furan derivatives	
Furfural	Furfuryl alcohol[2]
5-Methylfurfural	Furfuryl hydroxymethyl ketone[2]
5-Hydroxymethylfurfural	2-Hydroxyethyl-5-methyl-furan[2]
2-Acetylfuran	
2-Pentylfuran	
Sulphur containing derivates	
Thiophene	
Nitrogen containing derivatives	
2-Acetylpyrrole[2]	Methyl nicotinate[2]

Source: Klimes and Lamparsky (1976).
[1]Known constituent.
[2]Found only in methanolic extracts.
[3]Tentatively identified.

capture the true profile of the natural flavor, imitations must be similarly complex, but, at the same time, must be composed of synthetic chemicals the safety in use of which has been established and accepted under current legislation.

Imitation or artificial flavorings may, therefore, consist of:

(a) Natural materials (e.g., herbs and spices) and/or products derived directly from them (e.g., essential oils, oleoresins or extracts).

(b) Organic chemicals isolated from natural materials (e.g., citral from lemongrass oil).

(c) Synthetic organics identical with those found in nature, although not necessarily found in the named source of the flavor (e.g., vanillin in an imitation strawberry flavor).

(d) Synthetic organics not found in nature but of proven acceptability for use in foods (e.g., ethyl vanillin).

Whatever the nature of the chemicals used in creating an imitation flavor, the ultimate purpose is to reproduce as closely as possible, with the raw materials available, the aromatic effects of the natural flavoring in the end-product. In many cases a direct comparison can be made; but, in many more, the end-product is such that it would be unfeasable to manufacture had the natural flavor alone been available.

REGULATION OF FLAVORING SUBSTANCES

In many countries, the synthetic chemicals permitted for use in food flavorings are positively defined by legislation; in others, only those chemicals which may not be so used are regulated, allowing all others to be used as necessary to achieve the desired flavor profile. This difference in approach has lead to considerable diversity of legislation throughout the world and attempts to harmonize the regulations in the interests of international trade are making very slow progress.

In the United States, the Food and Drug Act of 1906 first defined materials which could be used by the food industry. This legislation was replaced by the Federal Food, Drug and Cosmetic Act in 1938, in which specific regulations were set out on the use of flavorings. These were amended in 1958 as a result of an investigation into the use of chemicals in food products. Under this amended Act, the Food and Drug Administration has issued lists of materials which are "generally recognized as safe" (GRAS) for use in foodstuffs, giving, in certain specific instances, limits in the concentration which may be present. In 1964, the Federal Register listed about 650 chemicals not generally recognized as safe, but permitted for use in foods as being "safe under good manufacturing practice." This initial legislation has been constantly reviewed both by federal legislators and the flavor industry itself and reference should be made to the Federal Register and the trade literature for up-to-date information on this very involved subject. The current status of flavoring materials has been published in a useful handbook by Fritzsche Brothers (1966). Updated listings under the U.S.A. (F.E.M.A.)., Council of Europe and U.K. classification systems were published by Ford and Cramer (1977). This is an on-going subject and substantial changes in the various classifications could well occur as more information on safety in usage becomes available. The published lists are intended as a guide and have no legal standing.

CLASSIFICATION OF FLAVORING CHEMICALS

Three classes of flavoring substances have been defined by the International Organization of the Flavor Industry (IOFI) (1973), namely:

(a) Natural flavors and flavoring substances: Preparations or substances acceptable for human consumption, obtained by physical processes from vegetable, sometimes animal, raw materials either in their natural state or processed for human consumption.

(b) Nature-identical flavoring substances: Substances chemically isolated from aromatic raw materials or obtained synthetically; they are chemically identical to substances present in natural products intended for human consumption, either processed or not.

(c) Artificial flavoring substances: Substances which have not yet been identified in natural products intended for human consumption, either processed or not.

Current legislative opinion favors the controlled usage of these materials by means of a restrictive list for natural and nature-identical flavoring substances and a positive list for artificial flavoring substances, defining precisely their usage levels in specific food products, beverages, etc. The acceptability of such chemicals would, of course, be dependent upon the establishment of their toxicity and safety in usage by internationally recognized test methods.

ORGANIC CHEMICALS USED IN FLAVORINGS

Aromatic compounds embracing the whole spectrum of organic chemistry are involved to a greater or lesser extent in the creation and formulation of imitation flavorings. The most important of these are listed in Table 7.6. The blending of such organics to form flavorings demands an exact knowledge of the sensory effects and interreactions which are likely to occur as well as an imaginative artistic flair and an excellent memory.

The relationship between structure and organoleptic attributes is not sufficiently defined to enable one to predict with any certainty the profile or flavoring effect of any chemical compound, although broad similarities do occur within certain groups and series of compounds. The organic chemist has to work on an empirical basis when deciding what molecular structural modification to make to a compound to achieve a particular aromatic effect, using whatever past knowledge is available as a guide. In the past, emphasis has been given to physical characters with little attention being paid to sensory attributes; fortunately, this attitude is changing as organic chemists realize the importance of establishing sensory data.

Within broad limits, it is found that the flavoring strength of a compound, as opposed to its profile, tends to increase as the molecular weight decreases and the vapor pressure increases. For this reason, chemicals having a branched chain substitution are likely to be of more value to the flavorist than the equivalent straight chain compound (e.g., iso-amyl butyrate is a more powerful flavoring chemical than n-amyl butyrate).

Chemicals are rarely so pure that they have a single aromatic profile; rather, they display a spectrum of attributes depending on the nature and quantity of trace impurities present as well as the

TABLE 7.6
CHEMICAL NATURE OF THE MAJOR AROMATIC COMPOUNDS USED IN FLAVORINGS

Hydrocarbons
Aliphatic
Aromatic
Cyclic terpenes (mono- and bicyclic terpenes)
Sesquiterpenes (mono-, bi- and tricyclic sesquiterpenes)
Alcohols
Aliphatic (saturated, unsaturated and terpenic alcohols)
Aromatic
Cyclic terpene alcohols
Sesquiterpene alcohols
Aldehydes
Aliphatic
Aromatic
Cyclic terpene aldehydes
Heterocyclic aldehydes
Acetals
Acids
Aliphatic
Aromatic
Anhydrides
Esters
Aliphatic
Aromatic
Terpene esters
Ethers
Heterocyclic compounds
Ketals
Ketones
Aliphatic
Aromatic
Cyclic terpene ketones
Ionones
Irones
Lactones
Oxides
Phenols
Phenol ethers
Compounds containing nitrogen and sulfur
Amines
Amino compounds
Imino compounds
Sulfides
Pyrazines

route taken in their synthesis. In some cases, it is, indeed, the odor of such trace impurities which may be detected rather than that of the material itself. Sometimes, such impurities become inseparably blended with that of the prime chemical entity and it is this complex odor pattern which is accepted as normal for the compound concerned. In some cases, a more chemically pure version would be of less value in either flavors or fragrances.

Even in cases where one chemical constituent is primarily responsible for the profile of a natural flavor, it is recognized that the equivalent synthetic compound, if used alone, does not reproduce the same complete and fully balanced effect as is achieved by using the natural flavoring. For instance, vanillin, the prime component of vanilla beans, when used as a single flavoring additive (for example, in an ice cream mix) will yield an acceptably flavored end-product but with nothing like the same full, rounded profile as that produced by the use of a natural vanilla extract prepared from vanilla beans or a blended flavoring composed of synthetic organics designed to imitate and replace the natural extract.

The organic chemist is able to assist the flavorist by producing alternative materials for the creation of close imitations of nature. For example, the modification of the vanillin molecule by changing the methoxy group to an ethoxy group to give ethyl vanillin is accompanied by a three-fold increase in the flavoring strength of the start material and a subtle modification of the flavor profile. The flavorist is thus able to use these different effects and by expert blending achieve a combination of attributes which, together, reproduce more closely the profile of the natural extract.

Flavoring aromatics may conveniently be classified on a traditional chemical basis, considering first the aliphatic and then the aromatic compounds and finally those of terpenoid structure. The groups concerned are:

Alcohols, acids, esters and lactones.
Aldehydes, ketones and ethers.
Terpene hydrocarbons.
Nitrogen and sulfur-containing compounds.
Complex molecules.

The inter-relationship between the various groups is given in Table 7.7.

TABLE 7.7
STRUCTURAL RELATIONSHIPS OF FLAVORING ORGANICS

Class of Compound	Distinguishing Systematic Suffix or Prefix	Functional Group	Typical Structure Aliphatic	Typical Structure Aromatic
Alcohols	-ol	—O—H	R·OH	$C_6H_5R \cdot OH$
Phenols	-ol	—O—H		C_6H_5OH
Carboxylic acids	-ic acid or -oic acid	$-\overset{\displaystyle O}{\overset{\Vert}{C}}-O-H$	R·COOH	C_6H_5COOH
Esters	-ate	$-\overset{\displaystyle O}{\overset{\Vert}{C}}-O-$	R·COO·R′	$C_6H_5 \cdot COO \cdot R$

TABLE 7.7 (continued)

Class of Compound	Distinguishing Systematic Suffix or Prefix	Functional Group	Typical Structure Aliphatic	Typical Structure Aromatic
Aldehydes	-al	$-C(=O)-H$	R·CHO	C_6H_5CHO
Ketones	-one	$-C(=O)-$	R·CO·R′	$C_6H_5CO \cdot R$
Ethers	-ether	—O—	R·O·R′	$C_6H_5O \cdot R$
Lactones	-lactone or -ide	$-C(=O)-O-$	R / \ O—CO	
Amines	-amine or amino	$-N(H)-H$	$R \cdot NH_2$	$C_6H_4(NH_2) \cdot R$
Nitro-compounds	nitro-	$-N(O)O$	$R \cdot NO_2$	
Sulfides	-ide	—S—	R·S·R′	
Disulfides	di-	—S—S—	R·S·S·R	
Thio-alcohols	-mercaptan	$-S-H$	R·SH	
Terpenoids	Structure may be open chain, closed chain or cyclic, saturated or unsaturated, based on units of isoprene:			

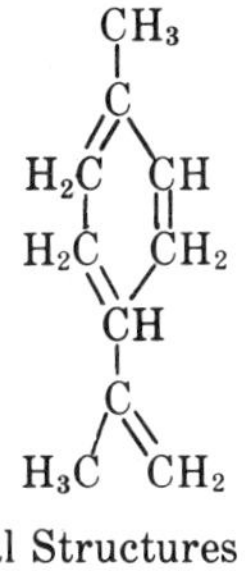

CH_3, C, H_2C, CH, H_2C, CH_2, CH, C, H_3C, CH_2

Typical Structures Include:

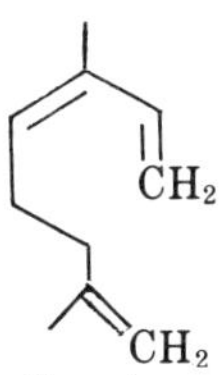

dipentene (CH_2, CH_2)

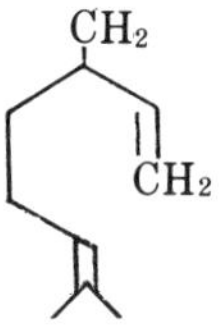

myrcene (CH_2, CH_2)

α-terpinene

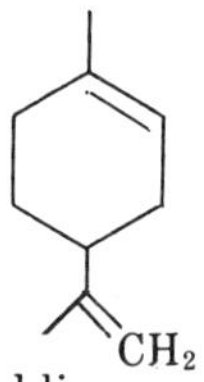

d-limonene (CH_2)

Δ^4-carene (CH_3)

camphane (CH_3, H_3C—CH_3)

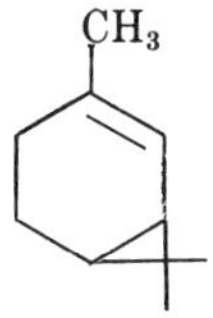

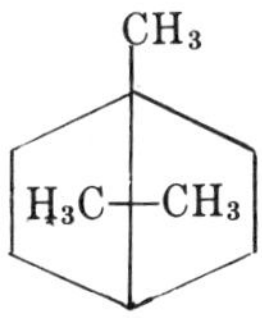

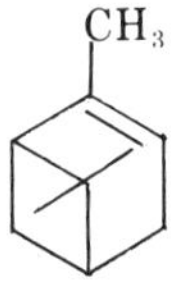

α-pinene (CH_3)

ALCOHOLS

Organic compounds containing one or more hydroxyl (–OH) groups are among the most important flavoring materials. In the aliphatic series, these include monohydric and unsaturated alcohols; in the aromatic series, alcohols in which the hydroxyl group is substituted in the side chain or phenols where it is directly on the benzene nucleus; as well as a whole range of alcohols of terpenoid structure. All the alcohols are distillable. Those of lower molecular weight (e.g., ethanol, isopropanol, etc.) are excellent solvents, for which purpose they are used in flavorings. As the molecular weight increases the compounds become more viscous and the higher boiling alcohols are waxy solids with almost no odor.

The lower members of the aliphatic series are decidedly sweet and spirituous, the pleasing notes being lost gradually as the molecular weight increases and replaced by an irritating character and a marked fattiness.

Unsaturated alcohols in which the double bond is near to the hydroxyl group have a harsh odor and are penetrating and irritating. Aromatic alcohols are generally more pleasant and have a rounded odor profile. The phenols have a generic similarity, although substitution of methyl groups on the benzene nucleus modifies and softens their odor impact. The substitution of two or more hydroxyl groups removes the odorous character of the compound in both the aliphatic and aromatic series [e.g., glycerol, $CH_2OHCHOHCH_2OH$ and resorcinol, $C_6H_4(OH)_2$].

The alcohols of importance in flavorings are given in Table 7.8.

CARBOXYLIC ACIDS

The saturated fatty acids (C-C_3) are generally unpleasant and have strongly pungent and irritating vapors. As the molecular weight increases (C_4-C_8) this character is replaced by a rancid, buttery/cheesy note which in turn softens and becomes more fatty/aldehydic. Members above C_{14} are waxy solids with little or no distinct odor. The unsaturated oleic series have characteristic odors, the lower members being pungent and acid whereas those of higher boiling point are slightly spicy, aldehydic and tallow-like. The aromatic acids do not have so marked a pattern of odors. All are faintly balsamic with either spicy or floral over-tones. The higher members of the series are odorless.

The presence of a hydroxyl group tends to suppress the odor [e.g., propionic acid CH_3CH_2COOH has a pungent odor whereas lactic acid $CH_3CH(OH)COOH$ is odorless]. The di- and tri-carboxylic acids are all odorless.

As a group, the organic acids impart a sour taste to a flavoring

TABLE 7.8
ALCOHOLS USED IN FLAVORINGS

Alcohol	Formula	Sensory Attributes
	Aliphatic Series	
Ethyl	CH_3CH_2OH	Sweetly spirituous
Allyl	CH_2CHCH_2OH	Irritating
Iso-propyl	$(CH_3)_2CHOH$	Pleasantly spirituous, slightly fruity
n-Butyl	$CH_3(CH_2)_3OH$	Fusil oil-like, irritating, dry
2-Butanol	$CH_3CH(OH)CH_2CH_3$	Sweetly fruity, vinous
Iso-butyl	$(CH_3)_2CHCH_2OH$	Unpleasantly harsh, sweet but irritating
n-Amyl	$CH_3(CH_2)_3CH_2OH$	Penetrating fusil oil-like, irritating
Iso-amyl	$(CH_3)_2CHCH_2CH_2OH$	Unpleasantly irritating but fruity on dilution
n-Hexyl (alcohol C6)	$CH_3(CH_2)_4CH_2OH$	Fruity, slightly fatty
2-Hexen-1-ol (leaf alcohol)	$CH_3(CH_2)_2CH{:}CHCH_2OH$	Powerful, fruity, green, sweet, leafy
3-Hexen-1-ol	$CH_3CH_2CH{:}CHCH_2CH_2OH$	Strongly herbaceous, leafy green
n-Heptyl (alcohol C7)	$CH_3(CH_2)_5CH_2OH$	Fatty, slightly herbaceous, pungent
3-Heptanol	$CH_3CH_2CH(OH)(CH_2)_3CH_3$	Strongly herbaceous, pungent; bitter taste
n-Octyl (caprylic alcohol) (alcohol C8)	$CH_3(CH_2)_6CH_2OH$	Strongly citrus-like, sweetly floral, herbaceous
2-Octanol	$CH_3(CH_2)_5CH(OH)CH_3$	Unpleasantly fatty
3-Octanol	$CH_3CH_2CH(OH)(CH_2)_4CH_3$	Sweetly nutty, green
3-Octanon-1-ol	$CH_3(CH_2)_4COCH_2CH_2OH$	Spicy with fruity, slightly green character
1-Octen-3-ol	$CH_3(CH_2)_4CH(OH)CH{:}CH_2$	Powerfully sweet, floral, herbaceous
n-Nonyl (pelargonic alcohol) (alcohol C9)	$CH_3(CH_2)_7CH_2OH$	Strong, fresh, rose-orange, slightly fatty
n-Decyl (alcohol C10)	$CH_3(CH_2)_8CH_2OH$	Sweet, freshly floral, orange-flower like, slightly fatty
n-Undecyl (alcohol C11)	$CH_3(CH_2)_9CH_2OH$	Floral, citrus-like, strongly fatty
Undecen-1-ol	$CH_2{:}CH(CH_2)_8CH_2OH$	Lemony, strongly fatty
n-Dodecyl (lauryl alcohol) (alcohol C12)	$CH_3(CH_2)_{10}CH_2OH$	Fatty, unpleasant, soapy; pleasantly floral on dilution
Myristyl	$CH_3(CH_2)_{12}CH_2OH$	Faintly coconut-like, slightly fatty
	Aromatic Series (Alcohol/Phenol)	
Benzyl	$C_6H_5CH_2OH$	Pleasantly sweet, slightly pungent
Phenyl ethyl (PEA)	$C_6H_5(CH_2)_2OH$	Sweet, rose-like, slightly fruity

(Continued)

TABLE 7.8 (Continued)

Alcohol	Formula	Sensory Attributes
Cinnamyl	$C_6H_5(CH_2)_2CH_2OH$	Sweet, warmly balsamic, floral
Anisyl	$OHCH_2C_6H_4OCH_3$	Sweetly floral, peach-like, balsamic
Eugenol	$CH_2{:}CHCH_2C_6H_3(OH)OCH_3$	Sharp, pungent, warmly spicy, clove-like
Iso-eugenol	$CH_3CH{:}CHC_6H_3(OH)OCH_3$	Warmly spicy, carnation-like, slightly woody
Guaiacol	$OHC_6H_4OCH_3$	Sweet, warmly phenolic
Phenol	C_6H_5OH	
p-Cresol	$CH_3C_6H_4OH$	Strongly phenolic, tar-like, smoky
Furfuryl	$C_4H_3OCH_2OH$	Irritating, warm, oily
α-Methylbenzyl	$CH_3CH(OH)C_6H_5$	Mildly floral, hyacinth-like
4-Phenyl-3-buten-2-ol	$C_6H_5CH{:}CHCH(OH)CH_3$	Strongly floral
1-Phenyl-3-methyl-3-pentanol	$C_6H_5CH_2CH_2C(C_2H_5)(CH_3)OH$	Warm, rose-like, fruity
1-Phenyl-1-propanol	$C_6H_5CH(OH)CH_2CH_3$	Floral, balsamic, honey-like
3-Phenyl-1-propanol (hydrocinnamyl alcohol)	$C_6H_5CH_2CH_2CH_2OH$	Sweetly floral, hyacinth-like, fruity (apricot-like)
Tetrahydrofurfuryl	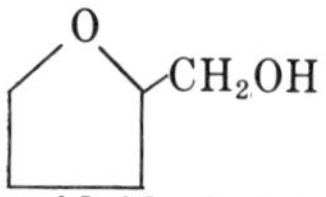 O; CH_2OH	Mildly warm, oily, sweetly nut-like on dilution
	Terpenoid Alcohols	
Citronellol (d-Citronellol)	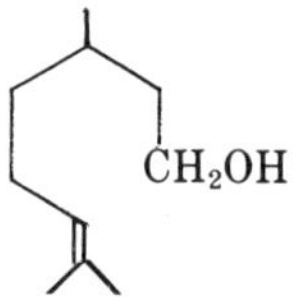 CH_2OH	Fresh, light, rose-like; bitter taste
Geraniol	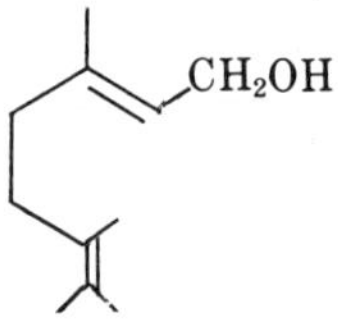 CH_2OH	Sweet, rose-like, warm, dry
Hydroxycitronellol	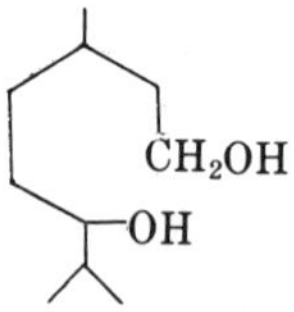 CH_2OH; OH	Sweetly floral, rose-like
Isopulegol	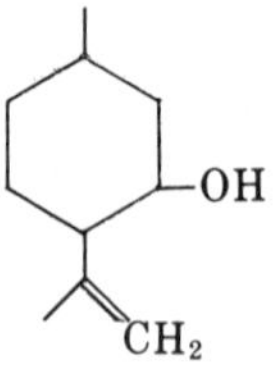 OH; CH_2	Sweetly minty with bitter herbaceous note

TABLE 7.8 (Continued)

Alcohol	Formula	Sensory Attributes
Linalool	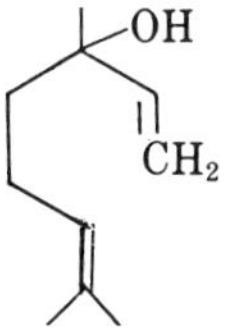 OH; CH_2	Light, sweet, freshly citrus, floral
Menthol (1-menthol)	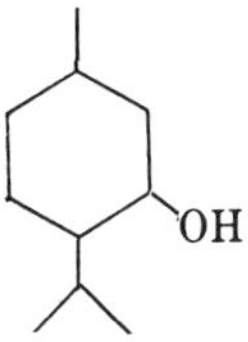 OH	Fresh, cooling, minty
Nerol	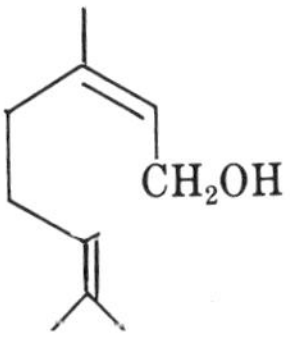 CH_2OH	Sweetly floral, rose-like, slightly fruity; bitter taste
Rhodinol (1-citronellol)	Indefinite composition (Arctander 1969)	Sweetly rose-like, slightly cooling, green
α-Terpineol	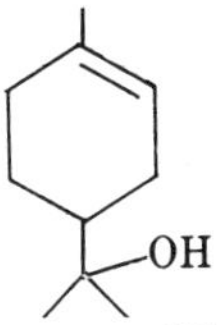 OH	Sweetly floral, lilac-like, fruity on dilution
β-Terpineol	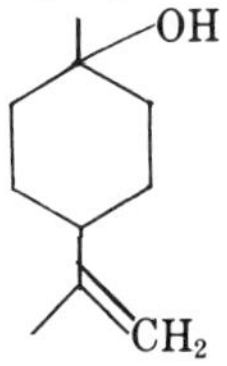 OH; CH_2	Dull, woody, slightly pungent
Tetrahydrolinalool	 OH	Sweet, strongly floral, citrus, slightly oily
Carvacrol	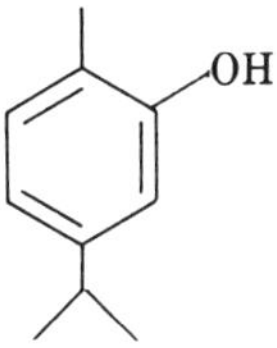 OH	Warmly pungent, smoky

(Continued)

TABLE 7.8 (Continued)

Alcohol	Formula	Sensory Attributes
Thymol	OH	Strongly phenolic, aromatic, sweet, spicy
	Cyclic Compounds	
Borneol	OH H CH_3—CH_3	Pungent, camphoraceous, minty; burning taste
Isoborneol	H OH CH_3—CH_3	Strongly camphoraceous, piney
Caryophyllene alcohol	CH_3 H_3C— OH	Warmly spicy, earthy, moss-like

composition. The di- and tri-carboxylic acids are particularly important in this respect as they are the natural acidulants of fresh fruits. Although not widely used per se in imitation flavorings, they are often incorporated into an end-product together with an imitation flavoring in order to reproduce a "natural" flavor character.

Most of the carboxylic acids are of little direct value in flavorings and are used primarily in the making of esters. The principal acids used for this purpose are given in Table 7.9.

ESTERS

Esters are organic compounds formed from carboxylic acids and alcohols by the elimination of water. The principal functional rela-

TABLE 7.9
CARBOXYLIC ACIDS USED IN FLAVORINGS

Acid	Formula	Sensory Attributes
Formic	$HCOOH$	Strongly pungent, irritating
Acetic	CH_3COOH	Pungent and irritating but pleasant on dilution
Propionic	CH_3CH_2COOH	Pungent, slightly cheesy
n-Butyric	$CH_3(CH_2)_2COOH$	Penetrating, rancid, butter-like, fatty
Iso-butyric	$(CH_3)_2CHCOOH$	Strongly butter-like, rancid, fatty
n-Valeric (pentanoic)	$CH_3(CH_2)_3COOH$	Cheesy, strongly rancid, unpleasantly pungent
Iso-valeric	$(CH_3)_2CHCH_2COOH$	Cheesy, flat, stale
n-Caproic (hexylic)	$CH_3(CH_2)_4COOH$	Acrid, rancid, sweaty
n-Heptylic (oenanthic)	$CH_3(CH_2)_5COOH$	Fatty aldehydic
n-Caprylic (octanoic)	$CH_3(CH_2)_6COOH$	Unpleasantly rancid, sweaty faintly fruity
n-Nonylic (pelargonic)	$CH_3(CH_2)_7COOH$	Strongly fatty aldehydic
n-Capric (decylic)	$CH_3(CH_2)_8COOH$	Unpleasantly rancid, fatty
n-Undecylic	$CH_3(CH_2)_9COOH$	Aldehydic
Lauric	$CH_3(CH_2)_{10}COOH$	Faintly waxy, coconut-like
Myristic	$CH_3(CH_2)_{12}COOH$	Nutmeg butter-like
Stearic	$CH_3(CH_2)_{16}COOH$	Faintly tallow-like, waxy
Unsaturated Acids		
Oleic	$CH_3(CH_2)_7CH{:}CH(CH_2)_7COOH$	Faint, tallow-like
Aromatic Acids		
Benzoic	C_6H_5COOH	Faintly balsamic; taste sweet-sour, slightly acrid
Phenyl acetic	$C_6H_5CH_2COOH$	Sweet, rose-like but unpleasant, honey-like on dilution
Cinnamic	$C_6H_5CH{:}CHCOOH$	Faintly balsamic, spicy; taste slightly fruity

tionships are given in Table 7.10. These compounds form the main armory of the flavor chemist and are of prime importance in the formulation of imitation flavors. Presently, the flavorist has over 200 esters at his disposal representing an enormous spectrum of profiles. Several attempts have been made to classify esters based either on their observed organoleptic characteristics or on their chemical structure. Some quite definite relationships exist but no really satisfactory and meaningful classification has yet been evolved. The one important attribute of many classes of esters is their fruitiness.

From broad similarities it is preferable to consider the esters according to their acid constituent (e.g., acetates, butyrates, ben-

TABLE 7.10
FUNCTIONAL RELATIONSHIPS AND DESIGNATIONS OF ALCOHOLS, ACIDS AND ESTERS

Alcohol		Acid		Ester Common Name	Systematic Name	Synonyms
Saturated Aliphatic Series						
Methyl	CH_3-	Formic	HCOO-	Formates		
Ethyl	C_2H_5-	**Acetic**	CH_3COO-	**Acetates**		
Propyl	C_3H_7-	Propionic	C_2H_5OO-	Propionates		
Iso-propyl	C_3H_7-	Iso-propionic	C_2H_5COO-	Iso-propionates		
n-Butyl	C_4H_9-	Butyric	C_3H_7COO-	Butyrates		
Iso-butyl	C_4H_9-	Iso-butyric	C_3H_7COO-	Iso-butyrates		
n-Amyl	C_5H_{11}-	Valeric	C_4H_9COO-	Valerates	Pentanoates	Valeria-nates
Iso-amyl	C_5H_{11}-	Iso-valeric	C_4H_9COO-	Iso-valerates		
n-Hexyl	C_6H_{13}-	Caproic	$C_5H_{11}COO$-	Caproates	Hexanoates	Hexylates
n-Heptyl	C_7H_{15}-	Heptylic	$C_6H_{13}COO$-	Heptylates	Heptanoates	Heptoates
n-Octyl (capryl)	C_8H_{17}-	Caprylic	$C_7H_{15}COO$-	Caprylates	Octanoates	Octylates
n-Nonyl	C_9H_{19}-	Pelargonic	$C_8H_{17}COO$-	Pelargonates	Nonanoates	Nonylates
n-Decyl	$C_{10}H_{21}$-	Decylic (capric)	$C_9H_{19}COO$-	Decylates	Decanoates	Caprates
n-Undecyl	$C_{11}H_{23}$-	Undecylenic	$C_{10}H_{21}COO$-	Undecylates	Undecanoates	
n-Dodecyl (lauryl)	$C_{12}H_{25}$-	Lauric	$C_{11}H_{23}CO$-	Laurates	Dodecanoates	
Myristyl	$C_{14}H_{29}$-	Myristic	$C_{13}H_{27}COO$-	Myristates		
Unsaturated Aliphatic Compounds						
Allyl	CH_2:$CHCH_2$-			Allyl esters		
		Allyl acetic	CH_2:$CHCH_2CH_2COO$-	Allyl acetates	4-Pentenoates	
		Acrylic	CH_2:$CHCOO$-	Acrylates	Propenoates	
		Propyl acrylic	$CH_3(CH_2)_2CH$:$CHOO$-	Propyl acrylates	2-Hexenaotes	
		Crotonic	CH_3CH:$CHCOO$-	Crotonates	*trans*-2-Butenoates	
		Tiglic	CH_3CH:$C(CH_3)COO$-	Tiglates	*trans*-2-Methylbutenoates	

		Undecylenic	$CH_2CH(CH_2)_8COO$-	Undecylenates	Undecenaotes	
		Acetoacetic	CH_3COCH_2COO-	Acetoacetates	3-Oxobutanoates	
		Pyruvic	CH_3COCOO-	Pyruvates	2-Oxopropanotes	
			$CH_3(CH_2)_4C\ CCOO$-	Heptyn carbonates	2-Octynoates	
			$CH_3(CH_2)_5C\ CCOO$-	Octyn carbonates	2-Nonynoates	
			$CH_3(CH_2)_7C\ CCOO$-	Decyn carbonates	Undecynoates	
			Hydroxy Compounds			
		Lactic	$CH_3CHOHCOO$-	Lactates	α-Hydroxy propionate	
			Aromatic Series			
Benzyl	C_6H_5-	Benzoic	C_6H_5COO-	Benzoates		
Phenyl ethyl	$C_6H_5C_2H_4$-	Phenylacetic	$C_6H_5CH_2COO$-	Phenylacetates Phenylethyl esters		-toluates
Phenyl propyl	$C_6H_5C_3H_6$-	Phenylpropionic	$C_6H_5CH_2(CH_2)_2COO$-	Phenylpropionates Phenylpropyl esters		
Cinnamyl	$C_6H_5CH{:}CHCH_2$-	Cinnamic	$C_6H_5CH{:}CHCHCOO$-	Cinnamates Cinnamyl esters		
Anisyl (*p*-methoxy benzyl)	$C_6H_4(OCH_3)CH_2$-	Anisic	$C_6H_4(OCH_3)CHCOO$-	Anisates Anisyl esters		*p*-Methoxy benzoates
		Anthranilic	$C_6H_4(NH_2)COO$-	Anthranilates	2-Amino-benzoates	
		Salicylic	$C_6H_4(OH)COO$-	Salicylates	2-Hydroxy-benzoates	
3-Phenylpropyl	$C_6H_5CH_2(CH_2)_2$-	Hydrocinnamic	$C_6H_5CH_2(CH_2)_2COO$-	Hydrocinnamtes Hydrocinnamyl esters	3-Phenylpropionates 3-Phenylpropyl esters	

Phenol	Formula	Ester
Eugenol	$C_6H_3(OCO\cdot R)(OCH_3)(CH_2CH{:}CH_2)$ [benzene ring: OCO·R; OCH$_3$; CH$_2$CH:CH$_2$]	Eugenyl esters

(Continued)

TABLE 7.10 (Continued)

Phenol	Formula	Ester
Iso-eugenol	 $CH{:}CHCH_3$	Iso-eugenyl esters
Terpenoid Alcohols	**Formula**	**Ester**
Borneol	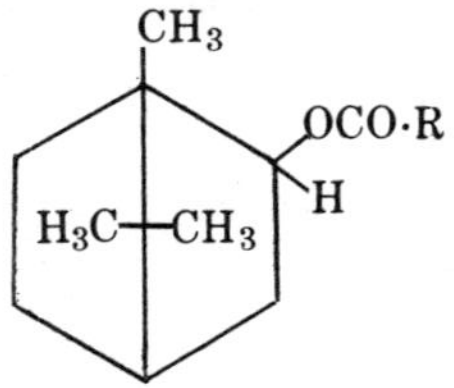 CH_3; $OCO{\cdot}R$; H; $H_3C—CH_3$	Bornyl esters
Iso-borneol	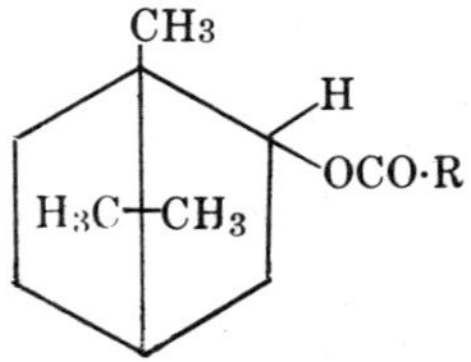 CH_3; H; $OCO{\cdot}R$; $H_3C—CH_3$	Iso-bornyl esters
Alcohol	**Formula**	**Ester**
Carveol	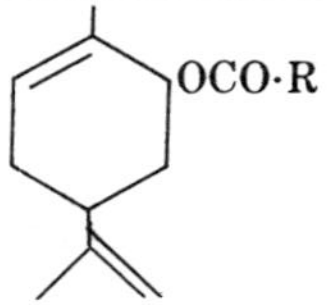 $OCO{\cdot}R$	Carvyl esters

Alcohol	Structure	Esters
Caryophyllene alcohol	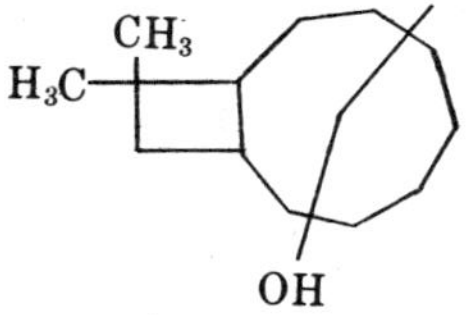 CH_3, H_3C, OH	Caryophyllene alcohol esters
Citronellol	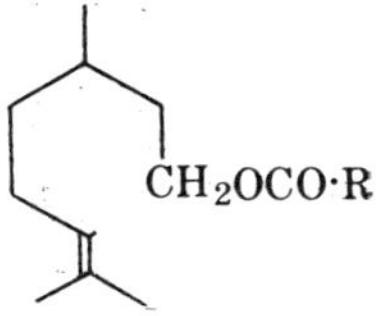 $CH_2OCO\cdot R$	Citronellyl esters
Geraneol	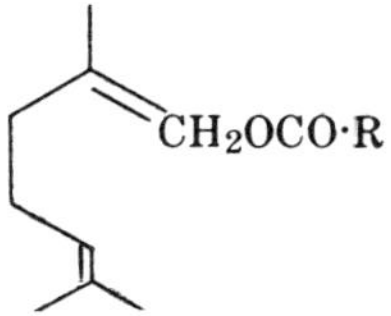 $CH_2OCO\cdot R$	Geranyl esters
Linalool	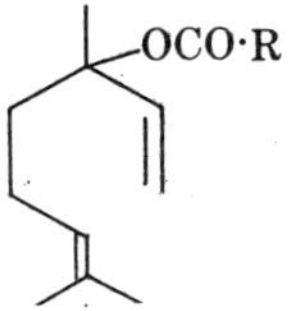 $OCO\cdot R$	Linalyl esters
1-Menthol	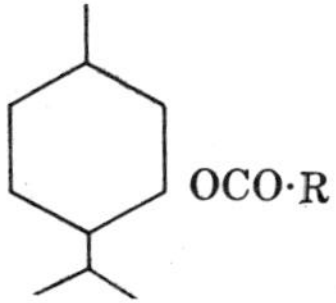 $OCO\cdot R$	Menthyl esters

(Continued)

TABLE 7.10 (Continued)

Terpenoid alcohols	Formula	Ester
Nerol	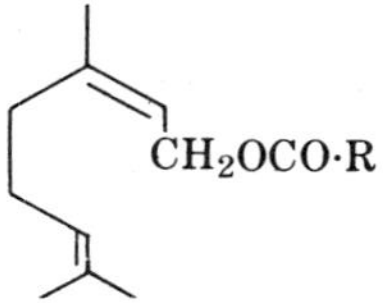 $CH_2OCO{\cdot}R$	Neryl esters
Rhodinol (1-citronellol)	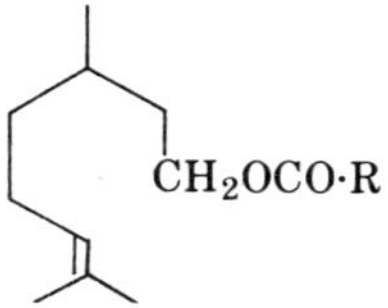 $CH_2OCO{\cdot}R$	Rhodinyl esters
Terpineol	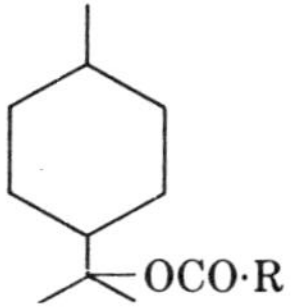 $OCO{\cdot}R$	Terpinyl esters

Cyclic Compounds	Formula	Ester
Cyclohexanol	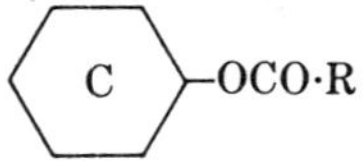 C $OCO{\cdot}R$	Cyclohexyl esters

zoates, etc.) rather than on their alcohol radical (e.g., methyl, ethyl, benzyl, etc.), although, as this moiety becomes larger and more unsaturated it increasingly determines the distinctive profile of the resulting ester. The profile of the constituent acid and alcohol has almost no relationship to that of the ester produced.

The principal esters used in flavor work are listed in Table 7.11 according to their acid radical. The profiles are of necessity short and reference should be made to one of the several standard texts (Actander 1969; Furia and Bellanca 1975) for a more detailed treatment of these materials.

LACTONES

Internal esterification, brought about by the loss of a molecule of water from the γ- or δ-hydroxyacids, results in a class of cyclic compounds called lactones. The five-membered ring system is readily formed and the resulting γ-lactones are very stable. As a group, they have strong and very distinctive odors and those of particular value in flavorings are listed in Table 7.12. None of the γ-lactones are found in nature but some of the large-chain lactones (e.g., ambrettolide or ω-6-hexadecenlactone) are of natural origin. These large-chain lactones have a distinctive musk-like odor.

The characteristic structure of the lactones is

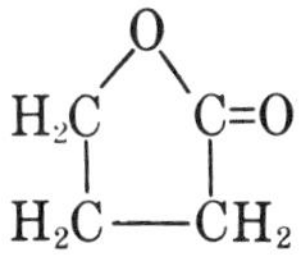

```
         O
       /   \
  H2C       C=O
   |        |
  H2C ——— CH2
```

γ-butyrolactone

ALDEHYDES

The aldehydes are characterized by the carbonyl radical and are intermediate between the alcohols and the acids. They are more reactive than the alcohols because of the presence of the double-bond linkage with oxygen.

Those of lower molecular weight are characterized by their unpleasant, sharply pungent odor and irritant effect in the nose. As the molecular weight increases, the odor profile gradually assumes a more pleasing fruity character, although all of them can be strongly penetrating in concentration and only display their true flavor quality on dilution. Aldehydes C_8 to C_{10} are markedly floral and very attractive on dilution; hence, why they are frequently used at low levels to introduce a fine floral nuance in many flavorings. The higher members of the series have only weak odors and are somewhat waxy.

TABLE 7.11
ESTERS USED IN FLAVORINGS

Ester	Sensory Attributes	
	Odor Type	Characteristic Notes
	The Formates	
Aliphatic Series		
Ethyl formate	Fruity	Pineapple
n-Propyl formate	Fruity	Rum, plum
Iso-propyl formate	Fruity, ethereal	Plum
n-Butyl formate	Fruity, ethereal	Plum
Iso-butyl formate	Fruity, ethereal	Plum
n-Amyl formate	Sweet, fruity	Vinous
Iso-amyl formate	Sweet, fruity	Black currant
n-Hexyl formate	Sweet, fruity	Apple
n-Heptyl formate	Sweet, fruity	Apple, orris root
n-Octyl formate	Sweet, fruity	Orange, rose
Aromatic Series		
Benzyl formate	Fruity, floral	Apricot, pineapple
Phenyl ethyl formate	Sweet, floral	Rose, hyacinth
Cinnamyl formate	Fruity, floral	Balsamic, herbaceous
Anisyl formate	Sweet, floral	Strawberry
Eugenyl formate	Sweet, spicy	Orris root
Iso-eugenyl formate	Sweet, spicy	Woody, orris root
α-Methyl benzyl formate	Sweet, floral	Woody
Terpenoids		
Bornyl formate	Fresh	Pine
Iso-bornyl formate	Fresh	Pine needles
Citronellyl formate	Fruity, floral	Jasmin, bergamot
Geranyl formate	Sweet, floral	Rose, leafy green
Linalyl formate	Fruity, floral	Sweet, rose, bergamot
Neryl formate	Floral	Herbaceous, rose, orange flower
Rhodinyl formate	Floral	Rose, leafy green
Terpinyl formate	Floral	Jasmin, bergamot
Cyclic Compounds		
Cyclohexyl formate	Fruity	Cherry
	The Acetates	
Aliphatic Series		
Methyl acetate	Fruity	Sweet, ethereal
Ethyl acetate	Fruity	Pineapple, vinous
n-Propyl acetate	Fruity	Pear, raspberry
Iso-propyl acetate	Fruity	Apple
n-Butyl acetate	Fruity	Pineapple
Iso-butyl acetate	Fruity, floral	Pear, hyacinth
Iso-amyl acetate	Fruity	Pear, banana
n-Hexyl acetate	Fruity	Pear, floral back-notes
n-Heptyl acetate	Fruity, floral	Pear, apricot
n-Octyl acetate	Fruity, floral	Peach, jasmin
n-Nonyl acetate	Sweet, floral	Orange, rose, honey
n-Decyl acetate	Sweet, floral	Orange, rose, pineapple
Undecenyl acetate	Sweet, floral	Rose
2-Ethylbutyl acetate	Fruity	Oily
2-Hexen-l-yl acetate	Fruity	Strongly sweet green
Aromatic Series		
Benzyl acetate	Floral, fruity	Jasmin, apricot, pineapple
Phenylethyl acetate	Floral	Rose, honey, raspberry

TABLE 7.11 (Continued)

Ester	Sensory Attributes	
	Odor Type	Characteristic Notes
3-Phenylpropyl acetate	Floral	Spicy, bitter/sweet
Cinnamyl acetate	Floral	Balsamic, jasmin, hyacinth
Anisyl acetate	Floral	Hawthorn blossom
Eugenyl acetate	Spicy	Clove
Iso-eugenyl acetate	Spicy, floral	Rose, carnation
α,α-Dimethylphenylethyl acetate	Floral, fruity	Pear
Ethyl acetoacetate	Fruity	Sweet, ethereal, rum
Ethyl α-toluate	Sweet	Honey
Guaiacyl acetate	Sweet	Slightly phenolic
Iso-amyl phenyl acetate	Sweet	Cocoa
Iso-propyl phenyl acetate	Floral	Rose, honey
α-methyl benzyl acetate	Green	Gardenia
Methylphenyl acetate	Floral	Jasmin, honey
2-Methyl-4-phenyl-2-butyl acetate	Floral	Jasmin, hyacinth, slightly rose
n-Octylphenyl acetate	Fruity Floral citrus	
Phenylethyl phenyl acetate	Floral	Heavy, sweet, balsamic, rose, honey
4-Phenyl-2-butyl acetate	Fruity	Mildly green, sweet
3-Phenyl propyl acetate	Floral	Spicy, bitter/sweet
Piperonyl acetate	Floral	Cherry
Propylphenyl acetate	Fruity	Mildly green, dull earthy
Terpenoids		
Bornyl acetate	Fresh	Pine
Iso-bornyl acetate	Fresh	Pine needles
Carvyl acetate	Fresh	Minty
Caryophyllene alcohol acetate	Fruity	Woody
Citronellyl acetate	Fruity, floral	Rose, apricot, slightly citrus
Geranyl acetate	Floral	Rose, lavender
Linalyl acetate	Floral	Bergamot, lavender
1-Menthyl acetate	Fresh	Minty, slightly rose *Flavor*: cooling
Neryl acetate	Floral	Sweet bergamot, rose, honey
Rhodinyl acetate	Floral	Rose
Terpinyl acetate	Floral	Lavender, bergamot
Cyclic Compounds		
Cyclohexyl acetate	Fruity	"Pear drops"
	The Propionates	
Aliphatic Series		
Methyl propionate	Fruity	Rum, black currant
Ethyl propionate	Fruity	Rum, pineapple
n-Propyl Propionate	Fruity	Apple, pineapple
Iso-propyl propionate	Fruity	Plum
n-Butyl propionate	Earthy	Sweet
Iso-butyl propionate	Fruity	Rum, pineapple
Iso-amyl propionate	Fruity	Apricot, pineapple
n-Hexyl propionate	Earthy	Sweet
n-Octyl propionate	Waxy	Myrtle berry, pineapple
n-Decyl propionate	Fatty, aldehydic	Cognac

(Continued)

TABLE 7.11 (Continued)

Ester	Sensory Attributes	
	Odor Type	Characteristic Notes
Aromatic Series		
Benzyl propionate	Fruity, floral	Peach, apricot, jasmin
Phenylethyl propionate	Fruity, floral	Fruity red rose, strawberry
Phenylpropyl propionate	Floral	Balsamic, rose, hyacinth
Cinnamyl propionate	Fruity, floral	Balsamic, spicy, woody, rose
Anisyl propionate	Fruity	Herbaceous, cherry
Ethyl-3-phenyl propionate	Floral	Rose, honey, hyacinth
Methyl-3-phenyl propionate	Floral, fruity	Fruity, balsamic, honey
α-Methyl benzyl propionate	Floral	Gardenia
3-Phenylpropyl propionate	Floral	Balsamic, hyacinth, jasmin
Terpenoid Compounds		
Iso-bornyl propionate	Fresh	Pine, turpentine
Carvyl propionate	Fresh	Minty, sweet, warm
Citronellyl propionate	Floral	Rose
Geranyl propionate	Floral, fruity	Rose, citrus
Linalyl propionate	Floral, fruity	Bergamot, lily
Neryl propionate	Floral	Jasmin, rose
Rhodinyl propionate	Floral	Rose, geranium
Terpinyl propionate	Floral	Lavender
Cyclic Compounds		
Cyclohexyl propionate	Fruity	Apple, banana
Allyl cyclohexane propionate	Fruity	Pineapple
The n-Butryrates and iso-Butryrates		
Aliphatic Series		
Methyl butyrate	Fruity	Apple
Ethyl butyrate	Fruity	Pineapple
n-Propyl butyrate	Fruity	Pineapple, banana
Iso-propyl butyrate	Fatty	Buttery, butyric acid-like
n-Butyl butyrate	Fruity	Apple, pineapple
Iso-butyl butyrate	Fruity	Apple, pineapple
n-Amyl butyrate	Sweet	Strongly penetrating
Iso-amyl butyrate	Fruity	Apricot, pineapple
n-Hexyl butyrate	Fruity	Apricot, pineapple
n-Heptyl butyrate	Fruity	Camomile, green tea, plum
n-Octyl butyrate	Herby	Green, citrus, parsley, melon
n-Nonyl butyrate	Fruity	Apricot, orange
n-Decyl butyrate	Fruity	Apricot
Ethyl-2-methyl butyrate	Fruity	Green apple, unripe fruit
Hexyl-2-methyl butyrate	Fruity	Green, strawberry
2-Methyl-allyl butyrate	Fruity	Ethereal, sharply penetrating, pineapple, plum
Aromatic Series		
Benzyl butyrate	Fruity	Apricot, pear, fatty
Phenyl ethyl butyrate	Floral	Rose, honey
Phenyl propyl butyrate	Fruity	Apricot, jasmin, woody
Cinnamyl butyrate	Fruity, floral	Balsamic, rose, citrus
Anisyl butyrate	Fruity, floral	Plum, rose
Ethyl-4-phenyl butyrate	Fruity	Sweet, plum, cooked-fruit note
α-Methylbenzyl butyrate	Fruity, floral	Apricot, jasmin
2-Phenylpropyl butyrate	Fruity	Plum, woody
Phenylethyl-2-methyl butyrate	Floral, fruity	Herbaceous, sweet, rose

TABLE 7.11 (Continued)

Ester	Sensory Attributes	
	Odor Type	Characteristic Notes
Terpenoid Compounds		
Citronellyl butyrate	Floral, fruity	Rose, balsamic, plum
Geranyl butyrate	Floral, fruity	Rose, apricot
Linalyl butyrate	Fruity	Bergamot, honey, buttery
Neryl butyrate	Floral	Leafy green, neroli-like, cocoa
Rhodinyl butyrate	Fruity, floral	Apple, rose, blackberry
Terpinyl butyrate	Fruity	Rosemary-like, balsamic, plum
Cyclic Compounds		
Cyclohexyl butyrate	Floral	Intensely sweet, fatty apricot
Allyl cyclohexane butyrate	Fruity	Pineapple
Iso-butyrates		
Butyl iso-butyrate	Fruity	Sweet, pineapple
Cinnamyl iso-butyrate	Fruity	Sweet, balsamic, apple, banana
Citronellyl iso-butyrate	Fruity, floral	Rose, apricot
α,α-Dimethylbenzyl iso-butyrate	Fruity	Sweet, peach, plum, slightly green
Ethyl iso-butyrate	Fruity	Apple
Geranyl iso-butyrate	Floral, fruity	Rose, apricot
Heptyl iso-butyrate	Fruity, floral	Orris root, rose, plum
Iso-amyl iso-butyrate	Fruity	Apricot, pineapple
Iso-butyl iso-butyrate	Fruity	Pineapple
Iso-propyl iso-butyrate	Fruity	Intensely ethereal
Linalyl iso-butyrate	Fruity, floral	Warmly green, lavender
α-Methyl benzyl iso-butyrate	Floral	Jasmin
2-Methyl-4-phenyl-2-butyl iso-butyrate	Fruity	Herbaceous, green tea
Neryl iso-butyrate	Floral, fruity	Rose, strawberry
Octyl iso-butyrate	Fruity	Fatty, parsley, fern-like
Phenyl ethyl iso-butyrate	Fruity	Pineapple, banana
2-Phenoxy iso-butyrate	Floral, fruity	Rose, honey, peach
2-Phenyl propyl iso-butyrate	Fruity	Sweetly woody
Piperonyl iso-butyrate	Fruity	Sweet, berry-like, plum
Propyl iso-butyrate	Fruity	Pineapple
Rhodinyl iso-butyrate	Fruity, floral	Pineapple, sweetly floral
Terpinyl iso-butyrate	Fruity, floral	Heavy, herbaceous
The Valerates (Valerianates) and iso-Valerates		
Aliphatic Series		
Methyl valerate	Fruity	Apple
Ethyl valerate	Fruity	Strongly apple
n-Propyl valerate	Fruity	Pineapple
n-Butyl valerate	Fruity	Apple, raspberry
n-Amyl valerate	Fruity	Ripe apple
n-Hexyl valerate	Fruity	Unripe apple
n-Heptyl valerate	Fruity	Over-ripe fruit
n-Octyl valerate	Fruity	Apple, aldehydic
Aromatic Series		
Benzyl valerate	Fruity, floral	Strongly fruity, musk-like
Phenyl ethyl valerate	Fruity, floral	Peach, rose
Phenyl propyl valerate	Fruity	Strawberry, raspberry
Cinnamyl valerate	Floral	Rose
Anisyl valerate	Floral	Sweetly heliotrope

(Continued)

TABLE 7.11 (Continued)

Ester	Sensory Attributes	
	Odor Type	Characteristic Notes
Terpenoid Compounds		
Citronellyl valerate	Floral	Herbaceous, rose, honey
Geranyl valerate	Floral, fruity	Rose, pineapple
Linalyl valerate	Floral, fruity	Lavender
Neryl valerate	Floral	Sweetly orange flower
Terpinyl valerate	Floral	Orange, sweetly balsamic
Iso-valerates		
Allyl iso-valerate	Fruity	Apple, cherry
Benzyl iso-valerate	Fruity	Apple, pineapple
Bornyl iso-valerate	Aromatic	Camphoraceous
Butyl iso-valerate	Fruity	Apple
Cinnamyl iso-valerate	Floral, fruity	Rose, apple
Cyclohexyl iso-valerate	Fruity	Apple, banana
Ethyl iso-valerate	Fruity	Sweet, apple
Geranyl iso-valerate	Floral, fruity	Rose, apricot
3-Hexenyl iso-valerate	Fruity	Strongly green, apple, buttery
Hexyl iso-valerate	Fruity	Over-ripe fruit, earthy
Iso-amyl iso-valerate	Fruity	Sweet, apple
Iso-bornyl iso-valerate	Herby	Herbaceous, woody, camphoraceous
Iso-propyl iso-valerate	Fruity	Strong, ethereal, apple
Linalyl iso-valerate	Fruity	Apple, peach
Menthyl iso-valerate	Herby	Sweetly herbaceous, balsamic, woody
2-Methylbutyl iso-valerate	Herby	Sweetly herbaceous, fruity
Methyl iso-valerate	Fruity	Strongly apple
Neryl iso-valerate	Fruity	Strongly bergamot
n-Octyl iso-valerate	Fruity, fatty	Apple, pineapple, fatty
Phenyl ethyl iso-valerate	Fruity, floral	Peach, apricot, rose
3-Phenyl-propyl iso-valerate	Fruity	Strawberry, raspberry, plum
n-Propyl iso-valerate	Fruity	Apple
Rhodinyl iso-valerate	Floral	Rich rose, cherry
Terpinyl iso-valerate	Floral	Sweetly floral, pine, citrus, apple back-notes
The Caproates (Hexanoates)		
Aliphatic Series		
Methyl caproate	Fruity	Pineapple, ether-like
Ethyl caproate	Fruity	Pineapple, banana
n-Propyl caproate	Fruity	Pineapple, blackberry
Iso-propyl caproate	Fruity	Pineapple, berry-like
n-Butyl caproate	Fruity	Pineapple
Iso-butyl caproate	Fruity	Apple, cocoa-like
n-Amyl caproate	Fruity	Banana, pineapple
Iso-amyl caproate	Fruity	Pineapple, fatty
n-Hexyl caproate	Herby	Herbaceous, grassy
n-Heptyl caproate	Green	Freshly cut grass
n-Octyl caproate	Fruity, green	Green parsley, banana
n-Nonyl caproate	Fruity	Nutty
Aromatic Series		
Benzyl caproate	Fruity, floral	Apricot, jasmin
Phenyl ethyl caproate	Floral, fruity	Rose, pineapple
Phenyl propyl caproate	Fruity	Peach, pineapple, green

TABLE 7.11 (Continued)

Ester	Sensory Attributes	
	Odor Type	Characteristic Notes
Terpenoid Compounds		
Geranyl caproate	Floral, fruity	Rose-geranium, pineapple
Linalyl caproate	Fruity	Warmly green, pineapple, pear
The Heptylates (Heptanoates)		
Aliphatic Series		
Methyl heptylate	Fruity	Orris root, gooseberry
Ethyl heptylate	Fruity	Cognac-like
n-Propyl heptylate	Fruity	Grape, ethereal
n-Butyl heptylate	Herby	Herbaceous, slightly floral, marigold
Iso-butyl heptylate	Herby	Herbaceous
n-Amyl heptylate	Fruity	Unripe fruit, banana
n-Hexyl heptylate	Green	Wet foliage
n-Heptyl heptylate		Green, fatty, vinous
n-Octyl heptylate	Fruity	Fatty
The Caprylates (Octanoates)		
Aliphatic Series		
Methyl caprylate	Fruity, winey	Orange, vinous
Ethyl caprylate	Fruity, winey	Apricot, vinous
n-Propyl caprylate	Fruity, winey	Plum, vinous
n-Butyl caprylate	Fatty	Buttery, ethereal
n-Amyl caprylate	Floral	Orris root, elderflower
Iso-amyl caprylate	Fruity	Sharp, apple, banana
n-Hexyl caprylate	Herby, fruity	Herbaceous, fresh vegetable
n-Heptyl caprylate	Green, fruity	Oily, slightly nutty
n-Octyl caprylate	Fatty	Oily, green tea, balsamic
n- Nonyl caprylate	Floral	Rose, musty, sweet
The Pelargonates (Nonanoates)		
Methyl pelargonate	Fruity, winey	Orange, vinous
Ethyl pelargonate	Fruity, winey	Cognac, rosy-fruity
n-Propyl pelargonate	Fruity	Melon, yeast-like
n-Butyl pelargonate	Fruity, floral	Apricot, rose
n-Amyl pelargonate	Fruity, winey	Apricot, rose, vinous
n-Hexyl pelargonate	Fruity, winey	Cognac-like
n-Heptyl pelargonate	Fruity, winey	Orange, rose, vinous
n-Nonyl pelargonate	Aldehydic	Orange, rose
The Decylates (Decanoates)		
Methyl decylate	Sweet, winey	Honey, vinous
Ethyl decylate	Fruity	Cognac, grape
n-Propyl decylate	Fruity	Grape
n-Butyl decylate	Fruity	Apricot
n-Amyl decylate	Fruity	Cognac-like
The Undecylates (Undecanoates)		
Methyl undecylate	Sweet	Honey
Ethyl undecylate	Sweet	Cognac, coconut-like
The Laurates (Dodecanoates)		
Methyl laurate	Fatty, floral	Weakly waxy, mimosa-like
Ethyl laurate	Fatty, floral	Weakly waxy, tuberose-like
n-Propyl laurate	Fatty	Weakly waxy, musty
n-Butyl laurate	Fatty	Weakly peanut butter-like
Iso-amyl laurate	Fatty	Very faintly oily

(Continued)

TABLE 7.11 (Continued)

Ester	Sensory Attributes	
	Odor Type	Characteristic Notes
	The Myristates	
Methyl myristate	Aromatic	Honey, orris root
Ethyl myristate	Aromatic	Mildly orris root-like
	The Benzoates	
Aliphatic Series		
Methyl benzoate	Fruity, floral	Aromatic, woody, floral
Ethyl benzoate	Fruity, floral	Aromatic, cananga-like, mild
n-Propyl benzoate	Balsamic	Nutty, sweet
Iso-propyl benzoate	Fruity	Sweet, berry-like, plum
n-Butyl benzoate	Balsamic	Amber-like
Iso-butyl benzoate	Floral	Leafy green, rose, geranium
n-Amyl benzoate	Balsamic	Amber-like
Iso-amyl benzoate	Fruity	Balsamic, slightly musk-like
n-Hexyl benzoate	Balsamic	Woody, green
n-Heptyl benzoate	Floral	Leafy green, sweetly floral
n-Octyl benzoate	Balsamic	Banana
Aromatic Series		
Benzyl benzoate	Balsamic	Almond
Phenyl ethyl benzoate	Floral	Sweet, rose, honey
Cinnamyl benzoate	Balsamic	Spicy
Eugenyl benzoate	Balsamic	Spicy, clove
Terpenoid Compounds		
Geranyl benzoate	Floral	Sweet, rose, cananga-like
Linalyl benzoate	Floral	Heavy, bergamot, lily
	The Phenylacetates	
Aliphatic Series		
Methyl phenylacetate	Floral	Jasmin, honey
Ethyl phenylacetate	Floral	Sweet, honey, rose, apricot
n-Propyl phenylacetate	Floral	Honey, rose, vinous
Iso-propyl phenylacetate	Floral	Rose, honey, vinous
n-Butyl phenylacetate	Floral	Balsamic, rose, honey
Iso-butyl phenylacetate	Sweetly fragrant	Musk-like, honey
n-Amyl phenylacetate	Floral	Rose, apple, honey
Iso-amyl phenylacetate	Sweetly fragrant	Rose
n-Hexyl phenylacetate	Fruity, floral	Fruity rose
n-Heptyl phenylacetate	Fruity	Citrus, lemongrass-like, verbena
n-Octyl phenylacetate	Fruity	Citrus, slightly fatty
Aromatic Series		
Benzyl phenylacetate	Floral	Jasmin, rose
Phenylethyl phenylacetate	Floral	Rose, hyacinth, balsamic, honey
Cinnamyl phenylacetate	Floral	Rich chrysanthemum-like, spicy, honey
Anisyl phenylacetate	Sweet, spicy	Anisic, honey
Eugenyl phenylacetate	Sweet, spicy	Clove, carnation
Iso-eugenyl phenylacetate	Sweet, spicy	Clove, carnation, honey
Terpenoid Compounds		
Citronellyl phenylacetate	Floral	Rose, verbena

TABLE 7.11 (Continued)

Ester	Sensory Attributes	
	Odor Type	Characteristic Notes
Geranyl phenylacetate	Floral	Sweet, rose
Linalyl phenylacetate	Floral	Lavender
	The Cinnamates	
Aliphatic Series		
Methyl cinnamate	Balsamic	Fruity, strawberry
Ethyl cinnamate	Balsamic	Sweet, peach, apricot, honey
n-Propyl cinnamate	Fruity	Peach, apricot, vinous
Iso-propyl cinnamate	Balsamic	Sweet, dry, amber
n-Butyl cinnamate	Balsamic	Cocoa-like
Iso-butyl cinnamate	Balsamic	Sweet, fruity
n-Amyl cinnamate	Balsamic	Sweetly cocoa-like
Iso-amyl cinnamate	Balsamic	Cocoa-like
n-Hexyl cinnamate	Balsamic	Warmly herbaceous, fatty
n-Heptyl cinnamate	Green	Leafy green, hyacinth
Aromatic Series		
Benzyl cinnamate	Balsamic	Spicy, honey
Phenyl ethyl cinnamate	Balsamic	Rose
Phenyl propyl cinnamate	Balsamic	Sweetly, floral-fruity
Cinnamyl cinnamate	Balsamic	Sweetly resinous
Terpenoid Compounds		
Linalyl cinnamate	Balsamic	Sweetly floral, lily
Terpinyl cinnamate	Balsamic	Sweet, muscatel-wine-like, heavy, fruity
	The Anthranilates (2-Amino Benzoates)	
Methyl anthranilate	Floral	Strongly orange-flower-like
Ethyl anthranilate	Floral	Weakly orange-flower-like
n-Butyl anthranilate	Fruity, floral	Sweet, weakly orange-flower-like, plum
Iso-butyl anthranilate	Floral	Faintly orange-flower-like
Cinnamyl anthranilate	Balsamic	Fruity, grape
Linalyl anthranilate	Citrus	Sweet, orange
Terpinyl anthranilate	Fruity	Harsh, orange

Unsaturation in the main radical increases the irritating acid character, this being further accentuated in the acetylenic compounds. The aromatic aldehydes are powerful and have very distinctive profiles depending on their complexity.

Because of their very pronounced profiles, some compounds are well known by names such as "aldehyde C_{16}" or "strawberry aldehyde," "aldehyde C_{14}" or "peach aldehyde." These materials are not in fact aldehydes but lactones; however, the names persist in common usage.

The principal aldehydes used in flavorings are listed in Table 7.13.

TABLE 7.12
LACTONES USED IN FLAVORINGS

Common Name	Systematic Name	Sensory Attributes
Ambrettolide	ω-6-Hexadecenlactone	Strongly musk-like
γ-Decalactone	4-Hydroxy-decanoic acid, γ-lactone	Fruity, peach-like, creamy coconut-like
δ-Decalactone	5-Hydroxy-decanoic acid, δ-lactone	Oily, peach-like
	4,4-Dibutyl-4-hydroxybutyric acid, γ-lactone	Coconut-like, oily
γ-Dodecalactone	4-Hydroxydodecanoic acid, γ-lactone	Peach-like, fatty, heavy with a suggestion of musk
δ-dodecalactone	5-Hydroxydodecanoic acid, δ-lactone	Strongly fruity, reminiscent of pear, peach and plum
γ-Heptalactone	4-Hydroxyheptanoic acid, γ-lactone	Sweet, nut-like
γ-Hexalactone	4-Hydroxyhexanoic acid, γ-lactone	Sweet, herbaceous, warm, coumarin-like.
S-Nonalactone (coconut aldehyde)	4-Hydroxynonanoic acid, γ-lactone	Strongly coconut-like, fatty, richly sweet
γ-Octalactone	4-Hydroxyoctanoic acid, γ-lactone	Fruity, coconut-like, sweet, slightly coumarin-like
ω-Pentadecalactone (exaltolide)	15-Hydroxy pentadecanoic acid, ω-lactone	Musk-like
γ-Undecalactone (peach aldehyde) (aldehyde C_{14})	4-Hydroxyundecanoic acid, γ-lactone	Strongly peach-like, unpleasantly strong, oily
Coumarin[1]	o-Hydroxycinnamic acid, lactone	Sweet, hay-like, reminiscent of vanilla

[1]In the United States and many other countries, coumarin is not permitted for use in foods.

TABLE 7.13
ALDEHYDES USED IN FLAVORINGS

Aldehyde	Synonyms	Formula	Sensory Attributes
Aliphatic Series			
Acetaldehyde		CH_3CHO	Ethereal, sharply pungent
Propionaldehyde	Propanal	CH_3CH_2CHO	Sweetly ethereal, irritating
Butyraldehyde	Butanal	$CH_3(CH_2)_2CHO$	Pungent, slightly fruity, green
Iso-butyraldehyde	Iso-butanal	$(CH_3)_2CHCHO$	Pungent, pleasantly fruity slightly ripe banana-like
n-Valeraldehyde	Pentanal	$CH_3(CH_2)_3CHO$	Dry, fruity, slightly nutty
Iso-valeraldehyde	3-Methyl butyraldehyde	$(CH_3)_2CHCH_2CHO$	Penetrating, pleasantly fruity, peach-like
2-Methyl butyraldehyde		$CH_3CH_2(CH_3)CHCHO$	Powerfully irritant, fermented fruit, roasted coffee-like
Caproic aldehyde	Hexanal Aldehyde C_6	$CH_3(CH_2)_4CHO$	Green, unripe apples, fatty
n-Heptanal	Enanthaldehyde Aldehyde C_7	$CH_3(CH_2)_5CHO$	Harshly pungent, unpleasantly fatty, fermented fruit
Capryl aldehyde	n-Octanal Aldehyde C_8	$CH_3(CH)_6CHO$	Pungent, fatty, sweetly citrus-like with honey-like back-note
2-Methyl-octanal		$CH_3(CH_2)_5(CH_3)CHCHO$	Sweetly floral, rose-like
Pelargonic aldehyde	n-Nonanal Aldehyde C_9	$CH_3(CH_2)_7CHO$	Strongly fatty, floral rose-like
Capric aldehyde	n-Decanal Decylic aldehyde	$CH_3(CH_2)_8CHO$	Strongly fatty, orange peel-like, floral on dilution
n-Undecanal	Undecylic aldehyde	$CH_3(CH_2)_9CHO$	Sweetly fatty, orange-like, floral, rosey on dilution
n-Dodecanal	Dodecylic aldehyde	$CH_3(CH_2)_{10}CHO$	Balsamic, freshly floral, lilac-like
Myristaldehyde	Tetradecanal	$CH_3(CH_2)_{12}CHO$	Strongly fatty, orris root-like
2-Hexenal	β-Propyl acrolein	$CH_3(CH_2)_2CH:CHCHO$	Green, leafy
cis 3-Hexenal	Leaf aldehyde	$CH_3CH_2CH:CHCH_2CHO$	Strongly herbaceous, green leafy
2-Ethyl-2-heptenal		$CH_3(CH_2)_3CH:C(C_2H_5)CHO$	Strongly green, oily
2, 6-Dimethyl-2-heptenal		$(CH_3)_2C:CH(CH_2)_2CH(CH_3)CHO$	Strongly green, oily, melon-skin-like, more fruity on dilution

(Continued)

TABLE 7.13 (Continued)

Aldehyde	Synonyms	Formula	Sensory Attributes
2-Dodecenal		$CH_3(CH_2)_8CH{:}CHCHO$	Strongly citrus, orange-like becoming tangerine-like on extreme dilution.
Aromatic Series			
Benzaldehyde		C_6H_5CHO	Powerful, sweet, almond-like
Phenyl acetaldehyde		$C_6H_5CH_2CHO$	Strongly penetrating, sweet, floral
Phenyl propionaldehyde		$C_6H_5CH_2CH_2CHO$	Strong, floral, warmly green
p-Tolylaldehyde	4-methylbenzaldehyde	$C_6H_4(CH_3)CHO$	Sweet, fruity, almond-like
Salicylic aldehyde	o-hydroxybenzaldehyde	$(OH)C_6H_4CHO$	Powerfully irritating, phenolic, in extreme dilution slightly nutty
Anisaldehyde	*p*-Methoxybenzaldehyde Aubepine	$C_6H_5(OCH_3)CHO$	Sweetly herbaceous, spicy
2-Methyl-3-tolyl propionaldehyde		$C_6H_4(CH_3)C_3H_7CHO$	Sweet, fruity, balsamic
Cyclamen aldehyde		$C_6H_4(C_3H_7)C_3H_6CHO$	Strongly floral
Cuminic aldehyde		$(CH_3)_2CHC_6H_4CH_2CHO$	Fruity, woody, strongly penetrating, spicy
Cuminyl acetaldehyde		$(CH_3)_2CHC_6H_4CH_2CH_2CHO$	Powerful, sweet, green, floral
Cinnamic aldehyde	3-Phenylpropanal	$C_6H_5CH{:}CHCHO$	Pungent, spicy, warmly sweet, balsamic, cinnamon-like
α-Menthyl cinnamic aldehyde		$C_6H_5CH{:}C(CH_3)CHO$	Warmly spicy, cinnamon-like
p-Menthyl cinnamic aldehyde		$CH_3C_6H_4CH{:}CHCHO$	Balsamic, spicy
α-Butyl cinnamic aldehyde		$C_6H_5CH{:}C(C_4H_9)CHO$	Floral, jasmin, lily-like
α-Amyl cinnamic aldehyde		$C_6H_5CH{:}C(C_5H_{11})CHO$	Strongly floral, jasmin-like
α-Hexyl cinnamic aldehyde		$C_6H_5CH{:}C(CHO)(CH_2)_5CH_3$	Floral, jasmin-like

Vanillin	3-Methoxy-4-hydroxy-benzaldehyde	CHO, OCH_3, OH	Very sweet, characteristically vanilla-like
Ethyl vanillin	3-Ethoxy-4-hydroxy-benzaldehyde	CHO, OCH_2CH_3, OH	Intensely vanilla-like, 2-4 times stronger than vanillin
Sulfur-containing Compounds			
Methional	3-Methyl thiopropion-aldehyde	$CH_3S(CH_2)_2CHO$	Powerful, onion-like, meaty
Terpenoid Aldehydes			
Citral	a-Geranial (*trans*) b-Neral (*cis*)	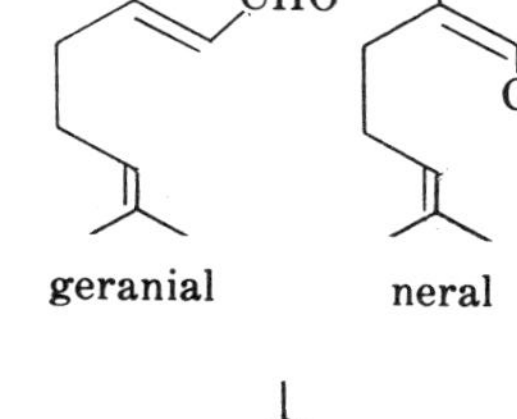	Powerful, lemon-like, bitter-sweet
Citronellal	d-Rhodinal	CHO	Fresh, green, citrus, slightly woody

(Continued)

TABLE 7.13 (Continued)

Aldehyde	Synonyms	Formula	Sensory Attributes
Hydroxycitronellal		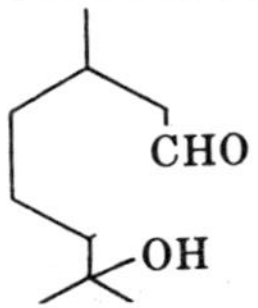 CHO OH	Sweetly floral, lily-of-the-valley-like
Heterocyclic Compounds			
Furfural		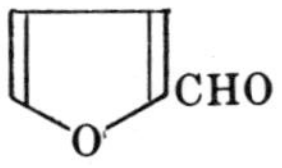 CHO O	Pungent, sweet, caramel-like, spicy
Piperonal	Heliotropine	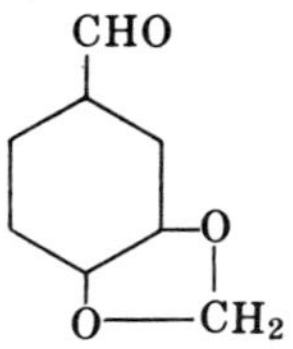 CHO O O—CH_2	Sweet, strongly floral, bitter-sweet taste

KETONES

The ketones are also distinguished by the carbonyl group but differ from aldehydes by having two substituent carbon radicals which may be the same or different

$$\underset{\text{acetone}}{H_3C-\overset{\overset{\displaystyle O}{\|}}{C}-CH_3} \qquad \underset{\text{methyl hexyl ketone}}{C_6H_{13}-\overset{\overset{\displaystyle O}{\|}}{C}-CH_3}$$

The diketones, containing the group $-\underset{\underset{\displaystyle O}{\|}}{C}-\underset{\underset{\displaystyle O}{\|}}{C}-$ are also valuable flavoring materials as they typically have a strong buttery character making them important components in imitation butter flavors. Ketones of low molecular weight are of little value as flavoring compounds but acetone does have a powerful and wide spectrum as a solvent. It is used in this capacity for the production of flavoring extracts and oleoresins from plant material. The first ketone of any importance as a flavoring material is C_7-methyl amyl ketone. This has a strong fruity character which persists through the next members of the series. Those of higher molecular weight have a marked floral character. The ketones display a wide range of aromatic effects most of which are pleasing. They are widely used in imitation fruit flavors. It is interesting to note that compounds in which the tertiary carbon atom is substituted have a distinctly camphoraceous profile.

The principal ketones used in flavorings are listed in Table 7.14.

IONONES

The ionones are ketones which are characterized by a strong violet-like odor and their structure and synthesis are discussed by Bedoukian (1967). They are an important group of aromatics, of greater use in perfumery than in flavoring, although trace quantities are used to give a pleasing nuance to flavors such as strawberry, raspberry and peach.

ACETALS

In the presence of a small amount of acid, the aldehydes condense with alcohols and glycols to form mixed ethers called acetals. The simplest of these is acetal, acetaldehyde diethyl acetal, which has a pleasantly nutty odor. Another example is cinnamaldehyde ethylene glycol acetal which has a warm, sweetly spicy profile reminiscent of cinnamon. The acetals are not widely used in flavorings although they are more stable than the aldehydes.

TABLE 7.14
KETONES USED IN FLAVORINGS

Ketone	Synonym	Formula	Sensory Attributes
Aliphatic Ketones			
2-Butanone	Ethyl methyl ketone	$CH_3 \cdot CO \cdot CH_2CH_3$	Acetone-like
2-Pentanone	Ethyl acetone	$CH_3 \cdot CO \cdot C_3H_7$	Acetone-like, vinous
2-Heptanone	Amyl methyl ketone	$CH_3 \cdot CO \cdot (CH_2)_4CH_3$	Fruity, spicy, banana-like
3-Heptanone	Ethyl butyl ketone	$C_2H_5 \cdot CO \cdot (CH_2)_3CH_3$	Fatty, green, fruity, melon-like
4-Heptanone	Dipropyl ketone	$CH_3(CH_2)_2 \cdot CO \cdot (CH_2)_2CH_3$	Pungent, fruity pineapple-like
2-Octanone	Methyl hexyl ketone	$CH_3 \cdot CO \cdot C_6H_{13}$	Green, unripe fruit, slightly camphoraceous
3-Octanone	Amyl ethyl ketone	$C_2H_5 \cdot CO \cdot C_5H_{11}$	Penetrating, fruity/floral
2-Nonanone	Methyl heptyl ketone	$CH_3 \cdot CO \cdot C_7H_{15}$	Herbaceous, fruity/floral, rose
2-Undecanone	Methyl nonyl ketone	$CH_3 \cdot CO \cdot C_{10}H_{21}$	Herbaceous, fruity, sweet, peach
3-Decen-2-one	Heptylidene acetone	$CH_3 \cdot CO \cdot CH{:}CH(CH_2)_5CH_3$	Fruity/floral, fatty, orris root-like, jasmin
5-Hydroxy-4-octanone		$CH_3(CH_2)_2CH(OH) \cdot CO \cdot (CH_2)_2CH_3$	Sweetly pungent, buttery
Diketones			
Diacetyl		$CH_3CO \cdot COCH_3$	Strongly penetrating, butter-like
Acetyl propionyl		$CH_3CO \cdot COC_2H_5$	Pungent, oily, butter-like
Acetyl nonyryl		$CH_3CO \cdot COC_8H_{17}$	Strong, creamy, coconut-like
Aromatic Ketones (listed alphabetically by systematic names)			
3-Benzyl-4-heptanone		$C_2H_5CH(CH_2C_6H_5) \cdot CO \cdot C_3H_7$ (structure: $C_2H_5CH \cdot CO \cdot C_3H_7$ with CH_2–benzene ring on CH)	Mildly herbaceous, sweet, plum-like, citrus

2,4-Dimethylacetophenone	$CH_3 \cdot CO$ (ring), CH_3, CH_3	Sweetly minty, floral
1,3-Diphenyl-2-propanone	$CH_2 \cdot CO \cdot CH_2$ (phenyl each side)	Fruity, almond-like
4-(*p*-Hydroxyphenyl)-2-butanone	$CH_3 \cdot CO \cdot CH_2CH_2$ (ring), OH	Sweetly fruity, cooked raspberry
4-(*p*-Methoxyphenyl)-2-butanone	$CH_3 \cdot CO \cdot CH_2CH_2$ (ring), OCH_3	Sweetly floral/fruity, cherry, raspberry

(Continued)

TABLE 7.14 (Continued)

Ketone	Synonym	Formula	Sensory Attributes
1-(4-Methoxyphenyl)-4-methyl-1-penten-3-one		$(CH_3)_2CH \cdot CO \cdot CH \cdot CH$–$C_6H_4$–$OCH_3$ (para)	Caramel-like, buttery, fruity
1-(*p*-Methoxyphenyl)-penten-3-one		$C_2H_5 \cdot CO \cdot CH \cdot CH$–$C_6H_4$–$OCH_3$ (para)	Sweet, buttery, fruity
1-(*p*-Methoxyphenyl)-2-propanone (anisic ketone)		$CH_3 \cdot CO \cdot CH_2$–C_6H_4–OCH_3 (para)	Mildly anisic
4-Methyl acetophenone		$CH_3 \cdot CO \cdot$–C_6H_4–CH_3	Sweet, fruity/floral, strawberry-like

3-Methyl-4-phenyl-3-buten-2-one	$CH_3 \cdot CO \cdot C(CH_3){:}CH$ 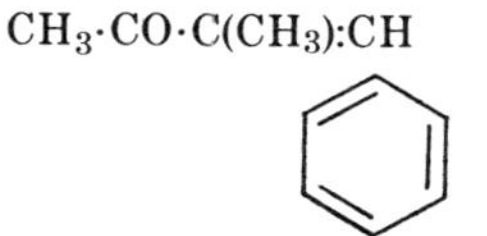	camphoraceous
4-Methyl-1-phenyl-2-pentanone	$(CH_3)_2CHCH_2 \cdot CO \cdot CH_2$ 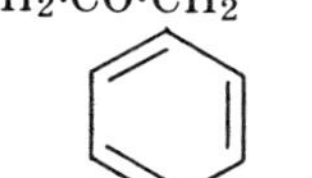	Sweet, spicy, woody back-notes
4-Phenyl-3-buten-2-one	$CH_3 \cdot CO \cdot CH{:}CH$ 	Sharply pungent, fruity, creamy, coumarin-like
Zingerone	$CH_3 \cdot CO \cdot CH_2CH_2$ 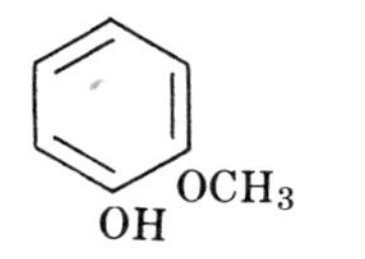	Strongly pungent, ginger-like
Terpenoid Ketones		
Camphor	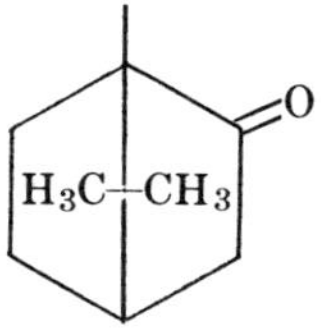	Characteristic warm minty, burning, penetrating

(Continued)

TABLE 7.14 (Continued)

Ketone	Synonym	Formula	Sensory Attributes
Carvone		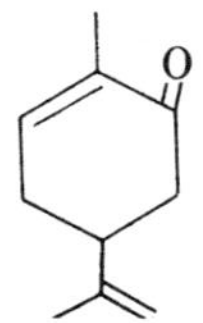	*1-Carvone*: spearmint-like *d-Carvone*: caraway-like
Dihydrocarvone		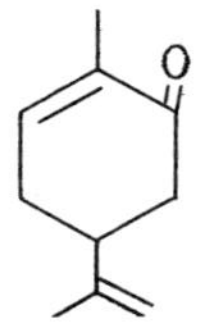	Strongly herbaceous, spearmint-like
Fenchone		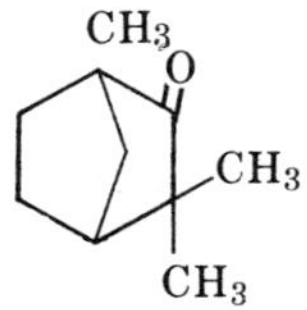	Sweetly camphoraceous, warm, bitter taste
Iso-pulegone		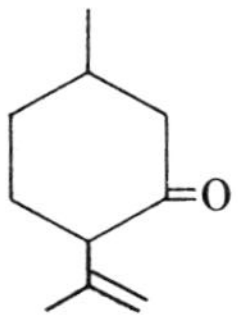	Strongly minty, woody, green

Menthone	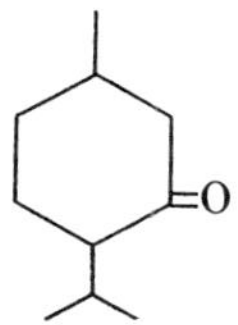	Fresh, cooling, minty, woody back-notes
d-Piperitone	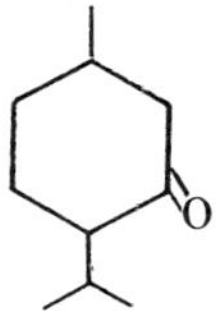	Sharply camphoraceous, minty
Pulegone	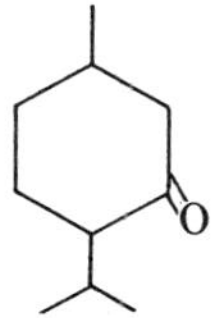	Strongly minty, camphoraceous, herbaceous
Irone and Ionones α-Ionone	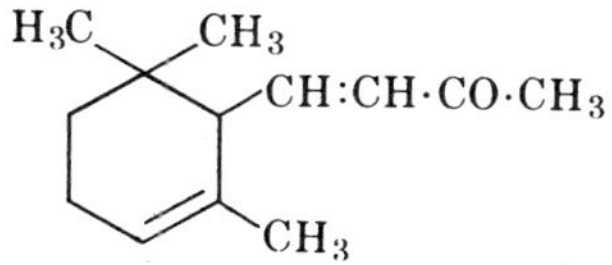 H_3C, CH_3, $CH{:}CH{\cdot}CO{\cdot}CH_3$, CH_3	Strongly and typically violet-like, sweet, warmly woody

(Continued)

TABLE 7.14 (Continued)

Ketone	Synonym	Formula	Sensory Attributes
β-Ionone		H_3C, CH_3, $CH{:}CH{\cdot}CO{\cdot}CH_3$, CH_3	Strongly violet-like, woody, green, less sweet than α-ionone
α-Irone		H_3C, CH_3, H_3C, $CH{:}CH{\cdot}CO{\cdot}CH_3$, CH_3	Characteristically violet-like, orris-like
α-Isomethylionone ("gamma methylionone")		H_3C, CH_3, $CH{:}C(CH_3){\cdot}CO{\cdot}CH_3$, CH_3	Sweetly violet-like, orris-like, woody (Profile depends on the commercial grade examined)

β-Isomethylionone	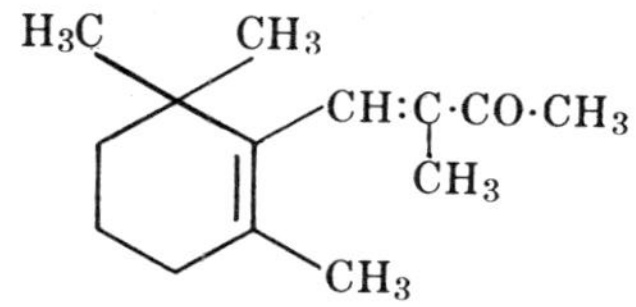 H_3C, CH_3, $CH:C(CH_3)\cdot CO\cdot CH_3$, CH_3	Warm, woody, lightly floral (Profile depends on the commercial grade examined)
α-Methylionone	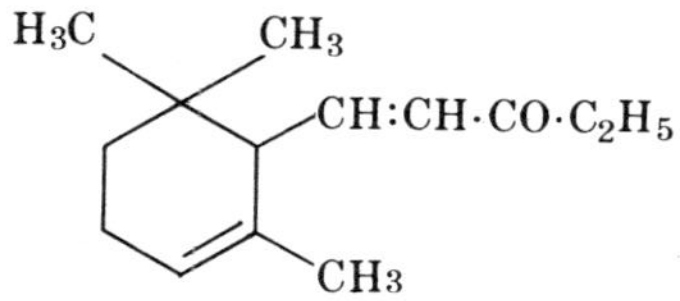 H_3C, CH_3, $CH:CH\cdot CO\cdot C_2H_5$, CH_3	Sweetly floral, similar to that of β-ionone but heavier
β-Methylionone	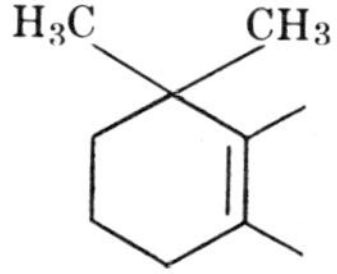 H_3C, CH_3	Warmly woody, leathery, similar to that of β-ionone but heavier.

TERPENOID HYDROCARBONS

The terpenes form a very distinctive group of organics which are of considerable importance in the chemistry of the essential oils and, hence, in flavor compounding. They are basically open-chain hydrocarbons of the general formula C_nH_{2n-4} where n may be 2, 3, or 4. They may be saturated or unsaturated, open-chain, closed-chain or cyclic in structure and the complex relationship between structure and odor has been the subject of much speculation over the past two decades. The terpene hydrocarbons act more as diluents or carriers than as flavor contributors, although their presence certainly makes a significant difference to the perception of an aromatic blend such as occurs naturally in the essential oils.

The terpenes listed in Table 7.15 have only a limited value as a source of flavor.

NITROGEN-CONTAINING COMPOUNDS—THE AMINES

From a flavor point of view the amines are of particular importance as they occur so widely in nature. The amines are substituted ammonia having the general formula:

NH_3	$R{\bullet}NH_2$	$RR'{\bullet}NH$	$RR'R''{\bullet}N$
ammonia	primary amine	secondary amine	tertiary amine

All amines of low molecular weight are powerfully aromatic but on dilution they have a decidedly ammoniacal odor which increases in "fishy" character in the secondary and teriary amines. These often-obnoxious odors are lost as the molecular weight increases. In the aromatic series, the presence of the amino group in the side chain results in a distinctive ammoniacal character but, when the substitution is directly on the nucleus, the odor is very feeble (e.g., anthranilic acid is odorless but the anthranilates all have strong orange-flower profiles).

Only trimethylamine is of limited use in flavorings. This compound is strongly ammoniacal and has an odor of herring brine.

SULFUR-CONTAINING COMPOUNDS

Most of the sulfur-containing compounds have strongly objectionable odors and many are associated with a marked bitterness. Some, however, are extremely sweet (e.g., saccharin).

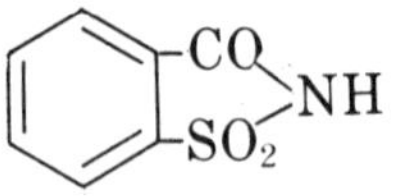

CO
NH
SO_2

saccharin

TABLE 7.15
TERPENE HYDROCARBONS USED IN FLAVORINGS

Terpene	Formula	Sensory Attributes
Monoterpenes		
Limonene (d-), (l-) and (dl-) (dipentene)		Sweetly fresh, clean, lemon-like
Terpinolene		Sweet, pine-like with citrus back-notes
Phellandrene (α-) and (β-)	α- β-	Freshly citrus-like with warm peppery notes
Carene (Δ^3-) and (Δ^4-)	Δ^4	Pleasant but nondescript
Pinene (α-) and (β-)	α- β-	Turpentine-like, dry woody, resinous
Sabinene		Woody herbaceous, warmly spicy

(Continued)

TABLE 7.15 (Continued)

Terpene	Formula	Sensory Attributes
Camphene (d-), (l-) and (dl-)	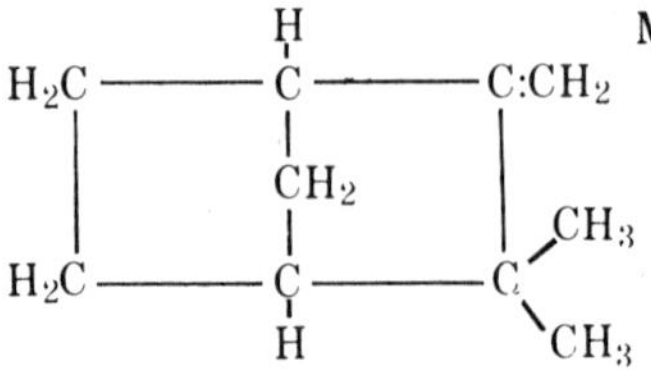 H H_2C —— C —— $C{:}CH_2$ CH$_2$ H_2C —— C —— C (CH_3, CH_3) H	Mildly camphoraceous
Sesquiterpenes Zingiberene		Warmly spicy, woody
Caryophyllene (α-) and (β-)	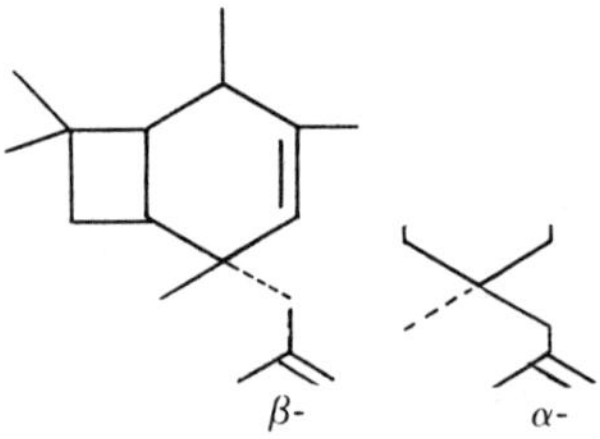 β- α-	

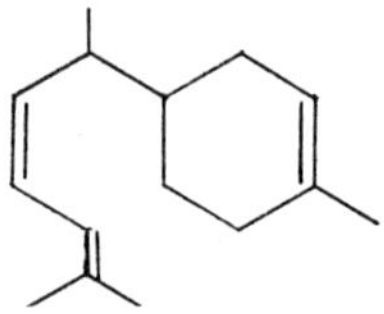

The sulfides play an important role in natural flavor chemistry as they often are responsible for the objectionable odors associated with rotting vegetation. Sulfides, disulfides and polysulfides are responsible for the characteristic odors of onion and garlic, although many of the subtle notes in the profiles of these vegetables are due to substituted thiols. Sulfur compounds are also present in roasted coffee and cocoa volatiles.

Sulfur and oxygen are chemically related and form similar compounds; for example

R•OH	R•SH	CH_2—CH_2 (O)	CH_3•S•CH_3
alcohols	mercaptans or alkane thiols	ethylene oxide	dimethyl sulfide

The -SH group is called mercapto-, thiol-, or sulfhydryl-. The compounds of most importance to the flavor chemist are given in Table 7.16. The number of such compounds included in the positive list of

TABLE 7.16
ORGANIC SULFUR COMPOUNDS

Compound	Formula	Sensory Attributes
Sulfides		
Dimethyl sulfide (methyl sulfide)	$(CH_3)_2 \cdot S$	Strongly objectionable, pungent, cabbage-like
Dimethyl disulfide	$CH_3 \cdot S \cdot S \cdot CH_3$	Strongly onion-like
Dispropyl disulfide (propyl disulfide)	$(CH_3CH_2CH_2)_2 \cdot S_2$	Very powerful, onion/garlic-like
Diallyl sulfide (allyl sulfide)	$(CH_2{:}CHCH_2)_2 \cdot S$	Powerful, very penetrating, garlic/radish-like
Diallyl disulfide	$(CH_2{:}CHCH_2)_2 \cdot S_2$	Objectionable, pungent, garlic-like
Allyl propyl disulfide	$CH_2{:}CHCH_2 \cdot S \cdot S \cdot C_3H_7$	Strongly garlic-like
Dibutyl sulfide	$(C_4H_9)_2 \cdot S$	Strongly repulsive, green, vegetable-like
Dibutyl disulfide	$(C_4H_9)_2 \cdot S_2$	Repulsive, strongly garlic-like
Dibenzyl disulfide	$C_6H_5CH_2 \cdot S \cdot S \cdot CH_2C_6H_5$	Strongly of burnt sugar, toast-like
Mercaptans		
Methyl mercaptan	CH_3SH	Objectionable, rotten-cabbage-like
n-Propyl mercaptan	C_3H_7SH	Strongly penetrating, sulfuraceous, cabbage-like
Allyl mercaptan (allyl thiol)	$CH_2{:}CHCH_2SH$	Penetrating, onion-like
n-Butyl mercaptan	C_4H_9SH	Powerful, sulfuraceous, cabbage-like
Amyl mercaptan	$C_5H_{11}SH$	Very penetrating, unpleasant, sulfuraceous, garlic-like
Benzyl mercaptan (benzyl thiol)	$C_6H_5CH_2SH$	Green, sulfuraceous, onion/leek-like
Isothiocyanates		
Allyl isothiocyanate (mustard oil, synthetic)	$CH_2{:}CHCH_2N{:}C{:}S$	Powerfully pungent, lachrymatory, characteristic of mustard seed.

flavoring materials permitted for use in the United States is strictly limited; many more have been reported as found in nature (Van Straten and deVrijer 1975).

THIOCYANTES

The isothiocyantes, containing the –N:C:S radical, are found in several pungent vegetables. The following are of interest:

Allyl isothiocyanate—produced by an enzymic reaction in black mustard seed. The synthetic product is often sold under the name "artificial mustard oil."

Acrinyl isothiocyanate—a nonvolatile compound produced enzymatically from white mustard seed.

Benzyl isothiocyanate—a component of garden cress.

Phenylethyl isothiocyanate
Phenyl propyl isothiocyanate } —responsible for the characteristic pungency of horseradish.

Only allyl isothiocyanate is included in the United States Federal Food and Drug regulations under Section 121.1164 as safe for use in foods.

PUNGENT COMPOUNDS

The chemicals so far considered have been included in flavorings because of their contribution to the odor element of the flavor profile. Several substances have a specific effect upon the chemoreceptors in the mouth giving rise to a sensation called pungency. Most of these are associated with the pungent spices and are incorporated into flavorings and seasonings in the form of the natural oleoresin, the strength of which is often stated in terms of the main pungent component.

The principal pungent chemicals are listed in Table 7.17. The relationship between pungency and molecular structure is discussed by Moncrieff (1967).

The above review of organics, which are widely used in the formulation of imitation flavors, must, of necessity, be limited. For the flavorist and expert in this specialist field, recourse must be made to the standard works of reference, given in the bibliography, in which all the permitted flavoring chemicals are listed and fully characterized. This section is aimed at those who require a basic knowledge of flavorings and flavor materials in order better to appreciate the creation and modification of flavorings for their particular purpose. The various tables are not exhaustive but do include those organics of most value in flavor work. The sensory attributes quoted are generally those observed in dilute solution and are intended as a guide to the trainee flavorist or any new-comer into this aspect of food technology. Inevitably, it is necessary for those involved in the development or modification of flavorings to acquire a first-hand knowledge of the odor and flavor profiles of those materials which are permitted for use in foodstuffs, to institute some system of re-

TABLE 7.17
ORGANIC CHEMICALS RESPONSIBLE FOR PUNGENCY

Compound		Present in
Capsaicin	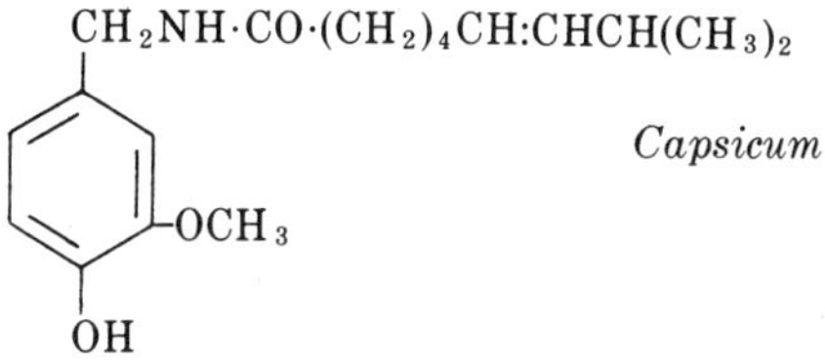 $CH_2NH{\cdot}CO{\cdot}(CH_2)_4CH{:}CHCH(CH_3)_2$ $-OCH_3$ OH 8-methyl-N-vanillyl-6-nonenamide	*Capsicum* sps
"Gingerol" (a mixture of homologous substances)	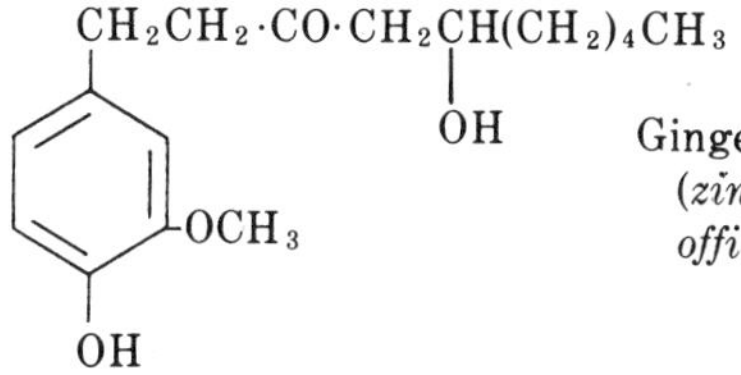 $CH_2CH_2{\cdot}CO{\cdot}CH_2CH(CH_2)_4CH_3$ OH $-OCH_3$ OH	Ginger (*zingiber officinale*, Roscoe)
Piperine	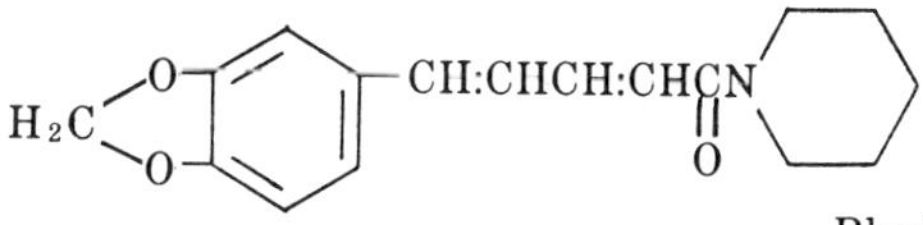 H_2C O O $-CH{:}CHCH{:}CHCN$ O 1-piperoyl piperidine	Black pepper (*piper nigrum*, L.)
Shogaol (part of *"gingerol")*	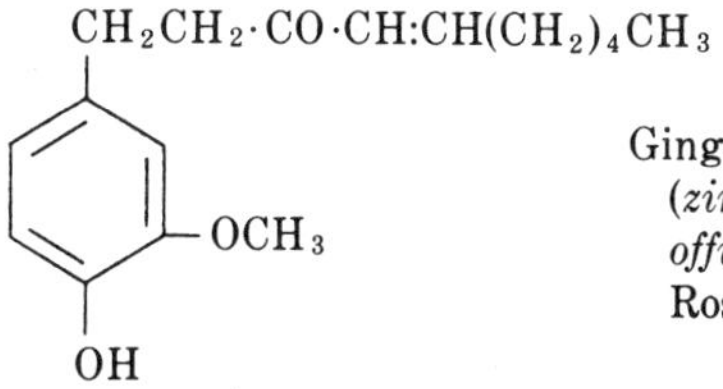 $CH_2CH_2{\cdot}CO{\cdot}CH{:}CH(CH_2)_4CH_3$ $-OCH_3$ OH 1-(4-hydroxy-3-methoxyphenyl)-4-decen-3-one	Ginger (*zingiber officinale*, Roscoe)
Zingerone	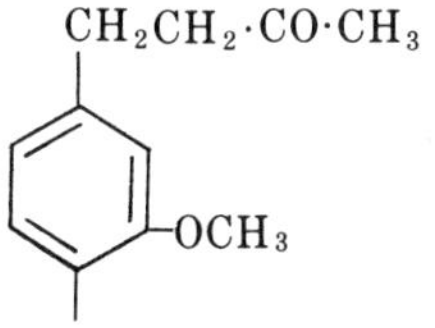 $CH_2CH_2{\cdot}CO{\cdot}CH_3$ $-OCH_3$ 4-(3-methoxy-4-hydroxyphenyl)-2-butanone	Allegedly in ginger, but this is doubted (Arctander 1969).

cording sensory impressions and to train a facility for mentally recalling these as the occasion demands. It is for this reason, that the various tables have been drawn up by chemical function rather than alphabetically as such relationships as do exist are then more evident and in consequence more likely to be remembered.

Much research has been directed to the problems of the relationship between chemical constitution and organoleptic attributes. Some degree of correlation exists between molecular weight, size and spatial configuration and perceived odor quality, but the relationship is somewhat tenuous and insufficiently precise to enable experts to formulate a really satisfactory theory of olfaction which will answer all cases or to predict the odor profile of any compound from a knowledge of its structure in other than the broadest terms. The problems of describing any odor or flavor effects are also the subject of much discussion and although considerable advances have been made by Harper *et al.* (1968), as yet no universally accepted vocabulary has been established.

So far, attention has been drawn to the nature and scope of the many natural and synthetic raw materials which are available to the flavorist and flavor technologist. The compounding of these into acceptable flavorings is both an art and a science, involving a precise knowledge of the attributes of the raw materials, how they will blend together, how stable they are likely to be under adverse manufacturing or storage conditions and whether or not they will provide the consumer with just the right flavor profile which will help to ensure maximum sales of the end-product. The nature of the flavoring will be determined to a large extent by that of the end-product. If this is liquid, then a liquid flavoring is likely to be most acceptable; if dry, then an equivalent powder flavor is called for. Processing conditions such as high temperatures, automated handling, the incorporation of air, etc., all have to be carefully considered in the selection of the most appropriate flavoring to be used. Three main types are generally available: (a) liquid flavorings, (b) flavoring emulsions, (c) dry powder flavorings.

LIQUID FLAVORINGS

Without doubt, flavorings in the form of a concentrated solution of natural and/or synthetic materials in a solvent permitted for use in foods, beverages, etc., are the most widely used throughout the various branches of these industries. To have a liquid flavoring which is uniform from batch to batch, is stable on storage prior to use, easy to handle on the factory floor, compatible with the other ingredients in the end-product and of a strength which will allow accurate measurement and ready incorporation directly into the product mix, all are of great convenience to most food processors.

The compounding of liquid flavorings is not particularly difficult but does require minute attention to the quality of the raw materials used and their accurate measurement into the bulk mix. A compounded flavor may consist of (a) natural flavoring materials: extracts, oleoresins, essential oils, fruit concentrates, distillates and isolates; (b) permitted or acceptable synthetic flavoring organics: nature-identical or not found in nature; (c) a solvent or carrier.

Each component has its part to play, whether this be as the prime contributor of the aromatic character, a product which intensifies flavor, a modifier to give the blended flavor a specific profile, a stabilizer or a solvent to enable the various solid and liquid constituents to be mixed into a uniform product having the desired flavoring profile and strength.

FLAVOR MANUFACTURE

The manufacture of bulk flavorings for use by the food and related industries is one for the expert. Many flavorists and those concerned with the use of flavorings in their end-products have little or no direct knowledge of what is involved in, say, the production of 500 gal. of a soluble orange flavor or 1000 litres of an imitation 10-fold vanilla flavor. The compounding floor of a modern manufacturing concern is fully equipped as befits the techniques involved. The plant used is generally fabricated in stainless-steel and, like the food processor which the industry serves, considerable attention is paid to hygiene and the elimination of contamination between one product and another.

The techniques and unit operations involved in flavor compounding are not particularly difficult or involved but, for success, they do require the following of certain procedures peculiar to the industry and demand considerable skill, patience and understanding on the part of the production team. The following aspects of production are of particular importance:

(a) Although ingredients may be either solid or liquid, compounding should be carried out by weighing, preferably in metric units only; formulations are designed to be in 1000 units by weight.

(b) There is always the temptation for the creative flavorist to take a short-cut and use an already-blended flavor to give to a new product a nuance which is already available in the earlier product. This system is certainly time-saving and convenient in the laboratory but in practice is fraught with difficulties, often causing considerable delays in the manufacturing schedule and an unnecessary strain on the costing system. The use of such practice is strongly to be discouraged.

(c) In the laboratory it is a common practice to use pre-made dilutions of many of the common aromatic components to enable minute quantities to be added with accuracy during the development of a flavor. In the final manufacturing formulation, the amounts derived by this process should have been translated back in terms of the active ingredient.

(d) The process techniques employed should be kept as simple and direct as possible involving the minimum of steps or stages at which mistakes can occur.

(e) In order to satisfy the high standards of hygiene required by the regulating authorities, all plant and utensils must be kept scrupulously clean; fully-trained operators should exercise the maximum precautions to prevent contamination from whatever source.

(f) A representative sample should be drawn and retained for an adequate period of time for re-examination in the event of any complaint. This should be in addition to any sample drawn for routine quality assurance.

For those not directly engaged in this work there is little to be gained by a detailed description of various techniques. Those so involved already know the paradoxical simplicity of the many individual methods used and the obvious complexity of the total operation in which there are so many products to be handled.

FLAVORING FORMULATIONS

It is obviously not practicable to discuss the detailed formulations of every type of imitation flavoring produced; in any case, the majority are closely guarded secrets of the flavor manufacturer concerned. Several authors (Merory 1968; Furia and Bellanca 1975) have already published a range of basic formulations which may be taken as a guide to those organics most likely to be present in any given flavoring. However, it should be appreciated that there is considerable artistry in the construction of imitation flavors and that the ultimate formulation is likely to bear little resemblance to any published list of components. For the creative flavorist such a formulation is of only marginal value. His training will already have given him a knowledge of what chemicals are most likely to be found in any particular flavor as well as the sensory attributes of chemicals which will provide any additional nuances required. For those not directly concerned with flavor development, the exact formulation can be of little interest; it is far more important to know how the finished flavoring stands up on processing, what sort of shelf-life it will have under given conditions, what effect it will give in the end-product and, if it is not satisfactory, what sort of modification is likely to be possible to meet a specific flavoring need. Once one is assured that the ingredients are safe in use and are those permitted by the current local legislation then a knowledge of the type of components likely to be present enables the technologist to discuss the subject with the experts with some degree of understanding of the problems involved. In flavoring matters, it is good to remember that sometimes a little knowledge can be a dangerous thing.

Table 7.18 sets out the most popular imitation flavorings and an indication of their major components. The proportions of these have not been given, as wide variations are likely to be encountered, depending on the exact profile achieved, the end-usage and the strength of the finished flavoring. Many flavor manufacturers include an additional code in their product title so as to identify the quality, flavor type and specific application of their flavorings. Such codes are an integral part of the title and should always be quoted when referring to a product so as to avoid misunderstanding. In addition to what may be called the "prime" imitation flavorings, most of the manufacturing houses produce flavorings designed to give specific effects in end-products (e.g., flavor for "Petit beurre" biscuits, American ice cream soda, "cola," "barbeque" chips, etc.). The range of available flavors and nuances of profile are almost limitless and are only restricted by what the consumer will accept.

SOLVENTS USED IN FLAVORINGS

In order to produce a uniform liquid flavoring, it is frequently necessary to use a solvent. In an earlier section, solvents were considered from the point of view of extraction and as such were completely removed from the final product. The solvents used in flavorings remain in the finished product and give rise to quite a different set of considerations. The solvents which are permitted for use for this purpose differ in the various countries and it is necessary to make reference to specific legislation to ensure compliance. The most important solvents used in flavorings are listed in Table 7.19.

The components used in flavorings may vary from crystalline solids, through oily or paste-like extracts to oily liquids; and it is the solvent which enables these to be combined into a homogeneous mixture by acting as a diluent and reducing the very strong aromatic effects to a level at which they can be incorporated directly into a food product or beverage. The following are the most important solvents available for this purpose.

TABLE 7.18
PRINCIPAL SYNTHETIC ORGANICS USED IN IMITATION FLAVORINGS

Imitation Fruit Flavors

Apple	
Acetaldehyde	Citronellyl iso-valerate
Allyl cyclohexane valerate	Decyl acetate
Allyl iso-valerate	Ethyl acetate
Allyl propionate	Ethyl oenanthate
Iso-amyl valerate	Ethyl valerate
Iso-amyl iso-valerate	Ethyl iso-valerate
Benzaldehyde	Geranyl iso-valerate
Benzyl iso-valerate	*cis* 3-Hexenyl-2-methyl butyrate
Butyl valerate	Methyl butyrate
Cinnamyl formate	Iso-propyl acetate
Cinnamyl iso-butyrate	Iso-propyl iso-valerate
Cinnamyl iso-valerate	Propyl iso-valerate
Apricot	
Allyl cinnamate	Dimethyl benzyl iso-butyrate
Allyl propionate	Ethyl acetate
Benzyl formate	Ethyl butyrate
Benzyl butyrate	Ethyl valerate
Benzyl propionate	Geranyl butyrate
Benzyl butyrate	Heptyl acetate
Butyl 2-decanoate	Heptyl propionate
Butyl propionate	2-Hexyl-4-acetotetrahydrofuran
Cinnamic acid	Methyl iso-butyrate
Citronellyl acetate	γ-Octalactone
Citronellyl butyrate	Phenyl ethyl dimethyl carbinol
γ-Deltalactone	Propyl butyrate
Decyl butyrate	
Banana	
Allyl heptylate	Ethyl heptanoate
Amyl heptanoate	Geranyl hexanoate
Amyl hexanoate	2-Heptanone
Bornyl acetate	Methyl butyrate
Cinnamyl iso-butyrate	Methyl 2-nonynoate
Cyclohexyl iso-valerate	Propyl butyrate
Dimethyl benzyl iso-butyrate	
Blackberry (Bramble)	
Anisaldehyde	Methyl anthranilate
Diphenyl ether	Methyl ionone
Ethyl eonanthate	Neryl iso-valerate
Ethyl vanillin	Propyl hexanoate
Heliotropin	Rhodinyl butyrate
Black Currant	
Iso-amyl formate	Linalyl propionate
Linalyl iso-butyrate	Methyl propionate

TABLE 7.18 (Continued)

Blueberry	
Iso-propyl hexanoate	Octyl propionate
Methyl ionone	Rhodinyl butyrate
Cherry	
Allyl benzoate	Ethyl butyrate
Allyl iso-valerate	Ethyl eonanthate
Anisyl acetate	Methyl benzyl propionate
Anisyl propionate	4-(*p*-Methoxyphenyl)-2-butanone
Benzaldehyde	Rhodinyl formate
Benzyl acetate	Rhodinyl iso-valerate
Cinnamaldehyde	Tolualdehyde
Cyclohexyl cinnamate	Vanillin
Cyclohexyl formate	
Grape	
Benzyl propionate	Ethyl caproate
Cinnamyl anthranilate	Methyl anthranilate
Cyclohexyl anthranilate	Octyl iso-butyrate
Dimethyl anthranilate	Propyl heptanoate
Melon	
Amyl butyrate	3-Heptanone
Amyl valerate	Methyl anisate
Benzyl benzoate	Methyl anthranilate
Benzyl cinnamate	Methyl o-methoxybenzoate
Cinnamic aldehyde	Octyl butyrate
2,6 Dimethyl-5-heptanal	Phenyl acetaldehyde
Ethyl formate	Vanillin
Ethyl valerate	
Peach	
Allyl cinnamate	Ethyl cinnamate
Allyl cyclohexane acetate	Ethyl valerate
Amyl acetate	Geraniol
Amyl butyrate	γ-Octalactone
Amyl formate	Methyl nonyl ketone
Amyl valerate	Musk ambrette
Anisyl propionate	Octyl acetate
Anisyl valerate	Phenyl ethyl alcohol
Benzaldehyde	Phenyl ethyl salicylate
Benzyl propionate	Phenyl ethyl iso-valerate
Citronellol	Rhodinyl iso-butyrate
γ-Decalactone	Vanillin
γ-Dodecalactone	
Pear	
Acetaldehyde	Dimethyl phenyl ethyl acetate
Amyl methyl ketone	Heptyl acetate
Iso-amyl acetate	Hexyl acetate

(Continued)

TABLE 7.18 (Continued)

Iso-amyl butyrate	6-methyl-5-hepten-2-one
Benzyl butyrate	Propyl acetate
Butyl butyrate	Iso-propyl benzyl carbinol
Pineapple	
Acetaldehyde	Cinnamyl acetate
Allyl cyclohexane acetate	Citral
Allyl cyclohexane butyrate	Decyl acetate
Allyl cyclohexane propionate	Ethyl butyrate
Allyl hexanoate	Ethyl cyclohexane propionate
Allyl α-ionone	Ethyl heptanoate
Allyl nonanoate	Ethyl propionate
Allyl phenoxyacetate	Ethyl iso-valerate
Amyl acetate	Hexyl iso-butyrate
Amyl butyrate	Iso-butyl iso-butyrate
Amyl hexanoate	Linalyl formate
Benzyl ethyl ether	Methyl haxanoate
Benzyl formate	Methyl 4-methyl valerate
Benzyl iso-valerate	Propyl iso-butyrate
Butyl hexanoate	Vanillin
Butyl iso-valerate	
Plum	
Iso-amyl formate	Guaiyl butyrate
Iso-amyl propionate	Heptyl butyrate
Anisyl butyrate	Heptyl formate
Bornyl acetate	Hexyl formate
Butyl anthranilate	Linalool
Butyl formate	Neryl propionate
Citronellyl butyrate	Phenyl ethyl cinnamate
Citronellyl formate	Phenyl propyl butyrate
Citronellyl propionate	Propyl formate
Dimethyl benzyl iso-butyrate	Iso-propyl formate
γ-Dodecalactone	Terpinyl butyrate
α-ethyl benzyl butyrate	
Raspberry	
Acetaldehyde	Ethyl valerate
Amyl cinnamate	*cis* 3-Hexen-1-ol
Benzyl salicylate	α-Ionone
Butyl valerate	Neryl acetate
Iso-butyl cinnamate	Phenyl ethyl acetate
Citral	Propyl acetate
Dimethyl anthranilate	Terpinyl acetate
Dimethyl sulfide	Vanillin
Ethyl acetate	
Red Currant	
Benzyl salicylate	Iso-butyl cinnamate
Guaiyl acetate	

TABLE 7.18 (Continued)

Strawberry	
Acetic acid	Ethyl valerate
Acetaldehyde	Ethyl 3-phenylglycidate
Amyl butyrate	Ethyl vanillin
Amyl formate	2-Hexen-1-ol
Iso-amyl salicylate	Hexyl 2-methyl butyrate
Anisyl formate	α-Ionone
Benzyl acetate	Methyl anthranilate
Benzyl iso-butyrate	Methyl amyl ketone
Benzyl salicylate	Methyl cinnamate
Cinnamyl iso-butyrate	4-Methyl acetophenone
Cinnamyl valerate	Methyl α-napthyl ketone
δ-Deltalactone	3-Methyl-3-phenyl glycidic acid ethyl ester (aldehyde C_{16})
Diacetyl	Methyl salicylate
Dipropyl ketone	β-Naphthyl methyl ether
Ethyl acetate	Phenyl ethyl benzoate
Ethyl butyrate	Neryl iso-butyrate
Ethyl benzoate	Vanillin
Ethyl lactate	
Imitation Citrus Fruit Flavors	
Grapefruit	
Ethyl butyrate	Linalool
Hexanal	Methyl mercapto pentanone
cis 3-Hexenol	Nootkatone
Lemon	
Amyl acetate	Geranyl acetate
Cinnamyl butyrate	Linalool
Citral	Limonene
Citral diethyl acetal	Methyl heptanone
Citronellal	Terpineol
Ethyl acetate	
Lime	
Amyl acetate	Ethyl acetate
Amyl butyrate	*cis* 3-Hexenol
p-Cymene	Terpineol
Dipentene	
Mandarin (Tangerine)	
Amyl butyrate	γ-Dodecalactone
Ethyl acetate	Methyl anthranilate
Ethyl anthranilate	Thymol
Bitter Orange	
Acetaldehyde	α-Ionone
Amyl acetate	Linalool
Amyl butyrate	Linalyl acetate
Ethyl acetate	*p*-Cymene
Dimethyl anthranilate	

(Continued)

TABLE 7.18 (Continued)

Sweet Orange	
Amyl acetate	2-Duodecanone
Amyl butyrate	Ethyl anthranilate
Cinnamyl butyrate	2-Hexen-1-ol
Cyclohexyl anthranilate	Linalyl anthranilate
Decanal	
Imitation Nut Flavors	
Almond	
Benzaldehyde	1,3 Diphenyl-1,2-propanone
Benzyl alcohol	Ethyl benzoate
Iso-amyl acetate	Ethyl formate
Cyclohexyl cinnamate	Vanillin
Coconut	
Amyl formate	Ethyl undecanoate
Benzaldehyde	Ethyl vanillin
Butyric acid	Methyl undecyl ketone
Caproic acid	Methyl nonanoate
4,4 Dibutyl-γ-butyrolactone	γ-Nonalactone
Ethyl acetate	γ-Octalactone
Ethyl heptylate	Vanillin
Hazelnut	
Cinnamic aldehyde	Maltol
Furfural	Vanillin
γ-Deltalactone	
γ-Octalactone	Various mercaptans
Peanut	
Butyl laurate	Vanillin
γ-Deltalactone	
γ-Octalactone	Various pyrazines
Walnut	
γ-Deltalactone	o-Methyl anisole
γ-Octalactone	Dimethyl resorcinol
Heptyl dimethyl acetal	Maltol
Imitation Dairy Flavors	
Butter	
Acetoin	γ-Dodecalactone
Benzilidene acetone	γ-Decalactone
Butyric acid	Ethyl butyrate
Butyl acetate	Ethyl formate
Butyl butyrate	Ethyl heptylate
Butyl iso-undecanoate	2,3 Hexane-dione
Iso-butyl acetate	Lactic acid
Caproic acid	1-(*p*-Methoxyphenyl), 1-penten-3-one
Caprylic acid	Triacetin
Capric acid	Valeric acid
Cinnamic aldehyde	Vanillin
Cinnamyl butyrate	

TABLE 7.18 (Continued)

Cream	
Similar to butter	
Cheese	
Butyl butyryl lactate	Lactic acid
Butyric acid	Methyl n-amyl ketone
Caproic acid	Iso-valeric acid
Ethyl butyrate	
Imitation Beverage Flavors	
Coffee	
Iso-amyl acetate	n-Heptyl mercaptan
Benzyl benzoate	n-Hexyl mercaptan
Citral	2-Methyl butyraldehyde
Diacetyl	Propyl heptylate
Ethyl acetate	Tetrahydrofururyl alcohol
Ethyl formate	Tolualdehyde
Ethyl nonanoate	
2-Ethyl furan	
Creme de Menthe	
Menthol	Menthyl acetate
Menthone	Menthyl iso-valerate
Cognac (Brandy)	
Acetaldehyde	Decanal dimethyl acetal
Isobutyl acetoacetate	Decyl acetate
Isobutyl butyrate	Decyl propionate
Isobutyl formate	Ethyl decanoate
Isobutyl 2-furan propionate	Ethyl heptanoate
Ethyl acetate	Ethyl propionate
Ethyl butyrate	Ethyl nonanoate
Ethyl formate	Hexanol
Ethyl eonanthate	Octanal dimethyl acetal
Ethyl pelargonate	Phenyl ethyl alcohol
Rum	
Acetic acid	Ethyl formate
Butyric acid	Ethyl propionate
Iso-butyl butyrate	Ethyl vanillin
Iso-butyl formate	Heliotropin
Ethyl acetate	Propyl formate
Ethyl butyrate	Rum ether (ethyl oxyhydrate)
Ethyl eonanthate	Vanillin
Whiskey	
Acetaldehyde	Ethyl benzoylacetate
Imitation Flavors for Specific Purposes	
Butterscotch	
Amyl acetate	Ethyl butyrate
Iso-amyl acetate	Ethyl acetate
Butyl acetate	Ethyl heptanoate

(Continued)

TABLE 7.18 (Continued)

Butyl butyrate	Ethyl nonanoate
Butyric acid	Ethyl vanillin
Capraldehyde	Valeric acid
Cinnamic aldehyde	Vanillin
Chocolate	
Amyl butyrate	Eugenol
Amyl phenyl acetate	Cinnamic aldehyde
Benzyl alcohol	Methyl (p-tert. butylphenyl) acetate
Benzyl butyrate	2-Methyl butyraldehyde
Ethyl acetate	Vanillin
Ethyl heptylate	Veratraldehyde
Ethyl nonanoate	
Honey	
Amyl acetate	Ethyl acetate
Amyl valerate	Ethyl cinnamate
Allyl phenoxyacetate	Ethyl hexanoate
Allyl phenyl acetate	Ethyl phenyl acetate
Anisyl phenyl acetate	Ethyl myristate
Iso-amyl acetate	Ethyl tolyl ether
Iso-amyl alcohol	Eugenyl iso-amyl ether
Benzyl cinnamate	Geranyl phenyl acetate
Benzyl iso-valerate	Methyl acetophenone
Benzyl phenyl acetate	Methyl phenyl acetate
Caprylic acid	Phenoxyacetic acid
Citronellol	Phenyl acetic acid
Citronellyl formate	Phenyl ethyl butyrate
Citronellyl butyrate	Phenyl ethyl propionate
Citronellyl iso-valerate	Iso-propyl acetophenone
Cinnamyl butyrate	Terpinyl acetate
Cinnamyl phenyl acetate	Terpinyl formate
Decyl acetate	
Peppermint	
2,4 Dimethyl acetophenone	Menthyl acetate
Menthol	Pulegone
Menthone	2-(*p*-Tolyl)-propionaldehyde
Rose	
Benzophenone	Citronellyl oxyacetaldehyde
Butyl phenyl acetate	Geraniol
Cinnamyl iso-valerate	Geranyl acetate
Citronellol	Geranyl formate
Citronellal	Nerol
Spearmint	
1-Carvone	1-Carvyl propionate
Carvyl acetate	Dihydrocarvyl acetate

TABLE 7.18 (Continued)

Vanilla	
Ethyl vanillin	Rum ether (ethyl oxyhydrate)
Heliotropin	Vanillin
3-Phenylpropionic acid	Veratraldehyde
Propenylguaethol	
Violet	
Heliotropin	α-Iso-methyl ionone
α-Ionone	Methyl heptine carbonate
β-Ionone	Methyl 2-nonanoate
α-Irone	

TABLE 7.19
SOLVENTS USED IN FOOD FLAVORINGS

Solvent	Boiling Point at 760 mm (°C)	Sp. Gr. 20/20°C	Flash Point	
			Open	Closed
Ethyl alcohol	78.4	0.7905	18°C (65°F)	14°C (57°F)
Isopropyl alcohol	82.3	0.7862	21°C (70°F)	13°C (56°F)
Ethyl acetate	77.1	0.899	−1°C (30°F)	−4°C (24°F)
Diethyl ether	34.4	0.714	−40°C (−40°F)	−31°C (−24°F)
Glycerol (glycerin)	290	1.260	177°C (350°F)	160°C (320°F)
Glyceryl monacetate (monacetin)	158	1.221		
Glyceryl diacetate (diacetin)	280	1.178		
Glyceryl triacetate	260	1.159	138°C (280°F)	146°C (295°F)
Propylene glycol	185	1.038	107°C (225°F)	99°C (210°F)

Source: Scheflan and Jacobs (1953).

Water

Water is certainly the cheapest of the solvents but is of limited applicability as so many aromatic chemicals are not soluble in it. It is of particular interest and value in the preparation of nonalcoholic flavoring emulsions, forming, as it does, the continuous phase. But even in such products, the aqueous medium is often a disadvantage as many of the essential oils tend to be unstable in the presence of water and readily acquire off-odors and off-flavors in such products.

Industrial water is an extremely variable commodity, the flavor of which can be significant due not only to the presence of dissolved organic matter and minerals, particularly sulfur-containing compounds, but also to the over-generous use of chlorine by the local water authorities. This intrinsic odor in public water supplies is quite sufficient to upset the flavor balance in a product, such as a soft drink, in which water is the major constituent. In the manufacture of flavorings, it is general practice to use distilled water made directly and not as a condensate of factory steam as this frequently carries a high level of "oily" odor. A preferred alternative is to use water which has been demineralized by passage through ion-exchange resins.

Ethyl Alcohol (Ethanol)

Ethyl alcohol is a wide-spectrum solvent which is eminently suitable for the production of food flavorings since most components are readily soluble in it. In most countries, however, ethanol attracts high revenue taxes which make products containing it relatively expensive. It also may involve the maintenance of accurate records of receipts and usage to enable the manufacturer to reclaim prepaid duties, as appropriate, or the running of a separate production unit under controlled bond conditions.

Ethanol is a colorless liquid which boils at 78.7°C; it has a pleasant odor and a burning taste. When distilled with water, it forms a constant boiling mixture at 78.15°C containing 95.6% ethanol. This is the strength of commercial alcohol which has to be further processed to make absolute alcohol. When the term "alcohol" is used in a formulation or process specification, in the absence of any other indication, it is taken to mean ethyl alcohol.

Commercial grades of alcohol are obtained from four main sources: (a) sugars—cane sugar, beet sugar, molasses, fruits, etc.; (b) starches—corn, potato, rice, etc.; (c) gases—principally, ethylene from petroleum refining; and (d) cellulose—sawdust, sulfite liquors, etc.

Commercial production results in an ethanol of 95% strength which complies with the specifications of various national pharmacopeias. The odor of alcohol is important to the flavor manufacturer and the most common source of objectionable off-notes are traces of fusel oil and certain aldehydes.

In order to comply with regulations, it is frequently necessary to quote the alcohol content of a product. This may be in terms of percentage of ethanol either by volume or by weight but, often, is stated as "proof spirit." The meaning of this term differs depending on the precise definition given in the appropriate legislation. In the United States, for example, it is twice the volume percentage of ethanol (e.g., 50% by volume ethanol is equivalent to 100 proof), whereas in the United Kingdom it is defined as "that which at the temperature of 51 degrees Fahrenheit weighs exactly twelve-thirteenths parts of an equal volume of distilled water." This is equivalent to 57.1% ethanol by volume. This is of particular importance where alcoholic products are being sold in different countries and where the alcohol content has to be stated on the label.

Ethanol, as a solvent for flavoring substances, has some disadvantages. It has a high vapor pressure and a relatively low boiling point, both of which tend to lead to excessive losses by evaporation. This can be a significant factor when one is manufacturing large volumes of flavorings containing highly aromatic components or is using alcoholic flavorings in food products involving high temperature processing (e.g., baked goods, sugar confectionery, etc.). Due allowance must also be made for the volume reduction which occurs when alcohol is diluted with water. Reference should be made to dilution tables which give the proportions of 95% alcohol and water required to produce a diluted alcohol of any given strength.

Iso-propyl Alcohol (Iso-propanol) (I.P.A.)

Iso-propyl alcohol is a colorless liquid having a characteristic harsh, spirituous odor. It boils at 82.4°C and is miscible with water in all proportions. In some countries, its use in food flavorings is restricted. In the United States, it may be used in accordance with good manufacturing practice (Regulation 121.1164; Federal Register, Vol 31, 77, April 21, 1966). In most countries, its use is not restricted as a flavoring solvent.

Flavorings made with iso-propyl alcohol tend to have a sharp odor over and above that of the aromatic components. This makes any evaluation of such flavorings very difficult and misleading by just smelling. These over-laying top-notes are generally lost when the product is used and do not affect the end-product.

Propylene Glycol (1,2 Propane-diol)

Propylene glycol is a colorless, practically odorless, hygroscopic, viscous liquid which boils at 189°C. It is miscible with water, alcohol and glycerol, as well as with many other organic solvents. In the United States, it is classified as a GRAS substance and finds almost universal acceptance as an alternative nonalcoholic solvent in flavorings. Unfortunately, its solvent power is not nearly as good as that of ethanol.

Glycerol (Glycerin)

Glycerol is a clear, colorless liquid which is very hygroscopic, odorless and nonvolatile. It boils at 290°C, usually with some decomposition. It has a sweetly warm taste. Glycerol is miscible with water, alcohol and propylene glycol but is rather a poor solvent for most organic chemicals. It is of particular value in the preparation of vanilla extracts where the content of water-soluble components is relatively high.

Triacetin (Glyceryl Triacetate)

Triacetin is the third of the series of acetins which are acetic esters of glycerol. The acetins are particularly useful in the preparation of nonalcoholic citrus flavorings.

Triacetin is a colorless, almost odorless, viscous liquid which is nonvolatile. It is not readily miscible with water but is with alcohol. Being listed as a GRAS substance, triacetin may be used without limitation but it has a bitter taste and this may be a definite limitation to its use in flavoring compositions.

Edible Oils

Odorless and tasteless fixed edible oils, either natural, fractionated or hydrogenated, (e.g., cottonseed oil, coconut oil fractions, hydrogenated palm-kernel oil) are widely used as solvents and carriers for flavorings, particularly those for use in the baking industry in which these products are also used as shortenings. Their solvent action is limited, particularly for hydrophilic materials; but, in spite of this, their ability to reduce the loss of volatiles during high temperature processing is significant. A disadvantage to their use is that the more unsaturated oils tend to oxidize on storage and the consequent development of rancid off-notes is both rapid and continuous. To reduce the onset of rancidity, it is necessary either to maintain the products in cold storage or to incorporate an antioxident such as butylated hydroxyanisole (BHA), although the use of such additives may be regulated.

Although generally regarded as bland, many of the vegetable fixed oils do have a slight flavor which is accentuated as they age. Many commercial grades have been processed to remove this character; but odor and color still remain the two most important attributes for quality assessment.

FLAVORING EMULSIONS

There are many occasions when an alcoholic flavoring is not suitable for a particular product or is too costly; in such cases, an acceptable alternative flavoring is necessary. This is often achieved by dispersing the aromatic flavoring materials in water as an emulsion using a suitable emulsifying agent and stabilizer. These latter materials may consist of the edible gums, such as gum acacia (Arabic), tragacanth, karaya, etc.; the starches; the alginates; methyl cellulose and its derivatives; or one of the mono- or di-glycerides of the fatty acids. The resulting emulsion may then be used in the same way as a solvent-based liquid flavoring.

Flavoring emulsions have several marked advantages over other liquid flavorings—they are much less expensive; they are generally less volatile and, hence, not as susceptible to losses during processing; and they can be prepared at a higher level of concentration making for economies in handling and usage. To offset these, however, emulsions are often prone to separation or creaming unless suitably stabilized; they generally require the addition of a preservative to reduce the growth of molds and other microorganisms which may give rise to unacceptable off-flavors; the presence of an excess of water may be deleterious to some of the more reactive organics present in the flavoring; and, as they do not contain any light volatile solvent such as alcohol, their overall odor effect tends to be dull.

The techniques and plant used in the manufacture of emulsions is fully discussed in standard texts on process engineering.

POWDER FLAVORS

There are many convenience food products and beverages (e.g., cake mixes, jelly crystals, soup mixes, instant breakfast drinks, etc.) in which the use of a liquid flavoring is technologically unacceptable. For such products, it is necessary to present the flavoring component in the form of a dry, free-flowing powder. This can be achieved in two ways:

(a) by dispersion of the flavoring materials onto an edible carrier or base (e.g., the "dry-soluble" or "plated" spice extractives, vanilla sugar, celery salt, etc.);

(b) by encapsulation in an edible water-soluble carrier or encapsulant which may be natural gums, modified starches or gelatin; (e.g., spray-dried flavors, micro-encapsulated flavors, etc.).

The method employed is determined by the nature of the flavoring materials used as well as that of the end-product. Powder flavors made by dispersion have an inherent disadvantage: the internal surface area is considerably greater and the volatile components are thereby more exposed to losses by evaporation or to oxidative

Courtesy of DEC International, Inc.

FIG. 7.16. A PILOT-SCALE SPRAY DRIER USED FOR THE PRODUCTION OF TRIAL BATCHES OF POWDERED FLAVORINGS

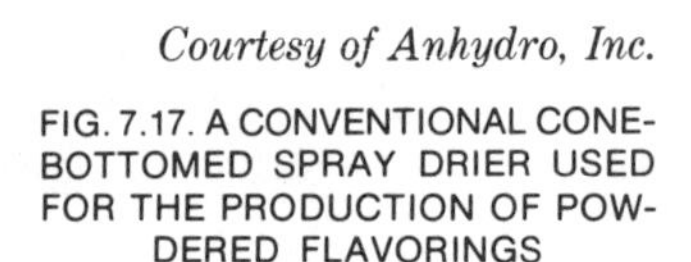

Courtesy of Anhydro, Inc.

FIG. 7.17. A CONVENTIONAL CONE-BOTTOMED SPRAY DRIER USED FOR THE PRODUCTION OF POWDERED FLAVORINGS

changes. The encapsulated products overcome this problem but have the disadvantage of being almost without impact until the capsule is ruptured, either physically or by solution in water, and its contents liberated.

DISPERSED FLAVORINGS

The easiest method of making a concentrated free-flowing powder flavor is to spread the aromatic components onto a suitable carrier or base which must be either edible or inert and permitted for

Courtesy of Anhydro, Inc.

FIG. 7.18. A FLAT-BOTTOMED SPRAY DRIER OF WIDE APPLICATION
The bottom of the drying chamber is continuously cleared by a rotating pneumatic collector.

incorporation into food products. The unit processes involved are fundamentally simple, consisting of little more than mechanical agitation to ensure intimate contact between the aromatic materials and the carrier. Mixing is continued until a uniform homogeneous mass results. After mixing the product is screened to ensure the correct particle size; to break up any agglomerates and ensure the absence of any hard lumps of undispersed flavoring materials (this is most likely to occur when dispersing sticky spice oleoresins); and to ensure absence of other extraneous matter.

The blending technique is not without problems. Almost any mixing operation involves a considerable and rapid movement of air within the system. This can lead to significant loss of volatiles unless adequate precautions are taken. In addition to differential losses by evaporation, there may be changes in the aromatic profile of the flavoring compound as a result of oxidative reactions. All flavors are altered or modified to some extent when exposed to the air; hence, the necessity to ensure that the mixing time is limited to the minimum required to produce uniformity. If mixing is unduly prolonged the temperature of the mass is likely to rise and may result in some adhesion between particles. At best, this can slow down any subsequent screening and, if bad, may make it impossible. The resulting product should be packed into well-filled and sealed containers as soon as possible after manufacture. The production of imitation powder flavors by this method is limited.

The method is extensively used for the production of so-called "dry-soluble" or "plated" spices and seasonings in which the spice oleoresin and/or essential oil is dispersed (usually at a rate which will give a product of equivalent flavoring strength to the natural spice) onto an edible carrier such as salt, dextrose, flour, MSG, etc. In this type of product, the active components of the spice extractive or essential oil tend to be protected to a large extent by the oily or resinous nonvolatile components which limits volatile losses or oxidative changes. If correctly manufactured, handled and stored prior to use, these dispersed spice products give to the food processor a uniform, standard product the advantages of which have already been discussed in an earlier chapter.

The free-flowing character of the dispersed spices will be determined by: (a) the nature and quality of the spice extracts, oleoresins or essential oils to be dispersed; (b) the quantity to be dispersed; (c) the nature and particle size of the dry carrier; (d) the method of dispersion and the time taken to ensure homogeneous mixing; and (e) the screen mesh size.

If the oleoresin is viscous, sticky or waxy (e.g., oleoresin nutmeg, extract of onion, etc.) and the carrier is coarse and not very absorbent (e.g., salt) the resulting product can end up looking like marzipan. If, however, the oleoresin is a mobile fluid (e.g., oleoresin celery seed) and the carrier is very absorbent (e.g., cracker meal or rusk) then the resulting product will be very dry and present no caking problems. One oleoresin which poses particular difficulties in dispersion is oleoresin black pepper. Depending on source, this oleoresin can vary in physical form between a thick, sticky, butter-like mass to a viscous fluid. In many cases, it is necessary to use a little heat to facilitate dispersion, although care must be taken to ensure that this is kept to a minimum; otherwise, significant losses of essential oil can result.

Where difficulty of caking is encountered the end-product can generally be made free-flowing by the addition of an anticaking agent in the final stages of the mixing cycle. Those of value for this purpose are listed in Table 7.20, although their use may be subject to legislative restrictions in certain classes of food product.

ENCAPSULATED FLAVORINGS

Several different techniques are now available for the economic production of dry encapsulated flavorings in which the essential aromatics are protected by a film of water-soluble gum, starch, gelatin or sugar. Depending on the technique employed, the resulting capsules will have a definite characteristic which determines their optimum usage. The methods involved are worthy of further consideration.

TABLE 7.20
ANTICAKING AGENTS USED IN DRY POWDER FLAVORINGS AND SEASONINGS

Carbohydrates
- Starch (edible grade)
- Lactose (milk sugar)
- Hydrolyzed cereal solids

Silica
- Silicon dioxide

Inorganic chemicals
- Calcium aluminium silicate
- Calcium phosphate, tribasic
- Calcium silicate
- Magnesium carbonate
- Magnesium oxide
- Magnesium silicate
- Magnesium stearate
- Sodium ferrocyanide, decahydrate (salt mfr, only)
- Sodium aluminium silicate

Microcrystalline cellulose
Modified food starches
Milk whey

NOTE: The above products are widely used to convey free-flowing properties to powder flavorings and blended seasonings. In many countries their use is governed by legislation and reference should be made to the appropriate regulations to determine the usage limits for specific applications.

Spray-Dried Flavors

In 1932, an English company, A. Boake, Roberts & Co. Ltd., (now an integral part of Bush Boake Allen Ltd.,) produced the first spray-dried flavor powders in which the aromatics were encapsulated and "locked in" by a thin film of gum acacia (Arabic). This was a great advance in flavor technology and today the majority of powder flavors used throughout the food and beverage industries are made by this technique.

The basic principle of spray-drying is the production of a dry powder by spraying an emulsion of the aromatic components into a stream of heated air or inert gas in a suitably designed plant. The water present in the emulsion is flash evaporated from the fine droplets leaving the flavoring materials encased in a shell of the encapsulant. The dried particles are carried away in the stream of air and separated in a cyclone collecting chamber. The great advantage of this method is that the rate of evaporation can be so adjusted that delicate, heat-sensitive substances may be dried satisfactorily. The same principle is used to dry dissolved solids (e.g., coffee extract and milk) where excessive heat would produce unacceptable changes in the end-product.

Some volatiles are inevitably lost during spray-drying. By trial and error, allowances are made in the formulation of the flavor concentrate used to prepare the emulsion so that the resulting dry powder has the desired flavor profile.

The plant used for this process is described in standard texts on process engineering.

The advantages of using spray-dried encapsulated flavorings are:

(a) They are nonvolatile since the flavoring materials are enclosed within an impervious shell which protects them from volatile losses and oxidative change.

(b) They are dry, free-flowing powders; nonhygroscopic; do not cake or form lumps; and are readily incorporated into food mixes to give a homogeneous dispersion of the flavoring.

(c) The flavoring strength and quality are retained on storage even at elevated temperatures over long periods. In the case of encapsulated citrus flavors based on citrus oils, some surface oxidation may occur under these conditions. Such products should not normally be stored other than under cool conditions.

(d) When mixed with water, the capsule breaks down and liberates the flavoring as a microscopic cloud.

(e) The particle size ranges from 10 to 200 μ with the majority of the roughly spherical particles being about 50 μ in diameter.

(f) The water activity is low (0.2–0.3).

Cold Microencapsulated Flavors

Although any process which produces capsules within the range of 1 to 200 μ can reasonably be called "microencapsulation," this term has come to be used more specifically to describe those capsules made by cold precipitation, coacervation or liquid phase separation techniques. The technology for the production of microcapsules by the application of thin polymeric coatings to small particles of solids, droplets of liquids, solutions or dispersions, was largely pioneered and developed by the National Cash Register Co., who own the majority of the patents for their manufacture and industrial application.

Basically, the technique involves the formation of three immiscible phases, comprising: (a) the continuous liquid (e.g., water); (b) the material to be encapsulated (e.g., flavoring compound); and (c) the coating material (e.g., gelatin).

These phases are mixed under very carefully controlled conditions which determine the deposition of the capsule wall. The technique has many applications outside the field of flavors (i.e., carbonless papers, slow-release drugs, etc.). In the case of flavors, the materials suitable for such encapsulation are strictly limited by the necessity to ensure their acceptance as food additives.

As the first stage of manufacture, the encapsulant, for example, gelatin, is dissolved or suspended as a sol in a liquid, usually water. The aromatic "core" material, which, if liquid, must be immiscible with water, is emulsified with the water phase; or if it is solid, is dispersed in the gelatin solution by gentle mechanical agitation. To form the capsules, the pH and temperature of the system must be changed so as to cause the encapsulant to come out of solution, or coacervate, and coagulate to form a wall around the emulsified droplets or suspended particles. At this stage, the cell wall is still liquid and requires hardening. This is usually done by heating, although desolvation and cross-linking techniques are also available to form the final microcapsules.

The amount of coating can be varied to control the level of the core material, which, in the case of flavorings, also determines the flavoring power of the end-product. At present, this method of encapsulation is expensive and has not achieved much acceptance by the food industry.

STABILIZED DRY FLAVORS

The Sunkist Co. of California has developed an alternative method of preparing powdered citrus flavors which, otherwise, would be prone to oxidation. In this technique, the flavoring compound is emulsified with molten sucrose or gelatin to form a homogeneous base. This melt is then extruded as fine rods into a cold liquid bath, which is usually alcohol. The formed rods are then further washed with alcohol to remove all the surface aromatics and are then dried and ground to the desired particle size. This product is not strictly encapsulated but the flavoring ingredients are locked into a matrix of long-chain molecules which has much the same effect as a continuous capsule wall. Very long shelf-life is claimed for the products which are particularly recommended for use in crystal beverage mixes and table desserts.

HEAT-RESISTANT FLAVORS

Although microencapsulated flavors have the advantage of stability when dry, they still suffer the same disadvantages as liquid flavors when mixed into a water-containing medium. Once the water soluble capsule wall is broken down and dissolved, the flavoring is released and its contents become vulnerable during any ensuing heat processing, such as baking or extrusion. During the early stages of baking, there is a considerable loss of water from the dough so that any volatile matter present is liable to be lost by steam-distillation. In the case of extrusion, the mass is heated by compression as it traverses the screw-press and experiences a sudden reduction in pressure as the product emerges from the die. This results in a rapid flashing-off of water and volatiles. Under such conditions, any unprotected flavoring inevitably suffers loss or degradation which, in turn, is reflected in the flavor of the end-product. To compensate for these losses and to obtain an adequate flavor level, necessitates the use of additional flavoring but even this may result in a significant and variable change in the overall flavor profile of the finished goods.

A range of heat-resistant capsules has recently been marketed by one flavor manufacturer. According to the published literature these consist of water-insoluble, multistage microcapsules which are designed to release their flavor content at a predetermined elevated temperature. The flavoring components in such products continue to be protected during any wet mixing, such as in dough production, and also throughout the early stages of baking. It is claimed that the use of these heat-resistant flavors results in a truer flavor balance which can be achieved with an economy of usage which is not possible with the equivalent liquid or spray-dried flavorings.

SPRAY-COOLED FLAVORS

A variant of the spray-drying process is known as "spray-cooling." The process involves the preparation of melt of a fatty material and the blended flavoring aromatics, followed by its atomization in a stream of chilled, inert gas. The minute fatty globules are instantly "set" and the entrapped flavor thereby protected. In view of the nature of the carrier, it may be desirable to add an antioxidant to delay the onset of rancidity on storage. Such fat-based flavors may be made stronger than is desirable for direct use and broken down to a usable strength by admixture with another carrier such as dextrose or finely powdered sucrose. It is claimed that such flavors undergo very little change during processing and are very stable on storage under cool conditions. The use of a fatty base must limit the use of these flavors to products which already contain fat as a normal ingredient (e.g., dry cake mixes, potato chips, etc.).

BIBLIOGRAPHY

ANON. 1976. Encyclopedia of food chemicals. Food in Canada *36,* No 4, 27–54.

ARCTANDER, S. 1969. Perfume and Flavor Chemicals, Vols I and II. Steffen Arctander, Montclair, N.J.

BEDOUKIAN, P. Z. 1967. Perfumery and Flavoring Synthetics, 2nd Edition. Elsevier Publishing Co., Amsterdam, London and New York.

BRENNAN, J. G., BUTTERS, J. R., COWELL, N. D. and LILY, A. E. V. 1969. Food Engineering Operations. Elsevier Publishing Co., Amsterdam, London and New York.

CHARM, S. E. 1971. The Fundamentals of Food Engineering. AVI Publishing Co., Westport, Conn.
EARLE, R. L. 1966. Unit Operations in Food Processing. Pergamon Press, Edinburgh.
FORD, R. A. and CRAMER, G. M. 1977. Reference list of flavoring substances in use in the United States. Perfum. Flavorist *2*, 1, 10–62.
FRITZSCHE BROTHERS. 1966. Fritzsche Guide to Flavor Ingredients as Classified under the Federal Food, Drug and Cosmetic Act. Fritzsche Brothers, New York.
FURIA, T. E. (Editor). 1973. Handbook of Food Additives, 2nd Edition. CRC Press, Cleveland.
FURIA, T. E. and BELLANCA, N. (Editors). 1975. Fenaroli's Handbook of Flavor Ingredients, 2nd Edition, Vols. I and II. CRC Press, Cleveland.
GRUNDSCHOBER, F. *et al.* 1975. Survey of worldwide use levels of artificial flavouring substances. Int. Flavours Food Additives *6*, 223–230.
HALL, C. W., FARRALL, A. N. and RIPPEN, A. L. 1971. Encyclopedia of Food Engineering. AVI Publishing Co., Westport, Conn.
HARPER, R., BATE-SMITH, E. C. and LAND, D. G. 1968. Odour Description and Odour Classification. A Multidisciplinary Examination. J. & A. Churchill, London.
IKEDA, R. M. and CROSBY, D. G. 1960. Chemicals and the Food Industry. Univ. Calif. Div. Agr. Sci. Manual *26*.
KLIMES, I. and LAMPARSKY, D. 1976. Vanilla volatiles—a comprehensive analysis. Int. Flavours Food Additives *7*, 272–273, 291.
MASTERS, K. 1972. Spray-drying. Leonard Hill Books, London.
MERORY, J. 1968. Food Flavourings: Composition, Manufacture and Use, 2nd Edition. AVI Publishing Co., Westport, Conn.
MONCRIEF, R. W. 1967. The Chemical Senses, 2nd Edition. Leonard Hill Books, London.
MONCRIEF, R. W. 1970. Odours. Wm. Heinemann Medical Books, London.
OSER, B. L. and FORD, R. A. 1975. Recent progress in the consideration of flavoring ingredients under the Food Additives Amendment. 9. GRAS substances. Food Technol. *29*, No. 8, 70–72. (Refs. to 8 previous reviews).
SCHEFLAN, L. and JACOBS, M. B. 1953. The Handbook of Solvents. D. Van Nostrand, Co., New York.
S.C.I. 1957. Molecular Structure and Organoleptic Quality. Society of Chemical Industry (London) Monogram *1*.
SLADE, F. H. 1967. Food Processing Plant, Vols. I and II. Leonard Hill Books, London.
Van STRATEN, S. and de VRIJER, F. 1975. Lists of Volatile Compounds in Food, 3rd Edition and Suppl. 1 to 6. Central Inst. Nutr. Food Res., Zeist, Netherlands, Rep. *R 4030*.
WEAST, R. C. (Editor). 1976–1977. Handbook of Chemistry and Physics, 57th Edition, CRC Press, Cleveland.

8

Creation and Development of Flavorings

In an editorial in *Food Technology* (Nov. 1968), under the heading "Five Faces of Flavor," the different viewpoints from which the subject of flavor could be regarded were described as follows:

> Flavor is difficult to define without recourse to the human sensing apparatus. To the food industry, flavor is the physician of its products. To the consumer, flavor is an experience, brief in duration but lasting in memory. To the flavorist, flavor is the stuff of his living. To the food researcher, flavor is still in many respects a romantic enigma. To the government, flavor is a regulatable additive in foods.

This eloquently emphasizes the multifaceted aspects of the flavor industry and determines the environment within which it must operate in the creation and development of flavorings.

In flavor manufacturing companies, this work is normally carried out by highly skilled flavorists whose activities are intermediate between the researcher who seeks to isolate, identify and synthesize those chemicals responsible for natural flavors, and the food technologist who must use the compounded flavorings to make his products a marketing success. From a position between the precise scientific and the pragmatic technological views of flavor, the flavorist must at once be a scientist, an artist in aromatic profiles and a technologist with an understanding of basic food processing. Success depends on the quality of interdisciplinary communication and collaboration and an ability to work with constraints which apply particularly to flavors. In addition, it is necessary to retain a freedom of thought and action to allow for an expression of creative talent so necessary in the formulation of imitation flavorings.

FLAVOR CREATION

All aromatic plant material depends for its characteristic odor and flavor on a highly complex mixture of organic chemicals. It is an essential part of the flavorist's training that he gain a complete knowledge of the odor and flavor profiles and the relative flavoring strength of all such chemicals as are permitted for use in food products. The perfumer has an even greater spectrum of compounds to handle as the same restrictions in use do not apply to fragrances. For the creation of a flavor which closely matches a natural counterpart, the flavorist must employ three prime attributes: namely, knowledge, practical experience and "flair."

Traditionally, imitation flavors were created by using chemicals the odor and flavor of which were reminiscent of the natural flavor to be matched—a process of mental association. This technique still plays a part. The selected flavor is first examined as a whole and its profile constructed by a panel of trained assessors; this becomes the target. The flavorist then critically examines the target flavor and profile and breaks it down into defined odor and flavor categories. Flavorists have their own systems for equating odor and flavor categories with specific chemicals and, as this is so subjective, little guidance can be given. It is better for the flavorist to establish his own "bench marks" and "anchor points" which are most meaningful to him. Using this knowledge, one can then create what may be called a caricature of the target flavor in which the main aromatic features are somewhat exaggerated. This initial product will be a very rough model and must be continuously modified by trial and error until a satisfactory match is achieved. Of necessity, this is a long and tedious process calling for considerable skill and patience. Today, instrumental methods of analysis, particularly gas-liquid chromatography coupled with mass spectrometry, give to the flavorist a basic formulation upon which to build the desired profile. The technique is far from perfect, as many components remain to be identified and synthesized economically. The final stages of flavor development still involve sensory assessment, for which there is no alternative.

The reasons for not achieving a satisfactory matching flavor using analytical data are many and include: differences in concentration and flavoring power between the natural flavor and the imitation version; instability of and interaction between flavoring chemicals, many of which are highly reactive; artifacts produced during the analysis; the influence of pH; the effect of the solvent used; the nature of the test medium; the method of sensory assessment; texture; etc. All of these factors play an important part in the reconstruction of a natural flavor. Generally, the principal reason for lack of success is the nonavailability of relatively few trace components which, in combination with those present in significant quantity, are responsible for the distinctive nuance of the flavor profile. The replacement of these chemicals calls for the skill of the flavorist in associating aromatic attributes. Unfortunately, it is all too often necessary to compromise because a theoretically attainable profile may not be practically feasible. Legal constraints on the acceptability of chemicals in foods, as well as a constant concern over the costs of flavorings, severely restrict the choice of components and, hence, circumscribe the extent to which a flavorist can push his creativity and ability to make a perfect match for a given natural flavor.

Flavor creation, then, is a process of trial and error, necessitating constant assessment of odor and flavor profiles of experimental blends, making, first, coarse and then, fine, adjustments until a satisfactory final profile is achieved. At this stage the following questions must be answered:

(a) Does the flavor taste as it should and is it instantly recognizable for what it claims to be?

(b) Is the profile well balanced?

(c) Does the flavor change on standing?

(d) Does the flavor produce other unacceptable mouth-feel effects?

(e) Does it have any after-taste?

(f) Is it of an acceptable strength for the end-use for which it was designed?

(g) Can its form be changed without changing the profile (i.e., can a powder version be produced)?

These, and many other questions, must be carefully considered and satisfactorily answered if a commercially viable flavoring is to be produced.

FLAVOR CHARACTERIZATION AS AN AID TO FLAVOR DEVELOPMENT

The following characteristics are apparent in a natural flavor:

(a) It is a complex mixture of organic chemicals embracing the whole spectrum of organic chemistry.

(b) The molecular structure of the components is related to their aromatic character.

(c) Quantitatively, the components may range from high percentages to traces only.

(d) Minor components often have a more significant effect on the profile than do the major components.

(e) Though, generally, only a minor constituent of foodstuffs, most flavorings exert a major effect upon its character.

(f) Many of the constituents are reactive and/or thermolabile, resulting in significant changes in profile depending upon circumstances.

All of these considerations make the characterization of specific flavoring components a difficult and tedious procedure. Gas-liquid chromatography was introduced as an analytical tool in 1952 and, for the first time, provided the researcher with an entirely new and highly sensitive technique for separating the components of natural flavors. The impact was considerable and resulted in a totally different concept of the nature of odor and flavor chemistry. During the past two decades, the basic technique and the instruments available have undergone considerable development and there is now an impressive array of published data on the chemical composition of almost everything that has an odor or a flavor. The conditions used for the isolation of aromatic components are now well understood and can be controlled to ensure a minimum of change in the system. In spite of these advances, the blind acceptance of analytical data rarely results in a first-class imitation of the natural flavor—there is still ample room for the creative flavorist to interpret these results into an integrated flavor response. Many nuances in a flavor profile cannot be described as "characteristic" of that flavor and a knowledge of what is and what is not an important attribute lies with the flavorist whose nose is still more sensitive than any instrumental technique as yet available.

INSTRUMENTAL METHODS

Reference should be made to the many excellent standard texts and to manufacturers' literature for descriptions of the highly sophisticated apparatus now available for flavor research. The following stages are of importance to the flavorist in gaining an understanding of the methods of establishing quantitatively and qualitatively the components responsible for the characteristic profiles of natural flavors and foodstuffs:

(a) Collection of a truly representative sample of the material under examination.

Many natural products display a wide individual variation in flavor character which will be reflected in the final results.

(b) Segregation of the aromatic components from other materials, usually inert cellulose, which may influence the results.

It is this stage which presents most problems due to the different physical nature of the raw materials. The following methods are widely used: (1) distillation, (2) solvent extraction, (3) adsorption of head-space vapors, (4) flash evaporation.

(c) Concentration suitable for injection into a chromatographic column.

Many aromatics are present naturally in very dilute solution. The volume of liquid which may be injected into a gas-liquid chromatographic column is severely limited (usually about 1 microliter) and yet has to be sufficient to give a positive response on the detector for even trace components.

(d) Running the gas-liquid chromatogram to a set program of time and temperature.

The pattern of the chromatogram is characteristic of the mixture under test but the form of the graphical presentation is influenced by the apparatus used, the method of detection, the rate of flow of the carrier gas, the nature of the column packing, etc., all of which should be stated on the published results.

(e) Examination of the individual components.

To enable the effluent gases to be assessed by sensory means, the stream can be split, part going to the detector and the remainder to a suitable smelling chamber, where an expert can assess and describe the sequence of odors as the fractions emerge from the column. The response is noted at the appropriate place on the chromatogram. For success, it is necessary to have an established vocabulary of well-understood descriptive terms.

(f) Identification of the components.

For any area of the chromatogram of particular interest from an odor point of view, the analysis can be re-run and the effluent gas introduced directly into a coupled fast-scanning mass spectrometer. The resulting spectrum enables the components to be characterized.

Other techniques are also available to the researcher: (1) Nuclear magnetic resonance (NMR) (2) X-ray defraction, (3) polarography, (4) ultraviolet spectometry, (5) infrared spectroscopy.

(g) Synthesis of the identified compounds.

The flavorist is often called upon to evaluate new synthetic chemicals, for, ultimately, it is the trained nose and palate which must be the arbiter of acceptability.

CONSTRAINTS ON FLAVOR DEVELOPMENT

The creation of a flavoring composition is not just a matter of putting together a number of aromatic chemicals or natural extracts in a solvent. The flavoring of processed foods, sugar and flour confectionery, baked goods, beverages, etc., involves many considerations outside those directly associated with the intrinsic profile of the flavoring used. Two major constraints determine what may be used in a flavoring intended for human consumption:

(a) The legislation governing the use of food additives in the country in which the product is to be consumed.

(b) The need to obtain a well-balanced flavor at an acceptable strength in the end-product despite adverse processing conditions.

LEGISLATIVE RESTRICTIONS ON FLAVORING MATERIALS

In most developed countries, the use of flavorings is controlled by legislation. Regulations differ widely, both in form and detailed content, making international trade in food products very difficult. With an ever-increasing consumption of factory processed foods and beverages, many of which depend for their acceptability on the use of imitation flavorings based on synthetic chemicals, it is not surprising that there is an increasing awareness, on the part of both the public and the legislators, of the possible dangers of ingesting chemicals in the diet. The arguments rage between those who defend the use of imitation flavorings and decry the assumption that "nature is safe" and those who declare, with equal conviction, that unless a flavor is entirely natural it cannot possibly be safe for use in foods. Of course, these are the extremes and the true situation is far less well-defined. There is now an enormous range of flavoring ingredients available and the uses to which they are put make it virtually impossible to exercise precise control over their application and daily input. Both the flavor industry and the food manufacturers are well aware of their responsibilities in this respect and fully collaborate in international restrictions designed to safeguard the health of the consumer.

Food additive legislation generally has two main aims: (a) to protect and safeguard the consuming public from any possible harmful effects of ingesting materials added to the natural diet to improve its character, color or flavor; and (b) to protect the consumer against deception.

The past decade has seen an enormous interchange of information and an increasing acceptance of the need for some internationally agreed standards for flavor legislation. This led to the formation, in 1969, of the International Organization of the Flavor Industry (IOFI), the members of which passed the following resolutions as a guideline for future action in dealing with government bodies responsible for drafting legislation on food additives, particularly flavoring materials:

(a) The flavoring of foods is necessary to give a satisfactory, and in the long-run, a sufficient diet.
(b) As foods containing flavorings are increasingly crossing international boundaries, the flavor industry supports harmonization of national flavor legislation.

(c) The basic principle of flavor legislation is to protect the public health, ensuring at the same time the industrial progress which is essential for the solution of problems of the future.
(d) Flavor and/or flavoring substances have odor and/or taste producing properties, three classes being distinguished:
  (1) natural flavors and flavoring substances,
  (2) nature-identical flavoring substances,
  (3) artificial flavoring substances not yet found in nature.
(e) The safety evaluation should be the same for natural and nature-identical flavoring substances provided the intake levels are of the same order of magnitude.
(f) Appropriate restrictions have to be imposed on the use of any natural and nature-identical flavoring substances which, under the proposed conditions of use, are harmful to the consumer's health.
(g) Artificial flavoring substances, as defined, should be subject to appropriate safety evaluation and only those approved for use in food should be legally permitted.
(h) Information to enable the authorities to judge whether the flavoring ingredients present a health hazard shall be made available by the industry to the appropriate health authority if requested.

These resolutions set out quite clearly the attitudes of the flavor industry to its responsibilities for the safety in use of flavoring products. The recommendations arising from them have still to result in firm legislation in any country or trading group. Reviews of the current status of national and international flavor legislation appear regularly in the trade journals and a review by Vodoz (1975) reveals that there is still much to be done to achieve any degree of harmonization. The responsibility of the creative flavorist to flavor legislation was the subject of the Littlejohn Memorial Lecture (Downer 1973).

Member bodies of IOFI which are working towards international harmonization include both government and industrial associations. These are listed in Table 8.1. The need for control is now almost universally accepted and regulations exist in all industrial countries having well-established food processing plants. There are differing opinions as to the nature and extent of the regulations necessary to ensure effective control. Three main approaches are available.

(a) Negative Lists of Substances

This type of legislation results in lists of substances which are specifically prohibited for use in foods or may be used only to a limited extent in particular food categories. The disadvantage of this system is that the regulatory authorities cannot be expected to

TABLE 8.1
MEMBER ORGANIZATIONS OF THE INTERNATIONAL ORGANIZATION OF THE FLAVOR INDUSTRY (IOFI)

Country	Organization
	Organization Headquarters International Organization of the Flavour Industry (IOFI) 8 Rue Charles-Humbert, CH-1205, GENÈVE.
Australia	Flavour and Perfume Compound Manufacturers Association of Australia, Box 3968 G.P.O., SYDNEY, 2001.
Austria	Fachverband der Nahrungs-und Genussmittelindustrie Oesterreichs, Zaunergasse 1-3 A-1037 WIEN.
Belgium	Groupement des Fabricants d'Essences, Huiles essentielles, Extraits, Produits chimiques, Aromatiques et Colorants (AROMA), 49 Square Marie-Louise, B-1040 BRUXELLES.
Brazil	Associaçao Brasileira des Industrias da Alimentaçao, Caixa postal, 8927, SAO-PAULO.
Canada	Flavour Manufacturers Association of Canada, Suite 10, 3625 Weston Road, WESTON, Ontario M9L 1V9.
Colombia	Asociacion Nacional de Industriales (ANDI), P.O. Box 4430, BOGOTA.
Denmark	Essens Fabrikant Foreningen, Grabrodretorv 16, DK-1154 COPENHAGEN K.
France	Syndicat National des Industries Aromatiques Alimentaires, 2 rue de Penthièvre, F-75 PARIS VIIe.
Germany	Verband der Deutschen Essenzenindustrie e.V., Neusser Strasse 104, D-5 KÖLN.
Italy	Associazione Nazionale dell'Industria Chimica—(ASCHIMICI), Via Fatebenefratelli 10, 1-20121 MILANO.
Japan	Japan Flavor and Fragrance Manufacturers' Association (JFFMA), 3F Nomura Bld., 3-2 Kodenma-Cho, Ninonbashi, Chuo-Ku, TOKYO.
Mexico	Asociacion Nacional de Fabricantes de Productos Aromaticos, A.C., Jose M. Rico No. 55, MEXICO, 12, D.F.
Netherlands	Vereniging van nederlandse aroma- en Reukstoffen-industrieën (NEA), Javastraat 2, 's - GRAVENHAGE.

(Continued)

TABLE 8.1 (Continued)

Country	Organization
Norway	Norske Aromaprodusenters Forening, (Association still to be instituted)
South Africa	The South African Association of Industrial Flavour and Fragrance Manufacturers, P.O. Box 4581, JOHANNESBURG, Transvaal.
Spain	Asociacion Espagnola de Fabricantes de Aromas para Alimentacion (AEFAA), San Bernardo, 23, MADRID, 8.
Sweden	Föreningen Svenska Aromtillverkare, Box 5501, S-114 85, STOCKHOLM.
Switzerland	Schweizerische Gesellschaft für Chemische Industrie (SGCI), Nordstrasse 15, CH-8035, ZURICH.
United Kingdom	The British Essence Manufacturers' Association (BEMA), 1-2 Castle Lane, Buckingham Gate, LONDON SW1E 6DN.
United States of America	Flavor and Extract Manufacturing Association of the United States (FEMA), 900 17th Street, N.W., WASHINGTON, D.C. 20006

know everything that is being used and, in consequence, potentially dangerous chemicals can be incorporated into flavorings without restriction.

(b) Positive Lists of Substances

As its name implies, this type of legislation entails the publication of lists of materials which may be used in foods or other products for human consumption. Everything else is specifically excluded. It is generally accepted that:

(1) The lists must, of necessity, be incomplete as many natural flavors contain chemicals which have yet to be identified. When, eventually, such components are characterized, their synthetic counterparts would not be listed nor could they be legally used even at the levels equivalent to those in which the chemical is found in nature.

(2) There must be precise criteria and universally acceptable methods of testing the safety in use of any material intended for inclusion in the accepted lists.

(3) There must be adequate methods for policing the system.

Too rigid an application of this system inevitably results in a disincentive to research into the nature of natural flavors and new flavoring chemicals, for even where these are nature-identical they cannot be used commercially.

(c) Mixed System of Substances

In a mixed system of control, there would be a restrictive list of natural and nature-identical substances and a positive list of synthetic chemicals not found in nature. Such a system avoids the major disadvantages of the more precisely defined negative and positive legislation. At the same time it does not inhibit research in the field of natural chemistry and the development of nature-identical chemicals for use in imitation flavorings. The system is adequately enforceable and yet sufficiently flexible to allow differing degrees of restriction to be applied, depending on the risks involved.

Anyone concerned with the creation, development and use of imitation flavorings must have a working knowledge of the current legislation governing these materials in the country in which the end-product is to be sold. This is not necessarily the same as that in which it was manufactured.

Most flavor laboratories have a system of color coding the bottles containing flavoring components so that the flavorist has a ready "eye guide" as to the legal status of the contents. This is strongly to be recommended. In addition to this, most flavorists maintain a simple card system of legislative data on all materials they are likely to require in the compounding of an imitation flavoring.

At the present time, FDA (U.S. Food and Drug Administration) and FEMA (Flavor and Extract Manufacturers' Association) are collaborating to produce a uniform, positive list of flavoring materials for use in the United States; the Council of Europe and BEMA are similarly active in Europe and the EEC countries; and IOFI is acting along similar lines on a universal basis. Progress is positive but necessarily slow as so many viewpoints have to be reconciled. This work will doubtless result in a very comprehensive bank of knowledge of flavoring substances and their safety in use. Hopefully, it will result in an international code for the guidance of all involved in the flavoring of foods and other products for human consumption.

GRAS SUBSTANCES

The technical committee of FEMA in the United States has, over the past 15 years, carried out a very comprehensive survey of materials currently in use or likely to be used in flavorings. All the relevent information was reviewed by an expert panel to determine whether or not the substances could be classified as "generally recognized as safe" (GRAS); those so classified are published in a

FEMA GRAS list. FDA also carried out a parallel investigation augmented by an extensive program of toxicological testing. This has resulted in an initial list of FDA-approved materials which may be used with safety in flavorings in the United States. Since the publication of the FDA GRAS list, it has undergone continuous re-assessment by qualified experts based on: (a) toxicological data, (b) biochemical and metabolic data, (c) occurrence in nature—qualitatively and quantitatively, (d) chemical structure and physical properties, (e) proposed level of use and maximum daily intake, (f) toxicological significance of constant ingestion, and (g) prior use in foods.

This is an on-going program. Amendments are given legal status by publication in the U.S. Federal Register under Section 121.3(2). The FDA published its criteria for including any chemical in the GRAS list in the U.S. Federal Register for 25th June, 1971 under the title, "Eligibility of Substances for Classification as Generally Recognized as Safe (GRAS) in food."

Both the FDA and FEMA GRAS lists enjoy wide recognition and form the basis for most decisions as to what may be used in a flavoring in the absence of specific restrictive legislation (Hall and Oser 1970; Wendt 1971–1977).

PROCESSING CONSTRAINTS ON FLAVOR DEVELOPMENT

The main constraint on the composition of any imitation flavoring is posed by the intrinsic volatility of its component chemicals. Any stage in food processing which results in volatile losses will inevitably determine the nature of the flavoring which can be used which will, in turn, dictate the technique to be used in its incorporation into the product and will control the ultimate flavor profile achieved.

The principal factors to be considered are: (a) temperature; (b) processing time; (c) pressure changes, positive and negative; (d) pneumatic conveyance and free air movement; (e) exposure to air; (f) permeability of packaging; (g) storage conditions; and (h) estimated optimum shelf-life.

Flavorings, be they natural, synthetic or a mixture of both, are composed of highly reactive chemicals which, though stable under recommended storage conditions, may reveal considerable reactivity under processing, particularly where this includes high-temperature retorting, low pH, aeration, etc. In some cases, the resultant chemical changes can give rise to quite unexpected flavors and/or unacceptable off-notes, either during the actual processing or slowly during ensuing storage.

Each branch of the food and related industries has its own particular processing problems with which the flavorist and the food technologist must be familiar, if a really successful flavoring is to be developed. Where the process is known to affect adversely the flavor stability, it is essential either that identical processing conditions be established on a laboratory scale or that a pilot batch of product be made in the factory. By this means, the flavoring can be more realistically evaluated and its performance under full production conditions can be established. As a very minimum, the following processing parameters must be reproduced as far as is possible: (a) temperature and time; (b) mixing sequence; (c) pH; (d) pressure; (e) throughput rate, particularly in relation to (a); (f) flavor addition; (g) packaging; and (h) storage, particularly if this involves subzero temperatures.

The use of specific flavorings in various end-products is discussed in a later chapter.

The flavor industry offers a comprehensive technical service to manufacturers in order to ensure that its products are correctly used to achieve optimum results. Unfortunately, the manufacturers most open to benefit by this expert knowledge are often those which are most secretive about their development projects and processing methods. Lack of information, as well as ill-defined requests for flavorings, are a cause of much costly, wasted technical effort. Very few flavorings are suitable for use in all types of food, confectionery or beverage products. There are so many divergent variables involved that only when precise information is available can the optimum flavoring be developed.

BIBLIOGRAPHY

Flavor Creation and Development

AYRES, J. and CLARK, W. R. E. 1975. Instrumental techniques in flavor research. Food Manuf. *50*, No. 1, 19–20, 22.

BEDNARCZYK, A. A. and KRAMER, A. 1971. Practical approach to flavor development. Food Technol. *25*, 24–26, 28, 32–33.

BRODERICK, J. A. 1972. Fruit flavor research. The practical flavorist vs. the basic researcher. Food Technol. *26*, No. 11, 37, 48.

BRODERICK, J. J. 1974. Banana—a feeling for nature. Flavour Ind. *5*, 284–285.

BRODERICK, J. J. 1975A. Blackberry—or a reasonable facsimile thereof. Int. Flavours Food Additives *6*, 41–44.

BRODERICK, J. J. 1975B. Cherry—common denominators. Int. Flavours Food Additives *6*, 107, 109.

BRODERICK, J. J. 1975C. Grape and preference. Int. Flavours Food Additives *6*, 171–172.

COOK, M. K. 1971. Flavour creation. I. A critical evaluation. II. The anatomy of flavour creation. Flavour Ind. *2*, 155–157.

DOWNEY, W. J. and EISERLE, R. J. 1970. Substitutes for natural flavors. J. Agric. Food Chem. *18*, 983–987.

FENAROLI, G. 1970. The influence of small amounts of some constituents on the sensory properties of aromas. Fr. Ses Parfums *12*, 423–425. (French)

FENAROLI, G. 1972. Flavoristics—science of flavor. Parfums, Cosmet. Savons Fr. *2*, 545–548. (French)

FRIJTERS, J. E. R. 1976. Sensory evaluation: A link between food research and food acceptance research. Lebensm. Wiss. Technol. *8*, 294–299.

GALETTO, W. G. and STAHL, W. H. 1974. An information acquisition and utilization system for the flavor chemist. J. Agric. Food Chem. *22*, 755–757.

LEES, R. 1974. Avoiding deterioration of stock and new flavourings. Confec. Prod. *40*, 403–404.

MERRITT, C., Jr. *et al.* 1974. A combined GLC-MS-computer system for the analysis of volatile components of foods. J. Agric. Food Chem. *22*, 750–755.

POLACK, E. H. 1970. The reconstruction of flavours. Flavour Ind. *1*, No. 7, 442–445.

POTTER, R. H. 1974. Gas chromatography—a flavourist's tool. Flavour Ind. *5*, 241–242.

SABINE, R. H. 1974. A synthetic flavour—naturally. Flavour Ind. *5*, 80–81.

SMITH, W. H. 1971. Towards developing good taste in flavors. Snack Foods *60*, Oct. 56–58.

STURM, W. and MANSFIELD, G. 1976. Tenacity and fixing of aromatic chemicals. Perfumer Flavorist *1*, No. 1, 6–10, 13–16.

TONG, S. T. 1975. Trace components in flavours. Int. Flavours Food Additives *6*, 350,355.

The Flavorist

BAUER, K. J. 1974. The flavourist as an internationalist. Flavour Ind. *5*, 239–240, 242.

BONICA, T. J. 1974. The flavourist as a processor. Flavour Ind. *5*, 172–173.

BRODERICK, J. J. 1972. The flavorist. Food Technol. *26*, No. 11, 37.

COWLEY, E. 1973. The training of a flavourist. Flavour Ind. *4*, 18–20.

DI GENOVA, J. 1974. The flavourist as an artist. Flavour Ind. *5*, 174–175.

FISCHETTI, F. 1974. The training of a flavour chemist—an organized programme. Flavour Ind. *5*, 166–168.

PERRY, P. 1974. The flavourist as a technical man. Flavour Ind. *5*, 171.

SABINE, R. H. 1973. A defence of the flavourist's art. Flavour Ind. *4*, No. 5, 213–215.

SHORE, H. 1974. The training of a flavourist—one on one. Flavour Ind. *5*, 165,168.

VOCK, M. H. 1974. The flavourist using the achievement of the organic chemist. Flavour Ind. *5*, 243–246.

WEISS, G. 1975. Communicating sensory objectives during prototype product development. Food Prod. Dev. *9*, No. 7, 34,36.

WESLEY, F. 1974. Flavour chemist uses salesmanship. Flavour Ind. *5*, 275–276.

WIENER, C. 1974. The flavourist as a biochemist. Flavour Ind. *5*, 237–238, 242.

Flavor Legislation

AMOORE, J. E. 1976. Synthetic flavors: A new approach to efficiency and safety. Chem. Senses Flavor *2*, 27–38.

ANON. 1970. Changes in food regulations. Food Technol. Aust. *22*, 630–631,633,635.

ANON. 1971A. International flavour legislation. Flavour Ind. *2*, 265.

ANON, 1971B. Flavour legislation—the present situation. Flavour Ind. *2*, 98–99.
ANON. 1974A. Natural flavor added to F.F. may be legally "artificial." Quick Frozen Foods *36*, No. 8, 38–39.
ANON. 1974B. Basic features of modern flavour regulations. Manuf. Confect. Mar., 28–33.
ANON. 1974C. Toxicology of flavours and food additives. Flavour Ind. *5*, 147.
ANON. 1975. Evolution of controls. Int. Flavours Food Additives *6*, No. 2, Editorial.
ANON. 1976. Flavourings in food—report of the review by the Food Additives and Contaminants Committee. Br. Food J. July–Aug., 112–114.
BLEDE, E. E. 1973A. Food regulations in Latin America. I. Food Prod. Dev. *7*, No. 7, 72–73.
BLEDE, E. E. 1973B. Food regulations in Latin America. II. Food. Prod. Dev. *7*, No. 8, 39–41.
BURKE, C. 1976. Flavourings legislation in the E.E.C. Int. Flavours Food Additives *7*, 265–267.
DOWNER, A. W. E. 1974. Social responsibility and the creative flavourist or bureaucracy versus the flavourist. (Littlejohn Memorial Lecture, London.) Flavour Ind. *5*, 71–76.
ELIAS, P. S. 1972. Food additives, toxicology and the E.E.C. Chem. Ind. *4*, 139–144.
ELIAS, P. S. 1973. The toxicological investigation of food additives. Proc. Soc. for Anal. Chem. *10*, No. 7, 173–175.
GERARDE, H. W. 1973A. Survey update—determining the safety of flavor chemicals. Food Prod. Dev. *7*, No. 3, 82,84.
GERARDE, H. W. 1973B. Survey update—FEMA expert panel. Flavour Ind. *4*, 162–163.
GERARDE, H. W. 1973C. The new product—is it toxic? Flavour Ind. *4*, 298–299, 308.
GRUNDSCHOBER, F. *et al.* 1975. Survey of worldwide use levels of artificial flavouring substances. Int. Flavours Food Additives *6*, 223–230.
GRUNOW, W. 1975. Toxicological evaluation of food additives. Ernaehr. Umsch. *22*, No. 4, B13–B16. (German)
HALL, R. L. 1960. Recent progress in the consideration of flavoring ingredients under the Food Additives Amendment. Food Technol. *14*, No. 10, 488.
HALL, R. L. 1975. GRAS—concept and application. Food Technol. *29*, No. 1, 48,50, 52–53.
HALL, R. L. and OSER, B. L. 1970. The safety of flavouring substances. Flavour Ind. *1*, 47–53.
HALL, R. L. and OSER, B. L. then OSER, B. L. and FORD, R. A. 1973–1976. Recent progress in the consideration of flavoring ingredients under the Food Additives Amendment. Food Technol. *15*, No. 12, 20, and *19*, No. 2, part 2, 151. *See also* GRAS substances. Food Technol. *24*, No. 5, 25–34; *26*, No. 5, 35–37; *27*, No. 1, 64–67; *27*, No. 11, 56–57; *28*, No. 9, 76–80; *29*, No. 8, 70–72.
HOOD, C. R. 1975. Food shortages may change flavor trends. Food Ind. South Africa Mar., 21,23,25.
JENKINS, J. J. 1975. International flavor legislation—a review. J. Assoc. Public Anal. *13*, No. 1, 1–13.
JONES, K. 1976. The present status of the flavour and food labelling regulations. Dragoco Rep. No. 3, 55–68.
KUIPER, L. 1962. Flavour legislation—a manufacturer's point of view. Food Manuf. June, 285–288.
LENANE, G. 1976. U.S.A. legislation on flavourings. Int. Flavours Food Additives *7*, 267–270.

MASON, M. E. 1975. Recent flavor developments. Act. Rep. Res. Dev. Assoc. Mil. Food Packag. Syst. *27*, No. 2, 71–82.
McNAMARA, S. H. 1975. Some legal aspects of providing new foods for hungry populations. Food Prod. Dev. *54*, No. 3, 57–60.
MIDDLEKAUFF, R. D. 1974. Legalities concerning food additives. Food Technol. *28*, No. 5, 42,44,46,48.
MOAN, G. le 1973. Flavoring compounds—toxicological problems posed by their use. Aliment. Vie *61*, 121–160. (French).
MONACELL, A. L. 1974. Flavour and patents. Flavour Ind. *5*, 164.
OSER, B. L. *et al.* 1965. Toxicological tests in flavoring matters. Food Cosmet. Toxicol. *3*, 563.
OSER, B. L. and FORD, R. A. 1973–1976. *See* above. Hall, R. L. and Oser, B. L.
PISANO, R. C. 1973. The American flavor industry—its growth and development and relationship with government. Flavour Ind. *4*, 384–386,388.
SAYER, M. D. 1976. Food development—innovation or stagnation. Food Technol. *30*, No. 8, 64,66–69.
SCHLEGER, W. 1971. Flavour research and food legislation. Flavour Ind. *2*, 413–415.
SCHMIDT, A. M. 1975. How the F.D.A. treats the benefit/risk question. Food Prod. Dev. *9*, No. 2, 78.
SCHRAMM, A. T. 1976. Politics vs. science in food additives regulation. Food Technol.. *30*, No. 8, 56–61
SPIHER, A. T. Jr. 1970. The GRAS list review. F.D.A. Papers, Dec. 1970/Jan. 1971. 12–13.
STOFBERG, J. 1975. Codex Alimentarius: The way to a world-wide food and flavor regulation? Cosmet. Perfum. *90*, No. 10, 71–74.
THOMPSON, D. R. 1973. GRAS: Present and future. Assoc. Food Drug Off. U.S. Q. Bull. *37*, No. 2, 141–146.
VODOZ, C. A. 1970. Intentional food additives in Europe. Food Technol *24*, No. 11, 42,44,46,48,50,52.
VODOZ, C. A. 1975. Flavour legislation—new developments and trends. Int. Flavours Food Additives *6*, 219–220,236.
VODOZ, C. A. 1977. Pregress toward the harmonization of international legislation for food flavourings. Inst. Food Sci. Technol. Conf. Proc. London *10*, No. 1, 40–47.
WENDT, A. S. 1971–1977. News from America. (A regular feature article dealing with aspects of current legislation in the U.S.A., flavoring topics, GRAS criteria, etc.) Flavour Ind now Int. Flavours Food Additives.
WODICKA, V. O. 1974. Regulatory problems with flavours. Flavour Ind. *5*, 293–294, 298.
W.H.O. 1974. Evaluation of certain food additives. World Health Organization Tech. Rep. Ser. *557*, Geneva, Switzerland.

Flavors in Processing

CABELLA, P. 1970. Heat is an important factor in effective flavor use. Candy Ind. *135*, No. 3, 25–26,36.
CLARK, K. J. 1970. Modern trends in flavoring and flavor creation for the soft drink industry. Flavour Ind. *1*, 388–389.
DOWNER, A. W. E. 1973. Application of flavours in the soft drink industry. Flavour Ind. *4*, 488–490.
DOWNEY, W. J. and EISERLE, R. J. 1970. Problems in the flavoring of fabricated foods. Food Technol. *24*, No. 11, 30,32,34,36.

FELLOWS, G. 1973. Confectionery flavours. Problems and factors affecting taste and choice of ingredients to protect or enhance flavours. Manuf. Confect. *52*, April, 70–78.

FRAZEN, K. K. and KINSELLA, J. E. 1975. Physicochemical aspects of food flavoring. Chem. Ind. No. 12, 505–509.

GORDON, J. and KIPLING, N. 1973. Flavour application in the dairy industry. Flavour Ind. *4*, 485–487,493.

GREMLI, H. *et al.* 1974. Some aspects of interaction between food and flavor components. Proc. 4th Int. Cong. Food Sci. Technol. *1d*, 9–11.

HEATH, H. B. 1970A. Flavourings in baked goods. I. Correct use for best results. Baking Ind. J. Mar., 20,24,26,30.

HEATH, H. B. 1970B. Flavourings in baked goods. II. Flavour influence on product acceptance. Baking Ind. J. Nov., 27–28,30.

LEES, R. 1973. Flavouring sugar confectionery. Confect. Prod. *39*, 488, 490–491.

LEWIS, J. A. 1973. The flavouring of dairy products. Dairy Ind. *38*, No. 6, 274–277.

MAIER, H. G. 1972. Fixation of volatile flavours in food products. Lebensm. Wiss. Technol. *5*, No. 1, 1–6. (German)

MEINER, A. 1973. Incorporating flavour and acid in continuous manufacturing systems. Manuf. Confect. *53*, 58–59.

OOSTERHUIS, P. C. 1976A. Systematic flavoring in confectionery practice. Int. Rev. Sugar Confect. *29*, No. 2, 43–47. (German)

OOSTERHUIS, P. C. 1976B. Systematic flavoring in confectionery practice. Int. Rev. Sugar Confect. *29*, No. 3, 78,80–81. (German)

PANNELL, R. J. H. 1973. Overcoming problems in flavouring ice cream. Dairy Ind. *38*, No. 6, 268–270.

PINTAURO, N. D. 1976. Food Flavoring Processes. Food Technol. Rev. Noyes Data Corp., Park Ridge, N.J.

SCHOONYOUNG, F. 1974. Attacking snack food problems through flavor application. Snack Food *63*, No. 10, 40–41,56.

TREVITT, M. R. A. 1973. Flavour application to tobacco. Flavour Ind. *4*, 491–493.

VAN EIJK, A. 1971. Flavourings in deposited articles. Adaptation of flavouring substances to the raw material and the end-product. Flavour Ind. *2*, 347–352.

VERREK, V. 1973. Some aspects of flavouring fat-based food products. Flavour Ind. *4*, 382–383.

Section III
Flavor Applications

9

Flavors in Food Processing

CRITERIA FOR THE USE OF FLAVORINGS

Three things may happen when flavoring is added to a product:

(a) The product achieves the same flavor as that of the added flavoring.

(b) It supplements or fortifies an existing flavor.

(c) It covers up or modifies some undesirable flavor character inherent to the product.

Having considered the raw materials from which flavorings are made and something of their creation and development, this section is concerned with the application of flavorings in specific groups of end-products intended for consumption. The use of flavorings in any type of consumer product, be it food, confectionery, soft drinks, ice cream, etc., is determined by three factors, namely: the acceptability to the consumer, the nature of the finished product as sold to the consumer and the processing conditions, storage and distribution which the product must undergo prior to ultimate sale.

When it comes to product acceptability, the consumers are restless; with very few exceptions, they look to the manufacturer to produce novel flavoring experiences, yet within the narrow confines of preconceived and often prejudiced views which may be quite irrational. Inevitably, the food manufacturer responds to these pressures in order to maintain or improve his share of the market; in consequence, the manufacturer develops a constant stream of new products based either on new concepts of presentation, new combinations of existing ingredients or new processing techniques. All of these make special demands on the flavor industry which, generally, is called upon to provide the food manufacturer with his flavoring needs. The flavorings used must reflect the changing public taste, which tends to be fickle and unpredictable. Flavorings must be compatible with the prime components of the end-product and be fully able to withstand the often stringent conditions imposed by the processing techniques employed. They must be stable, so that, from the point of view of the flavor, the product sold to the consumer, even after many months of storage, shall be as fresh and acceptable as when it leaves the factory. On top of all this, flavorings must comply with the increasing load of legislative restrictions which govern both domestic and international trade in these products.

Very few flavorings are suitable for all possible applications. The processing conditions encountered in the various branches of the user industries often present unique problems in flavor use. This has resulted in the establishment, over the past two decades in particular, of an unusually close working relationship between food and flavor manufacturers. It is now well-appreciated that development teams should include a flavor technologist. If this is not practicable, the development team should have available a flavorist as a consultant during the life of the project.

In the past, much product development was carried out under a cloak of secrecy and to some extent this still holds true. But, with today's facilities and the uniform capabilities of the major manufacturing companies, the initial advantage of launching a revolutionary new product on the market is comparatively short-lived. The supermarkets are full of duplicated product lines which appear within a remarkably short time of the initial launch of an obviously successful new product. There are now very few occasions when an entirely new product becomes the unassailable brand leader safe from competition. In order to optimize the development project's chances of producing a new product having a high level of acceptablity and attraction, the need for full collaboration between the various interested disciplines involved is obvious. If adequate and common-sense safeguards are made, this generally can be achieved without loss of security of information or market advantage.

Many food manufacturers still demand samples of flavorings from all the normal competitive supply houses, generally without indicating the end-use; then, by a process of trial and error, they eliminate those of little or no interest and proceed with the development using those flavorings which have been found to be most acceptable to the development team. Frequently, this does not represent the best possible flavoring for the product in mind, merely the best of what has been submitted. Some of the larger food processors maintain a flavor library from which samples may be called to satisfy the outline requirements of any project. This is eminently satisfactory so long as the samples are regularly reviewed and renewed as necessary. There are many ways in which flavorings may be chosen but none is better than the direct involvement of a flavor technologist who is an expert in this particular branch of product development and appraisal. Where a flavor manufacturer's staff is called upon to collaborate directly in this way, it is fully appreciated that this places them in a particularly priviledged position; but this method of giving a technical service is well established and the integrity of the flavor industry is not in doubt. New product development is now very costly and likely to become increasingly so. In consequence, one cannot blame any company for adopting a protective attitude for fear that its secrets may be made available to the competition. The flavor houses, themselves, are similarly very security-minded; however, the efficacy of their technical associations with the user industries over many years is attested to by the continuing success of their combined efforts.

Before specifically reviewing the industries concerned, certain general considerations are worthy of comment.

The Finished Product

Many so-called "new products" are not, in fact, so new. Very few are entirely new, that is, nothing similar has already been available in any market. In many instances, the product is merely "new" to the particular manufacturer who is either trying to imitate an existing brand leader, or product available in another market, or is new in the sense that it is an extension to an existing range. The most popular way of revitalizing a sagging demand is to offer a product labelled "new formulation." This description may involve a change in the flavor profile or a totally new presentation aimed at improving sales interest without losing the impetus and goodwill of the product brand name. Whatever the nature of the "new" concept, it is advantageous to consider the problems of flavor and involve a flavor technologist at a very early stage of the project planning. This will then allow sufficient time to create and evaluate an entirely new flavoring, should this be necessary. All too often a product is developed or modified without regard to this criterion and, in consequence, the flavoring problem is stated too late in the program giving inadequate time to carry through the necessary evaluations and storage trials.

The Process

The flavor technologist, or flavorist and food technologist working together, must be able to produce a facsimile of the end-product under laboratory conditions. This is the only sure way of establishing the acceptability of a flavoring in the finished product short of carrying out an expensive pilot or full production batch which could possibly be a failure. Alternative media may give a valuable indication of the ultimate flavor profile but nothing replaces assessment under actual formulation and processing conditions. For this reason, the general evaluation and assessment of a range of flavor samples by tasting under quite arbitrary conditions can lead to misleading results and the possible rejection of flavorings designed for and performing best under stated processing conditions. Where the processing is likely to affect the flavor profile or flavor stability due to inherent physical or chemical interactions, it is essential that identical raw materials and processing conditions be reproduced in the test laboratory or kitchen so that critical data can be established and evaluated for each stage of the process so far as flavoring is concerned. This is of particular importance when the new product is based on an untried technology where the outcome of each stage is uncertain and must be established experimentally. Again, close interdisciplinary collaboration gives the best chance of success.

Unit Operations as They Affect Flavor

When it comes to a consideration of the unit processes involved there are certain conditions which may significantly affect the resulting flavor profile. The most important of these are:

Time and Temperature.—Most flavoring materials are to some extent thermolabile. At elevated temperatures they may be lost to the system through evaporation or may change due to chemical interactions with other constituents. A knowledge of the temperatures to which the flavoring will be exposed and the dwell-times involved is most important in deciding the nature of the flavoring, particularly the best solvent to be used in its preparation.

The Mixing Sequence.—in many products, it is necessary to add the flavoring at the beginning of the process which exposes it to all the conditions encountered in processing; in other products, it may be added towards the end of the process thus avoiding potentially damaging processing conditions. Wherever possible, the flavoring should be exposed to the minimum of treatment consistent with uniformity in the end-product.

Open or Closed System.—Obviously, the handling of flavorings in an open system (e.g., blending in an open vat) is likely to result in greater losses than in a closed system (e.g., retorting in cans). Every precaution should be taken to minimize exposure by using covered vessels.

pH.—Many flavorings contain ingredients which are sensitive to changes in pH. It is essential that this particular condition be carefully reproduced during any development stage.

Pressure.—Both positive and negative pressure changes are likely to endanger added flavoring by altering the relative concentration of components in the headspace. Vacuum filling may result in the more volatile components of the added flavoring being preferentially lost to the system which results in an unbalanced profile in the end-product.

Contact with Air.—This is of particular concern in products which are aerated (e.g., ice cream, marshmallow, etc.). High-speed mixing can result in considerable volatile losses, but more importantly, can produce conditions conducive to oxidation. The pneumatic conveyance of powdered products containing flavorings (e.g., soup mixes, instant dessert powders, etc.) may also result in a significant loss of volatiles unless encapsulated products are employed.

Each product has its own technology so that few generalizations can be made about the behavior of flavorings. For this reason, it is necessary to consider products by groups in which the processing conditions are broadly the same. General flavor problems encountered during product development are frequently the subject of articles in the appropriate trade journals.

10

Meat, Poultry and Fish Products

This group includes all types of meat, meat by-products, poultry and fish, sold to the consumer either fresh for domestic preparation and cooking or processed into dried, semidried, fermented, deep-frozen or canned which may be eaten directly either cold or after reheating as appropriate. The opportunities for added flavorings (in this group they are usually called seasonings) are almost limitless. These items form the basis of what may be classified as savory foods in which the prime taste adjunct is salt and in which additional flavoring effects are achieved by the use of herbs and spices. Flavor enhancers such as MSG, the ribonucleotides and various processed materials such as hydrolyzed vegetable protein and yeast extract, which have a marked meaty character, are also widely used in compounded seasoning mixtures (Lyall 1965).

SEASONINGS

Generally, herbs and spices, or products derived from them (e.g., essential oils or oleoresins), are blended together to form seasonings; rarely does the flavoring effect of one herb or spice suffice to make a well-rounded flavor profile. Domestically, all too often only pepper and salt suffice to give an adequate savor, although the housewife is certainly becoming more venturesome in the use of herbs and spices. A seasoning should enhance the natural flavors present; maybe modify them a little to suit individual palates; but never swamp them. A few national dishes (e.g., goulash, chili con carne, etc.) tend to disprove this contention but, generally, seasonings should be used primarily to add zest and interest to a dish to make it appetizing to the consumer. There are no set rules for using spices; indeed, there is little precise information on this elusive art but most domestic recipes advise the nature and amount of spices to be used to suit aver-

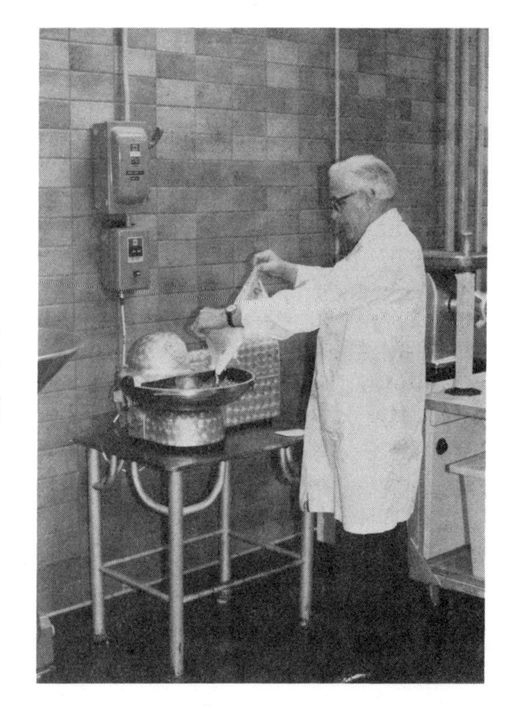

Courtesy of Griffith Laboratories, Inc.

FIG. 10.1 THE EXPERIMENTAL EVALUATION OF A SEASONING IN A CHOPPING FOR SAUSAGES

age tastes. There is now a very flourishing trade in packaged herbs and spices as well as preblended seasonings for specific product types. Skill in blending seasonings is only built up by practical experience as these valuable flavoring adjuncts do not all have the same flavoring power—some are excessively strong and overpowering; others are extremely weak and as readily suppressed.

The spices available to the meat processor range from the whole and milled natural products to a variety of processed spices made from essential oils and oleoresins. The whole spices are basically the most stable for any lengthy storage. Unfortunately, whole spices do not yield their flavor to food products and where they are used (e.g., in certain sausages, paté, etc.) the effect is only obtained when the spice is crushed between the teeth during mastication. For the large majority of meat products, it is necessary to employ finely milled spices in which condition their stability is considerably reduced. Up to about 20 years ago, virtually all seasonings used in meat processing were composed of ground spices; today, the industry is more technologically based and many traditional techniques have been modified. Increasingly, seasonings are being formulated with processed spice products, the nature of which has already been discussed in an earlier chapter.

Seasonings may include:

- Ground spices
- Spice essential oils and/or oleoresins, usually plated onto an edible carrier
- Encapsulated spice extractives
- Flavor enhancers
- Preservatives (where permitted)
- Food colors (where permitted)
- Imitation flavorings (e.g., smoke) (where permitted)
- Salt, dextrose, sucrose, etc.
- Milk whey, sodium caseinate
- Hydrolyzed vegetable protein, yeast extract, etc.

In most countries, the regulations governing this meat product group are more severe than for other foodstuffs. But as long as the individual constituents are permitted for use in meat products, there are no limits to what may be included in a seasoning formulation. Formerly, the composition of added seasonings was a closely guarded secret but times have changed. With high labor costs, most meat packers are prepared to have their individual formulations compounded for them by one of the specialist supply houses. For convenience, many of these are in individual unit-packs, sufficient to provide the seasoning necessary for one batch of end-product. Such a system has the advantage that it removes from the factory floor the onus of weighing and mixing errors and provides a ready control over the added seasoning without the need to employ specialist staff.

The stability of spices depends on the storage conditions used. It is desirable that stocks of seasonings be turned over at least every two months. They should be stored in well-sealed containers in a cool place away from direct heat and sunlight. Unit containers should preferably be made of a laminated material to minimize losses on storage. Once opened, seasonings stored in bulk containers (e.g., 50-kilo keg) should be used rapidly or any unused part transferred to a smaller container and resealed. Encapsulated spices and liquid spice seasonings (e.g., Soluspices) do not suffer from this disadvantage but should, nevertheless, be stored in a cool place.

FORMULATIONS AND PROCESS TECHNOLOGY

Sausages

Technically, a sausage is an emulsion having water and protein as the continuous phase and fat and solid fibers as the disperse phase.

The muscle protein, myosin, is the prime emulsifier and, to be effective, this must be released from the muscle-fiber cells. This is achieved by the combined action of salt and the physical breakdown induced by chopping or mincing. The addition of certain phosphates aids water retention and emulsion stability as well as reducing shrinkage during cooking. The use of such additives is generally the subject of legislative control.

The appearance and texture of a sausage is determined, not only by its constituents, but by the degree of comminution employed in its manufacture. The texture may be coarse to creamy smooth, depending upon the traditional characteristics of a particular type of sausage (Komarik *et al.* 1974).

The formulation of sausages is very variable and is dictated by the need to satisfy a wide spectrum of tastes and traditions. In many countries, the formulation is controlled so that any detailed formulae given here would be inappropriate. The following comments must be regarded as generalizations which may or may not apply in specific instances. The items covered by regulations usually include:

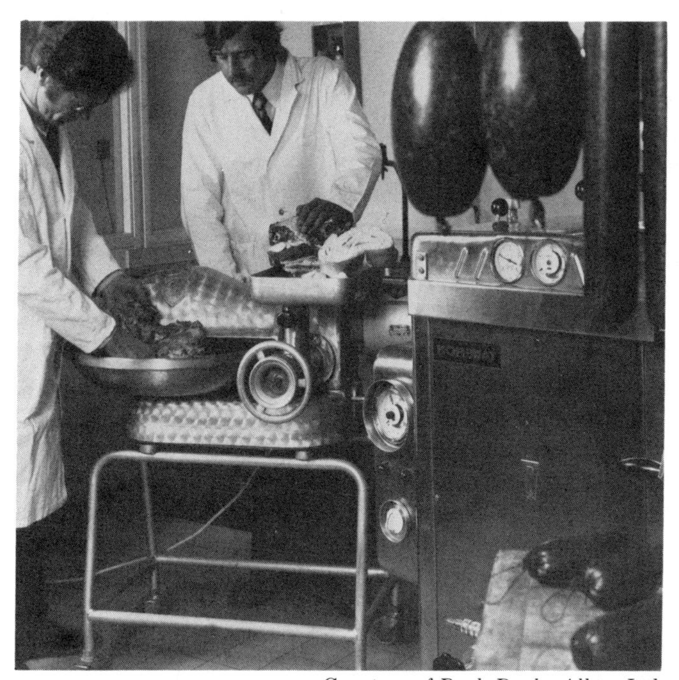

Courtesy of Bush Boake Allen, Ltd.

FIG. 10.2. THE PILOT SCALE PRODUCTION OF DRIED SAUSAGES IS PART OF THE EVALUATION OF SEASONING COMPOSITIONS

Courtesy of Bush Boake Allen, Ltd.

FIG. 10.3. SEASONINGS MUST BE EVALUATED UNDER THE SAME PROCESSING CONDITIONS AS USED IN THE FACTORY
Here sausages are undergoing a smoking/cooking process.

(a) meat content by type and weight; (b) fat content; (c) nonmeat protein; (d) fillers; (e) additives (e.g., permitted preservatives, colors, phosphates, flavorings); and (f) water content.

Within the regulations, sausages may contain the following ingredients and be processed as outlined.

Fresh Sausages (e.g., Breakfast Sausage).—May contain meat, fat, filler (bread, rice, rusk, cracker meal, soya flour, cereal flours), binder (bread, rusk, soya products, milk powder, flours), spices and other seasonings, coloring, flavor enhancers (MSG, hydrolyzed vegetable protein, yeast autolysates), salt, phosphates, preservatives (sulfur dioxide) and water.

Method.—Preminced, chilled lean meat (pork or beef) is placed in a bowl-chopper with the seasoning and other additives. Chopping is commenced and the filler, presoaked with iced water, is added. Chopping is continued until the desired texture is achieved; the cubed fat is added and chopping continued as necessary. The mix is transferred to a filling machine; prewashed and dried casings are run onto the nozzles and the sausages are filled. Linking may be done mechanically or by hand. The resulting strings of sausages are hung on conveyor hooks and passed through a cooling unit which simultaneously dries the surface. The sausages are weighed off, wrapped and stored at 1–2°C.

Where a bowl-chopper is not available, a mincing machine fitted with a 3/16 in. (5 mm) plate for the meat and a 3/8 in (10 mm) plate for the fat may be used.

Skinless sausages are made with the same mixture as above and are filled into 40–50 ft long cellulose casings which are linked in the usual way. Immersion for 3 min in water at 65°C (150°F) partially cooks the outer layer of the sausage and immediate cooling in iced water sets the coagulated proteins and fixes the shape. The skin is then removed either mechanically or by hand. A continuous process which eliminates the need for skins is now widely used; however, the temperatures encountered during the process may affect the added seasoning involving a higher initial usage rate to compensate for losses.

Cooked Sausages (e.g., Frankfurters, Wieners, Etc.).—May consist of beef, pork and other filler meats, cereal, nonfat milk solids, glucose syrup solids, dextrose, salt, sodium nitrate and nitrite (as permitted), sodium ascorbate, spices and other seasonings, phosphates and water.

Method.—Wide variations occur throughout the industry (Gerrard 1969). Frankfurters are usually prepared from beef and/or pork trimmings. The meat is minced and the phosphates, salt, filler, sugar, cure, seasoning and water all added. Chopping is continued as necessary and, just before completion, any sodium ascorbate is added. The temperature must be maintained at below 15°C (60°F) throughout the chopping period. The mix is then vacuumized to remove air and is then filled into casings. The filled sausages are held at 3°C (36°F) to cure, followed by drying at 55°–60°C (132°–140°F) for 10 to 20 min. They are then smoked at 74°C (165°F) with 75–80% RH for 1¼ to 1½ hr. At the end of the smoking period the sausage is cooked, either by raising the temperature of the smokehouse or by hot-water sprays, so that the internal temperature is 65°C (150°F). The sausages are then cooled to 5°C (40°F) by cold-water sprays or cold air. The product is then hung to dry.

An alternative patented method is used in the United States. This involves the incorporation of glucosedeltalactone which, during the smoking, is hydrolyzed to gluconic acid, thus reducing the pH and accelerating the curing stage. The smoking is carried out at a higher temperature (95°–150°C) (200°–300°F) for only 20 min. A subsequent cooking stage is not necessary. The product is claimed to be more succulent and have a better flavor.

Fermented Sausages (e.g., Salami, Etc.)—Ingredients may clude meat, fat, spices and other seasonings, salt, sugar, sodium nitrate.

Method.—The technology of making fermented sausages has undergone considerable development in recent years but the traditional methods are still widely used. In this, the ingredients are mixed and chopped to the desired texture (often very traditional), held for 2–3 days at 3°C (36°F) to allow curing to take place and then stuffed into punctured natural casings. The resulting sausages are then held at 21°–24°C (70°–75°F) with 75–80% RH for 2 to 10 days to initiate fermentation. If required, the sausages may be smoked at 32°C (90°F), then dried at 7°–13°C (44°–55°F) with 70–72% RH for 10 to 19 days, depending on the size and type of sausage. This last stage is critical and requires that it be carried out using 15–25 air changes per hour. The moisture loss should be in the order of 40% of the original weight.

A modern development, in which the bacterial growth is initiated and controlled, is gaining wide acceptance. In this, the cultures of specific organisms are mixed with the ingredients prior to chopping and the resulting sausages are allowed to ferment, under carefully specified conditions, for only 16 hr. The resulting products are claimed to have a more consistent and acceptable flavor.

Burgers (e.g., Beefburgers, Hamburgers, Etc.).—These usually include meat, fat, filler, binder, onion, spices and other seasonings, flavor enhancers, salt and water.

Method.—These types of product are extremely popular and can be made with the simplest of equipment necessary to produce a uniform mix and a coarse texture. Fresh lean meat is desirable; it is difficult to produce a good quality product if a high percentage of filler is used. Where the addition of onion is appropriate, it is preferable to incorporate dehydrated onion powder or a liquid onion flavoring rather than fresh onions which produce too loose a texture.

Cured Meats

The curing process is a long-established method of preservation which results in products having a very acceptable color, texture and flavor. The technology involved is outside the scope of this book. For those who wish to get a better insight into the numerous methods employed, reference should be made to the AVI Food Products Formulary Series, Volume 1. Meats, Poultry, Fish, Shellfish (Komarik *et al.* 1974). Basically, the preservatives used are salt and nitrate/nitrite, although smoking is often a supplementary form of preservation and flavor modifier. The bulk of curing is confined to bacon and ham, but the principles are of much wider application. The use of nitrate/nitrite as a curing agent is of long standing but recently has been the subject of regulatory discussion. In the United States, proposed regulations have been published in the Federal

Register, November 11, 1975, which would: ban the use of nitrate in all cured meats and poultry products except fermented sausages and dry-cured prime meats; lower the maximum permitted nitrite in cured meat and poultry products to 156 ppm and at the same time establish a maximum level for residual nitrite in specified products. It is also proposed that the use of nitrate/nitrite be prohibited in foods intended for babies. These proposals are part of the continuing debate on the safety of nitrates in cured products.

There is very little call for the use of added flavorings, although some success has been achieved by the use of seasoned brines and by the tumbling technique for the flavoring of hams. Honey and maple flavors are of particular interest in this respect. Additional flavor notes may be introduced by the use of special woods as a source of smoke; maple, hickory and oak produce characteristic profiles in the end-product (Gerrard 1969; Kramlick *et al.* 1973).

Canned Meat Products

The range of products in this group is almost limitless as the main source of flavor is the added gravy used with the meat, which itself can be appropriately seasoned. The composition of these products is defined and controlled in some countries, particularly with regard to meat content and product description.

The main ingredients encountered are: prepared meat (either fresh or preroasted, in cubes, pieces, slices, or ground); textured vegetable proteins (where these are permitted); gravy or sauce comprised of flours, salt, sugar, flavor enhancers, tomatoes, onions, herbs and spices, caramel color; jelly comprised of gelatine, agar, salt and water; cereals usually of rice, noodles, dumplings, etc.

Depending on the regulations, imitation meat flavors may be incorporated into certain types of canned meats.

The production of corned beef is somewhat special as it results in a 35% shrinkage of the original meat and yields a liquor by-product which may be used as "beef broth" or concentrated as a so-called "meat extract." The curing medium for corned beef is salt, nitrate/nitrite (as permitted), sugar, glucose syrup solids and MSG.

Canned meat products have their own particular processing characteristics and problems so that generalizations could be misleading. The following products illustrate the main factors to be considered when creating a seasoning for this group.

Canned Stewed Steak in Gravy.—The clean deboned meat is prepared to whatever size is desired for the product (from 1¼ in. cubes to ¼ in. strips to coarsely ground). The can is filled with the meat, usually automatically by weight, and the premade gravy added. The can is double-seamed and vacuum packed prior to sterilization for 30 min at 107°C (224°F) followed by 1½ hr at 115°C (238°F) or longer depending on the can size.

Canned Whole Chicken or Chicken Breasts.—Approximately 2-lb birds are used for canning. These are fully prepared (cleaned, drawn and well washed to remove all blood) and handpacked into cans containing half of the required liquor (a seasoned chicken broth or plain brine). After packing, the remainder of the liquor is added, the cans sealed and processed at 115°C (238°F) for 75 min followed by rapid cooling to 40°C (104°F).

Canned Corned Beef.—Cleaned, deboned meat is passed through a rotary disc cutter, sieved (with recycling) until uniform, then subjected to open steam heating in broth for 25 min. After cooking, the meat and broth are separated, the latter being used to make a concentrated "meat extract." The heated beef is cured to 4°C (40°F) for 12–14 days with salt and nitrate/nitrite mix and is then blended and filled, by means of a rotary can stuffer, into cans which already contain a quantity of gelatine solution. The cans are vacuum sealed and processed at 115°C (240°F) for 1½–2½ hr, depending on the size of the cans, followed by rapid cooling to 40°C (104°F).

Canned Pork Luncheon Meat.—There is an enormous range of luncheon meats which consist of chopped, cured meats with added seasoning and sometimes cereal. Individual formulations are given by Komarik *et al.* (1974), the following processing method being common to them all:

The lean meat is chopped, together with the seasoning, cure, cereal and water, until a coarse texture is obtained. The fat, cut into strips, is added and chopping continued until a smooth, fine texture and even color are obtained. The mix is then vacuum treated to remove all air (or the chopping may be carried out in a vacuum bowl-chopper) and is then filled out into cans. Fully automatic, high speed fillers are now in general use for this stage. The filled cans are vacuum sealed and processed at 110°C (230°F) for a sufficient time to ensure sterilization. The cans are cooled rapidly, under pressure in the case of large cans to ensure that they retain their correct shape.

Meat Pies

There are two main varieties of meat pie: (a) made with a short-paste crust and filled with various meats plus gravy; usually sold in a foil tray and eaten hot (e.g., steak and kidney pie, chicken pie, etc.); and (b) made with a hard-paste crust filled with comminuted meat often surrounded by jelly and usually eaten cold (e.g., pork pie, Grosvenor pie, etc.)

These products come in all shapes and sizes, those intended for reheating usually being distributed under refrigeration, those for eating cold being sold within a short period of manufacture. Meat pies are the subject of legislation in many countries, the regulations defining minimum meat content and designation.

The following illustrate the processing conditions involved.

Steak and Kidney Pie.—Pastry: fat, flour, salt and water. Filling: meat, kidney, onion, flour, seasoning and caramel color. The short-pastry dough, known as "cold paste" is made conventionally or by a hot high-speed method. It is then quickly sheeted and cut to form bases and tops. The filling meat is prepared to the desired size and deposited into the preformed pie base. The gravy is premade, cooked to thicken uniformly, rapidly cooled and metered onto the meat portions. The pies are topped and baked at 232°C (450°F) for 25 mins. They are then cooled to 15°C (60°F) and either wrapped or boxed prior to freezing.

Pork Pie.—The traditional hand-raised pork pie, which is so popular in the United Kingdom has largely been replaced by automatic pie formers.

Pastry: fat, flour, salt and water. Filling: meat and fat (about 50:50), binder, filler, gelatine, seasoning. In certain types of pies eggs or cheese are included.

The pastry dough, known as "hot" or "boiled" paste, is made using boiling water and is passed either to a cutter to produce the tops or to a divider to give the bottoms which are later pressed to form the walls of the pie. The filling is deposited into the shells and the tops placed in position. The filler is prepared from selected, cured meat which is minced with any filler, binder, seasoning and fat to give the desired texture. After the pies are formed, the tops are glazed by spraying with gelatine solution. Pies are then baked at 205°C (400°C) for about 40 min to 1½ hr, depending on the size of the pie. The cooked pies pass through a cooling tunnel and jelly (10% gelatine) is injected between the casing and the shrunken meat filling. The pies are finally cooled to 15°C (60°F) prior to cellophane wrapping.

Fish Products

The processing of fish involves a specialist technology and reference should be made to one of the standard texts for specific details of processing methods. The following products give an indication of areas where seasonings or flavorings may be applied.

Smoked Fish.—The smoking of fish is traditional and the process has remained largely unchanged over the years (Tressler and Lemon 1951). The source of smoke depends upon the wood available in the main processing regions; these give to the product a very distinctive flavor. Liquid smoke flavorings have a limited use and some experimentation has been done on their use in the smoking of kippers, but none of the methods reproduces the aroma and flavor obtained by direct smoking. An essential part of the process is the drying which takes place at the same time; this inhibits bacterial growth but does little to reduce mold growth.

Kippers are traditionally smoked by exposure for 3½ hr at 27°C (80°F), during which time the water content is reduced by some 12%, the flesh attaining a rich smoke flavor and an overall golden-brown color.

Canned Herring.—Frozen herrings are first thawed, headed, gutted and filleted. The fillets are washed well in brine and steam-cooked for 10 min, lightly dried and filled into flat cans containing a flavored sauce. The cans are lidded, vacuum sealed and retorted for 60 min at 115°C (240°F).

The sauce used is based on a simple white sauce which is stable under these conditions to which is added a seasoning or flavoring. The most important flavors used include tomato with light herbs and spices such as dill and fennel.

Cured Herring.—A variety of methods are used for pickling fish. In most cases the fish is dressed, headed, gutted and washed well prior to immersion in the pickling liquor. This usually is a brine/vinegar mix to which may be added whole spices such as bay leaves, allspice, pepper, clove and chili pepper. The fish is allowed to marinate until the curing liquor has penetrated the flesh—curing times varying from 3 to 7 days. The cured product is then removed, packed into brine and is ready for distribution, usually under refrigeration.

Comminuted Fish Products (e.g., Fish Cakes, Fish Sausages, etc.).—The processing of whole fish pieces (e.g., fish sticks) results in a considerable weight of trimmings which are perfectly edible but not salable as such. These are generally processed into fish cakes or other products in which the fish is comminuted. Fish cakes consist of shredded fish admixed with potato or other starchy material and bound together with a little egg; fish sausages have a composition not unlike that for meat sausage the fish being chilled, chopped with salt and then mixed with starch in a slurry, seasoned and chopped to a fine emulsion before filling into casings. The filled sausages are then cooked at 100°C (210°F) for 30 min, rapidly cooled to 2°C (35°F) for distribution. In both types of product, seasoning may readily be incorporated. For fish cakes, this is normally limited to pepper and parsley whereas with fish sausages a wider range of spices, including nutmeg, mace and ginger, as well as garlic and onion, may be used to give a definite character to the end-product.

Fish sausages based on fish protein concentrates, which have little intrinsic flavor, usually contain hydrolyzed vegetable protein as well as imitation meat flavorings in order to give them an acceptable flavor.

FACTORS AFFECTING THE USE OF SEASONINGS

The basic technology involved in meat processing is highly complex and often specific to a particular product line, but certain aspects emerge as significant in determining the nature of any added seasonings. The most important factors are:

(a) The nature of the raw materials.

(b) The pretreatment of the raw materials, particularly the use of salt and curing mixtures.

(c) The stage and method of incorporating the seasoning.

(d) The degree and nature of any comminution stage: open bowl-chopper, vacuum bowl-chopper, mincing.

(e) Any postmixing treatment, particularly cooking, smoking, drying, retorting.

(f) Temperature and times involved at each stage, particularly any open-system heat processing or abnormally high temperatures encountered in some automated processes.

(g) The nature and level of any added preservatives.

(h) Method of packaging, particularly if this involves vacuum packing.

(i) Post production handling and storage, particularly if this involves refrigeration.

(j) Domestic preparation prior to eating.

All of the above will have an effect upon the flavor of the product as consumed and any assessment of a seasoning must include the critical factors in any trial production runs.

EVALUATION OF SEASONING BLENDS

For the initial stages of the development of a seasoning blend, the often-complicated preparation of the final product is frequently unjustified. In such cases, use may be made of a neutral medium. The following have been found satisfactory for this purpose:

Neutral Soup

This should be made from a base consisting of salt (18%), sugar (22%) and corn starch (60%). The base can be premixed and is convenient. For use, weigh out 4.5 g of the mixture into a 250-ml beaker, mix into a smooth cream with a few milliliters of cold water and add boiling water to make 100 ml of soup; stir well and bring to the boil to thicken. The seasoning mix is added to this base, stirred well to ensure even distribution and is then ready for evaluation. For tasting, the temperature should be about 60°C (140°F).

Reconstituted Mashed Potato

Make up in accordance with the manufacturer's instructions but omit any added butter. This medium is of value for the assessment of seasonings containing onion and garlic.

Thick White Sauce

A traditional thick white sauce, made with equal quantities of fat and flour and using milk as the liquid, makes a very good basis for the evaluation of curry powders and other highly piquant seasonings.

It does not matter which base is used so long as the test conditions are reproducible. The final testing should always be carried out in the end-product itself and any final adjustments made as a result of a panel evaluation.

SEASONINGS

Unlike flavors which have a specific profile, seasonings can be blended to give a wide spectrum of effects. Before one can consider the blending process it is necessary to have a clear idea of the nature of the meat itself as its intrinsic flavor and texture greatly affect what herbs and spices are compatible to give the desired profile in the end-product. Meat has five main constituents: fibrous tissue; fat; connective tissue; fibrous capsules; and sheaths and bone. The proportion of each present, particularly of fat, largely determines overall flavor of the raw materials which, in turn, influences the ultimate cooked flavor of the product. It is obvious that the variables are enormous and that the blending of seasonings, as well as the amount used, cannot be subjected to hard and fast rules. Robert Carrier in his book, *The Great Dishes of the World*, aptly states that the correct spice combination is the one that tastes right to you. He suggests that it is wise to be selective and, unless following a tested recipe, not to combine too many at one time. A spice blend which is suitable for one product may be quite unsuitable for another made from similar ingredients but processed differently, just as a seasoning which is suitable for pork is most unlikely to be as attractive in a beef-based product.

Some of the considerations to be taken into account in the building of a seasoning have already been discussed in an earlier chapter. In

Courtesy of Griffith Laboratories, Inc.

FIG. 10.4. BULK BLENDING OF SEASONINGS FOR THE MEAT INDUSTRY

Fig. 7.9 the gradation of intrinsic flavor of the meat, poultry or fish in the formulation is balanced against a similar grading of the flavoring power of culinary herbs and spices. This has been used to build up a collection of basic seasoning formulations which are given in Table 10.1. These are, of course, only of value as a guide and considerable variations are possible by simple readjustment of the relative proportions of the various spices used.

When it comes to blending herbs and spices, they are not unlike colors—some go well together, others clash. In deciding what spices to use, select first those which are complementary to the particular

TABLE 10.1
SPICE FORMULATIONS FOR MEAT PRODUCTS

	Sausages (Fresh, Dry and Semidry)											Canned		Meats					Other Meat Products	
Ground Spices or Equivalent Spice Extractives on Salt/Dextrose Base, % w/w	Fresh Pork (UK)	Fresh Pork (USA)	Fresh Beef	Beef (Kosher-style)	Bockwurst	Saveloy	Bologna	Chorizo	Frankfurter	Liver	Salami	Pork Luncheon Meat	Meat Loaf	Mock Chicken	Stewed Beef in Gravy	Meat Balls	Spaghetti in Meat Sauce	Ham	Pork Pie (UK)	Hamburger Mix
Allspice	10						5							5			50		35	35
Bay laurel																	3		5	
Cayenne (capsicum)	1		5			0.5		0.8								1	2			

Cinnamon/cassia						3											10		
Celery	2													50	8	3			10
Clove																	10		10
Coriander	10						10	1.5		10		3							
Dill seed																	1		
Garlic powder											5			5	1.5	10			
Ginger		12	10	12	30	6	5	1.5	10	2.5	30	5	5			0.5	5		
Mace	5		10		20	2.5			10			2	10				0.5	10	
Marjoram										7.5									
Mustard flour							10												
Nutmeg	10	10	20	9		2.5	10		5	5		5	20						
Onion powder					15										62	75			5
Oregano								0.7									10		
Paprika								88							15				
Black or White pepper	45	40	55	40	35	85	60	7.5	65	75	65	85	60	40	13	7	8.5	15	15
Sage	10	26		26						10			5			3			25
Thyme	5	12		13		0.5									0.5	0.5		35	

meat to bring out its particular flavor character. Then modify this basic flavor by the use of much smaller quantities of other herbs and spices to emphasize a particular flavor note. In this respect, the spices may also be classified from their principal effect and a useful guide is given in Table 10.2.

TABLE 10.2
FLAVORING EFFECTS OF HERBS AND SPICES

Flavor Character	Herb or Spice
Pungent	Capsicum; black and white pepper; ginger; mustard; horseradish.
Fruity	Anise; caraway; cardamom; coriander; cumin; dill; fennel.
Phenolic	Clove; pimento (allspice); cinnamon leaf; West Indian bay.
Bitter	Nutmeg; mace.
Sweetly aromatic	Sweet bay laurel; rosemary; sweet basil; tarragon (estragon); sweet marjoram; parsley herb; dill weed; Spanish sage.
Strongly aromatic, spicy	Celery; fenugreek; lovage; cinnamon; cassia; saffron.
Heavy, overpowering	Thyme; origanum; savory; Dalmatian sage.

The use of the correct proportion of salt is most important. This significantly influences the effect of added herbs and spices and its level in the end-product should be established before the seasoning formulation is finalized.

PROTEIN EXTENDERS AND STRUCTURED ANALOGS

The problems of feeding an ever-increasing population has been the subject of intense study. Many organizations associated with this program have stressed the urgent need to provide more food from traditional sources and develop nontraditional sources consistent with nutritional needs. Progress in this field has been rapid. Currently, several sources of nonconventional proteins have been proven satisfactory from a nutritional point of view. All require considerable modification before they are acceptable as human food. The problems involved include both texture and flavor since the prime materials are usually without form and require "texturization." These proteins have either an unpleasant flavor or are insipid and require the addition of flavorings to make them edible.

The following are the present commercially viable protein sources for structured protein foods: (a) soy bean, cottonseed and groundnut proteins, (b) fish protein concentrates, (c) leaf protein concentrates, (d) single-cell proteins.

In all cases, the concentrated proteinaceous material is a whitish, amorphous powder having almost no odor, a distinctly bitter and often unpleasantly mealy flavor, and an objectionable after-taste. In some commercial grades, these characters have been considerably reduced and reference should be made to the manufacturer's specification sheets for detailed claims in this respect.

Considerable research has gone into the texturizing of these protein foods. In the case of soy protein, this may now be converted into a fibre-like structure, either by extrusion or spinning, which has a chewing characteristic not unlike that of meat.

The flavoring of such products has to take into account the need for: (a) provision of a prime flavor, either to enable the product to be used on its own or with other traditional protein foods; (b) a "cover" to disguise the basically unpleasant flavor inherent to the product; and (c) a seasoning blend that will make the end-product attractive to the consumer.

The range of products in which vegetable proteins may be used, either as a powder or in the textured form, is wide and embraces soups, sauces, puddings, burgers, meat extenders, cookies, extruded sweet and savoury snack foods and even sugar confectionery items. The formulations for many such products associated with meat are given by Komarik *et al.* (1974). In the case of the textured soy proteins, these may be converted into meat analogs to be used as the main protein source; or, admixed with natural meat, poultry or fish as an extender depending on the current legislation governing the use of these materials in meat products. Analogs and extended meat products are finding a growing market as dietary supplements, particularly for institutional feeding where costs of providing an adequate nutritional intake are high. Their use, of course, is to some extent dictated by the relative costs compared with natural meat.

Most simulated meat products are based on the use of textured vegetable protein fibers bound together with a mixture of hydrolyzed vegetable protein, monosodium glutamate or other flavor enhancers, caramel coloring, imitation meat flavorings, fats, binding agents and seasoning. Where such products are cooked by broiling or roasting and do not come into contact with water (e.g., burgers) the flavor retention is reasonable; but where they are incorporated into a product having gravy or sauce, then the added flavorings tend to be leached off the fibers into the liquor leaving the textured protein somewhat tasteless when eaten. In some cases, it is necessary to ensure that the gravy is itself well flavored and rather thicker than usual so as to cover this deficiency.

Meat Flavorings

The chemistry of meat flavors is very complex (Cole and Lawrie 1975) and is based largely on the Maillard reaction between amino acids and sugars which occurs naturally in meat tissue. With meats, one expects the difference in flavor which results from different methods of cooking and the degree of heat that is used, e.g., roast beef is totally different from boiled beef. Such flexibility in end-product flavors cannot, as yet, be achieved with any of the structured proteins. Where these are used, it is necessary to impose the selected flavor by the use of appropriate imitation flavorings (e.g., roast beef, fried chicken, etc.). Several of these imitation flavorings are now available. Their correct application makes it possible to fabricate products which closely imitate natural meat in both flavor and texture. In such simulated meat products the flavor profile is, of course, fixed and determined by that of the added flavoring. One cannot modify the end-product flavor to any great extent by changing the method of preparation.

To date, the approach to flavoring meat analogs has been largely classical with the flavorist attempting to reproduce an observed aromatic effect using permitted flavoring chemicals. Imitation meat flavors of the future are more likely to be based on established reactions similar to those which take place naturally when meat is cooked. Already, there are a large number of patents taken out in this field; but rather than producing the end-flavor, it may be advantageous to create blends of precursors which would be intrinsically tasteless and only yield the appropriate flavor on cooking. Unfortunately, the Maillard reaction is complex and tends to be open-ended; the nature of the final mixture of aromatic chemicals is determined by the conditions and duration of the reaction and the nature of the amino acids and sugars involved (see Van Den Ouweland *et al.* 1978). Herein lies the main problem for the flavor technologist—how to achieve a stable and consistent flavor which will maintain its characteristic profile, not only during the pre-use storage but right through processing up to the time of consumption of the end-product. By reacting together selected amino acids and sugars, a range of meat-like flavors can be produced with profiles similar to beef, pork and chicken. Innumerable patents have been registered covering almost all possible combinations of reactants and processing conditions; but however successful the resultant flavoring may be, its use is still limited and in no way reproduces the flexibility of flavor effect which can be achieved by cooking natural meats.

The problem of applying flavoring to textured meat-like fibers has yet to be satisfactorily solved. The ultimate aim is to produce a fiber carrying the precursors similar to those found in meat. Unfortunately, the heat created during the extrusion also induces chemical changes in the precursors and, as yet, this difficulty has to be overcome. Because textured vegetable protein has the tendency to retain a greater proportion of both fat and water than meat with which it is mixed, two methods of flavoring have been evolved:

(a) Oil-soluble flavors may be sprayed onto the extruded fibers which will absorb up to 130 times their dry weight. Such flavors are retained much more satisfactorily than water-soluble ones. However, their profiles are limited because so many of the "meaty" components are not readily soluble in fixed oils or fats.

(b) Oil-in-water emulsions are sprayed onto the fibers. This system overcomes the flavor profile problem as the water-soluble "meaty" notes can be incorporated into the continuous phase. This method is preferred as it displays relatively good flavor absorption and retention during cooking.

BIBLIOGRAPHY

Meat and Fish Processing

An annotated bibliography, covering books issued between 1967 and 1974, is published by International Food Information Service, Commonwealth Agricultural Bureaux, Farnham Royal, Slough, SL2 3BN, England. The following are the major texts.

BRANDERBURG, W. and KRAMER, H. 1972. Industrial Fish Processing, Izdatel'stvo Pishchevaya promyshlennosi, Moscow, U.S.S.R. (Russian)

DANILOV, M. M. 1969. Handbook of Food Products. Meat and Meat Products. Israel Program for Scientific Translations, Jerusalem, Israel.

FISHER, J. 1961. Pork and Pork Products. Barrie and Rockliff, London.

GERRARD, F. 1969. Sausage and Small Goods Production, 5th Edition. Leonard Hill Books, London.

GERRARD, F. 1971. Meat Technology, 4th Edition. Leonard Hill Books, London.

GRAU, R. 1969. Meat and Meat Products, 2nd Edition. Verlag Paul Parey, Berlin, W. Germany. (German)

KOMARIK, S., TRESSLER, D. K. and LONG, L. 1974. Food Products Formulary, Vol. I. Meats, Poultry, Fish, Shellfish. AVI Publishing Co., Westport, Conn.

KARMAS, E. 1970. Fresh Meat Processing. Food Processing Review, Noyes Data Corporation, Park Ridge, New Jersey.

KRAMLICH, W. E., PEARSON, A. M. and TAUBER, F. W. 1973. Processed Meats. AVI Publishing Co., Westport, Conn.

KREUZER, R. (Editor). 1971. Fish Inspection and Quality Control. Fishing News (Books), London. (Papers presented at the F.A.O. Congress, Halifax, Novia Scotia, Canada.)

LEVIE, A. 1970. The Meat Handbook, 3rd Edition. AVI Publishing Co., Westport, Conn.

PESLE, O. and SCHWIERZINA, A. 1970. Quality of Fish and Fish Products. VEB Fachbuchverlag, Leipzig, E. Germany. (German)
PRICE, J. F. and SCHWEIGERT, B. S. 1971. The Science of Meat and Meat Products. W. H. Freeman & Co., San Francisco, Calif.
TRESSLER, D. K. T. and LEMON, J. M. 1951. Marine Products of Commerce, 2nd Edition. Reinhold Publishing Co., New York.

Flavors and Meat Processing

ANON. 1971. Meat flavour—a review. Flavour Ind. *2*, 212–215.
ANON. 1975. Processing and flavouring of proteins. Proc. 3rd Flavour Application Symp. Br. Soc. Flavourists, London. Int. Flavours Food Additives *6*, 275–282.
ANDERSON, R. L. and WARNER, K. 1976. Acid sensitive soy proteins affect flavor. J. Food Sci. *41*, 293–296.
BRECLAN, E. W. and DAWSON, L. E. 1970. Smoke-flavored chicken rolls. J. Food Sci. *35*, 379–382.
CHOU, I. C. and BRATZLER, L. J. 1970. Effect of sodium nitrite on flavor of cured pork. J. Food Sci. *35*, 668–670.
COLE, D. J. A. and LAWRIE, R. A. 1975. Meat. Proceedings of 21st Easter School in Agricultural Science, University of Nottingham, 1974. (American Edition) AVI Publishing Co., Westport, Conn.
DOUGAN, J. and HOWARD, G. E. 1975. Some flavoring constituents of fermented fish sauces. J. Sci. Food Agric. *26*, 887–894.
DWIVEDI, B. K. 1975. Meat flavor. CRC Crit. Rev. Food Technol. *5*, 487–535.
FISCHETTI, F. 1975. Flavoring textured soy proteins. Food Prod. Dev. *9*, No. 6, 64.
FRANZEN, K. L. and KINSELLA, J. E. 1974. Parameters affecting the binding of volatile flavor components in model food systems. I. Proteins. J. Agric. Food Chem. *22*, 675–678.
GOOSENS, A. E. 1974. Protein food—its flavours and off-flavours. Flavour Ind. *5*, 273–274, 276.
GORBATOV, V. M. 1971. Liquid smokes for use in cured meats. Food Technol. *25*, No. 11, 71–77.
GORDON, A. 1972. Meat and poultry flavours. Flavour Ind. *3*, 445–446,448.
GRACE, J. 1974. The use of degraded proteins in foodstuffs for nutrition, flavour and flavour enhancement. Food Technol. Aust. *26*, No. 2, 60–61, 63–64,67.
HASHIDA, W. 1974. Flavour potentiation in meat analogues. Food Trade Rev. *44*, No. 1, 21–32.
HEATH, H. B. 1965. Spicing meat products. Food Manuf. *40*, No. 1, 52–54,56.
HILL, M. H. 1973. Compounded seasonings for meat products. Flavour Ind. *4*, 164–168.
HIRAI, C. *et al.* 1973. Isolation and identification of volatile flavor compounds in boiled beef. J. Food Sci. *38*, 393–397.
HOERSCH, T. M. 1970. Synthetic meat flavor. Can. Pat. 831,472. Jan. 6.
LAWRIE, R. A. 1970. Variation of flavour in meat. Flavour Ind. *1*, 591–594.
LIEBICH, H. M. *et al.* 1972. Volatile components in roast beef. J. Agric. Food Chem. *20*, 96–99.
LYALL, N. 1965. Some Savoury Food Products, Food Trade Press, London.
MARBROUK, A. F. 1973. Preparation of aqueous beef flavor precursor concentrate by selective ultra-filtration. J. Agric. Food Chem. *21*, 942–947.
MAGA, J. A. 1973. A review of flavor investigations associated with soy products, raw soya beans, defatted flakes, flours and isolates. J. Agric. Food Chem. *21*, 864–867.

MAY, C. G. 1974. An introduction to synthetic meat flavours. Food Trade Rev. *44*, No. 1, 7, 9–10, 12–13.

NIXON, L. N. and JOHNSON, C. B. 1975. Volatile medium chain fatty acids and mutton flavour. J. Agric. Food Chem. *23*, 495–498.

PERSSON, T. and VON SYDOW, E. 1973. Aroma of canned beef. GLC and MS analysis of the volatiles. J. Food Sci. *38*, 377–385.

PERSSON, T. and VON SYDOW, E. 1974. Aroma of canned beef. Application of regression models relating to sensory and chemical data. J. Food Sci. *39*, 537–541.

PERSSON, T. *et al.* 1973. Aroma of canned beef. Models for correlation of instrumental and sensory data. J. Food Sci. *38* 682–689.

PETERSON, R. J. *et al.* 1975. Changes in volatile flavor compounds during retorting of canned beef stew. J. Food Sci. *40*, 948–954.

PRENDERGAST, K. 1974. Protein hydolysate—a review. Food Trade Rev. *44*, 14, 16–21.

RAO, CH.S. *et al.* 1975. Effects of polyphosphates on the flavor volatiles of poultry meats. J. Food Sci. *40*, 847–849.

RAO, CH.S. *et al.* 1976. Effects of polyphosphates on carbonyl volatiles of poultry meats. J. Food Sci. *41*, 241–243.

ROZIER, J. 1970. The flavor of meat and pork products. Fr. Ses Parfums *13*, No. 69, 244–251. (French)

ROZIER, J. 1971. The flavor of meat and meat products. Riv. Ital. Essenze, Profumi Piante Off. *53*, Mar., 17–26. (Italian)

SATO, K. and HERRING, H. 1973. Chemistry of warmed-over flavor in cooked meats. Food Prod. Dev. *7*, No. 9, 78,80,82,84.

SATO, K. *et al.* 1973. The inhibition of warmed-over flavor in cooked meats. J. Food Sci. *38*, 398–403.

TILGNER, D. J. and DAUN, H. 1970. Antioxidative and sensory properties of curing smokes obtained by three basic smoke generating methods. Lebensm. Wiss. Technol. *3*, No. 5, 77–82. (German)

VAN DEN OUWELAND, G. A. M., OLSMAN, H., and PEER, H. G. 1978. Challenges in Meat Flavour Research. *In* Agricultural and Food Chemistry. R. Teranishi (Editor). AVI Publishing Co., Westport, Conn.

WASSERMAN, A. E. 1972. Thermally produced flavor components in the aroma of meat and poultry. J. Agric. Food Chem. *20*, 737–741.

WASSERMAN, A. E. and TALLEY, F. 1972. The effect of sodium nitrite on the flavor of frankfurters. J. Food. Sci. *37*, 536–538.

WILSON, R. A. 1975. A review of thermally produced imitation meat flavors. J. Agric. Food Chem. *23*, 1032–1037.

WILSON, R. A. and KATZ, I. 1972. Review of literature on chicken flavor. J. Agric. Food Chem. *20*, 741–747.

WILSON, R. A. and KATZ, I. 1974. Synthetic meat flavours. Flavour Ind. *5*, 30–35,38.

ZIEMBA, Z. and MALKKI, Y. 1971. Changes in odor components of canned beef due to processing. Lebensm. Wiss. Technol. *4*, 118–122. (German)

11

Baked Goods and Bakery Products

This important branch of the food industry embraces such widely different products as: bread and rolls; sweet yeast dough products; biscuits, cookies and crackers; pies and pastries; cakes; and breakfast cereals.

All of these are based on cereal flour admixed as necessary with sugar, eggs, milk, shortenings, leavening agents or yeast and flavoring materials. For those interested in the formulation of a wide range of bakery products and the precise techniques used in their manufacture, reference should be made to Sultan (1976) or Matz (1972). In the United Kingdon the term "biscuit" is used to include sweet, plain and savory varieties of sheet-baked products; in the United States, the terms "biscuit" and "cookie" are synonymous, although "biscuit" is also used to describe products of the fresh bread, cake or roll manufacturer. The term "cracker" is more widely used in the United States where it originated. To avoid unnecessary duplication of words, the term "biscuit" will be used throughout this section to include all varieties.

FERMENTATION EFFECTS ON FLAVOR

Baked goods rise as a result of yeast fermentation. The chemical changes which take place during prebaking fermentation may have far-reaching effects upon the resultant flavor of the end-product, as well as upon its acceptability. The need to control fermentation conditions often dictates the nature of any added flavorings, as these may impair or even inhibit the growth of the yeast cells. Other factors involved include the nature of the flour and the cereal from which it originated, the degree of water present, the amount and nature of any added sweeteners, shortenings and emulsifiers, as well

Courtesy of Bush Boake Allen, Ltd.

FIG. 11.1 AN EXPERIMENTAL BAKERY FOR THE DEVELOPMENT OF FLOUR CONFECTIONERY AND BAKED GOODS AND THE EVALUATION OF FLAVORINGS

as the incorporation of milk, eggs, etc. Because of the complex interrelationships involved, flavorings for this group of products can only be assessed effectively in an end-product made under conditions which parallel those encountered on the manufacturing scale, particularly as this concerns temperature and time.

The bakery industry, in its many aspects, covers an enormous range of specialist products, many of which involve elaborate plant and individual techniques. Generalization on the use of flavorings is of little practical value and where new products are being developed for which the technology is not already well established, the application of flavoring is best carried out by trial and error on a laboratory or pilot plant scale. Taking baked goods as a whole, there are four ways in which flavor can be incorporated into the product: (a) mixing into the dough prior to baking, (b) spraying onto the surface as the product emerges from the oven, (c) dusting onto the surface after oiling, (d) applying as a cream filling, glaze or coating.

Courtesy of Bush Boake Allen, Ltd.

FIG. 11.2. A LAYOUT OF AN EXPERIMENTAL LABORATORY IN WHICH FLAVORINGS CAN BE EVALUATED UNDER FACTORY CONDITIONS

BAKED-IN FLAVORS

Many baked products, particularly cakes, depend for their flavor on natural materials added to the batter (e.g., cocoa powder, chocolate, butter, and eggs) and the balancing of these in the original recipe dictates the nature and flavor of the product. The process of baking, involving as it does high temperatures, imposes considerable contraints on the use of flavorings prior to baking and hence the number suitable for this method of application is limited. In most cases, baked goods result from a multistage process which has the following steps: preparation of a wet dough, rising of the dough by yeast fermentation prior to baking or chemically during baking, molding or sheeting to give product form, baking at an appropriate temperature for sufficient time to cook the flour, drying and/or cooling, packaging. In the production of biscuits, for example, the dough is rolled to the desired thickness (about ½ in.), cut into appropriate shapes and sheeted for baking in a traveling oven. The conditions for baking are dictated by the formulation of the biscuit dough, depending on the amount of sugar, fat and water present. The temperature attained is generally within the range of 177°–260°C (350–500°F); the moisture content is reduced to about 3%. The initial stages of baking inevitably result in the loss of water as steam. This produces conditions which are ideal for volatilization of any aromatic components of added flavorings. In practice, it is found that the use of fat-based or oil-based flavorings reduces losses. Even with liquid flavorings, it is advisable to use a solvent such as propylene glycol or triacetin and to admix the flavoring with any fat present in the formulation. This significantly reduces volatile losses. This approach to reducing volatile losses has been further developed to produce a range of heat-resistant flavors.

Flavorings may be affected by the pH of the system. Most baked goods can only be made using an alkaline dough and it has been found that practically all flavoring materials are either significantly altered or even destroyed by the combination of high temperature and a high pH. Unfortunately, the practical range of pH is very limited so that little adjustment is feasible in order to achieve better flavor retention. Losses arising from this cause can only be compensated for by an increased usage of the flavoring.

In cases where a solvent may be counter-indicated, the flavorings are best added as emulsions, although care is necessary to ensure that any foam formation is not destroyed. Any emulsion should be added with gentle stirring which is just sufficient to ensure uniform distribution in the mix.

Under these conditions the flavors which best withstand baking and have a high level of popularity include: almond, butter, butterscotch, caraway, cardamom, cinnamon, ginger, honey, maple, nut flavors, vanilla.

There are also certain blended flavorings suitable for specialist products such as "petit beurre," "Marie" and "speculaas" biscuits. This does not, of course, exhaust the list of natural or imitation flavorings, but their effective usage level may be disproportionately high in order to achieve an acceptably flavored product. The advice of the flavor manufacturer should be sought on the correct usage level to give optimum results.

HEAT-RESISTANT FLAVORS

In many types of product, the use of encapsulated flavors effectively overcomes volatile losses during processing. But in the case of baked goods, where a wet dough has to be prepared, their efficacy is cancelled out as the capsule wall, being water-soluble, breaks down as the dough is mixed and releases the enclosed flavor. A range of multistage encapsulated flavorings under the trade name RESALOK[1]

[1]RESALOK is a registered trade-name of Bush Boake Allen Ltd., London.

has been patented by one flavor manufacturer. These products consist of encapsulated flavorings in which the final outer-coating of the capsule is water-insoluble. This enables the flavoring to be incorporated into the wet dough, the capsule remaining intact. The flavor is, in consequence, fully protected not only through this initial mixing stage but also during that stage of the baking process when water is being lost. The capsules are designed to release their flavor content only at a pre-determined elevated temperature at which the capsules burst. As the losses are reduced to a minimum, this type of heat-resistant flavoring shows a marked economic advantage over traditional liquid flavorings as the usage rate can be considerably reduced to take advantage of the greater flavor retention in the end-product.

SAVORY FLAVORS

Cracker biscuits form the basis for most savory flavorings. The formulation for these may vary but most of this type of product is made by mixing the flour and yeast with some of the water and shortening called for in the formulation. A small quantity of diastatic malt may be added as necessary. The dough is allowed to fermen at 25°–33°C (85°–90°F) from 16 to 20 hr. The resulting sponge is then mixed with the remaining ingredients, including any flavoring materials, and is allowed to rise for a further 4 hr. It is then rolled to the desired thickness and cut prior to baking in a continuous band oven.

Savory flavors, based on cheese together with onion, garlic, celery and other herbs and spices as appropriate, can be applied in the dough, although the exact conditions for fermentation may have to be established experimentally. It is well known that certain spices (e.g., thyme, sage, allspice, etc.) inhibit the action of yeast during the fermentation period and, in consequence, little or no oven-springing takes place during the baking sequence. The resulting biscuits are dense, hard and "flinty" and are barely edible. For this reason, care must be taken when incorporating spices or spice essential oils into yeast-raised doughs. The difficulty does not apply to chemically-raised doughs or proteinase-conditioned crackers.

Where flavors are to be added to the dough, it is advisable that the addition be made at as late a stage as possible, generally when the initially-raised sponge is remixed just prior to forming. Uniform mixing is important and it may be advantageous to blend the flavoring with other liquid ingredients to aid dispersion. The long mixing times employed in the preparation of certain doughs using a slow-speed two- or three-spindle vertical tub mixer results in little loss of

Courtesy of Weston Foods, Ltd.

FIG. 11.3. BLENDING OF FLAVORED DOUGH FOR CHOCOLATE COOKIES

Courtesy of Weston Foods, Ltd.

FIG. 11.4. A BAND OVEN USED FOR THE CONTINUOUS PRODUCTION OF BISCUITS AND COOKIES

flavoring materials. When, however, high-speed mixers are used, the relatively high temperatures induced in the dough by friction can result in significant loss of volatiles. In such cases, it is not unusual to increase the usage levels of the added flavoring by as much as 20% to compensate for the losses.

SPRAYED-ON FLAVORS

This is a method of applying flavorings to baked goods after the cooking stage is completed. It is usually restricted to thin biscuits since their regular shape is particularly suited to ensure uniform spread of the flavoring. The flavorings used are oil-based and for this purpose are mixed with either hydrogenated vegetable oil or a high grade corn oil.

Hot biscuits emerge from the moving band oven and are automatically transferred to a stainless steel wire mesh conveyor. This carries the biscuits through an oil spray chamber during which the top surface is given a thin film of the flavored oil. This occurs at a temperature of 40°–50°C (104°–122°F) and the biscuits have an open texture and are very absorbent. The spraying method requires an efficient hood over the dispensing apparatus in order to remove the excess oily mist. If this is not done, it settles and coalesces in the chamber causing a messy interior and possible contamination in other parts of the plant.

An alternative method which creates less mist and is easier to control is achieved by dispensing the oil onto rapidly rotating wire brushes which direct droplets of the flavor onto the surfaces of the biscuits.

Although these techniques are relatively simple to operate as part of a continuous production line, they do present certain problems:

(a) It is necessary to clean the plant very thoroughly when changing from one flavor to another. Many flavorings, particularly those such as onion and garlic, tend to cling and a generous time allowance must be made to ensure complete freedom from residual unwanted odors.

(b) The flavors most suitable for this method of application tend to fade in intensity, due, possibly, to a lack of available aromatics which become entrapped by the nonvolatile fatty base.

DUSTED-ON FLAVORS

This is, in a way, an extension of the sprayed-on method in that the flavor is added after the baking process. In the case of crackers, the application takes place in two stages:

(a) Coating the cracker with oil.

(b) Then dusting the oiled surface with a powder flavoring. Sev-

Courtesy of Bush Boake Allen, Ltd.

FIG. 11.5. EXPERIMENTAL SURFACE DUSTING OF CRACKERS WITH POWDERED FLAVORING

eral products made by this technique are available but, as the method applies more directly to the flavoring of snack lines, it will be considered in detail in a later section dealing with this product group.

In both the sprayed-on and dusted-on operations, it is necessary to ensure that no parts of the part coming into contact with the oil or the oiled biscuits should be fabricated in copper. Even traces as low as 5 ppm of this metal catalyse autoxidation of the fixed oil giving rise to premature rancidity and the development of unacceptable off-odors and off-flavors in the end-product. Iron contamination may do much the same thing and the material of construction must be a good grade of stainless steel.

FLAVORS IN FILLINGS AND COATINGS

The fillings used in baked goods may conveniently be classified as: dairy cream, synthetic cream, buttercream, fondant, custard, and glazes.

Dairy and Synthetic Cream

Either heavy whipping cream containing 40% butterfat or light cream containing 18–20% butterfat is used as the basis for natural cream fillings and toppings. Cream-based fillings present few problems in manufacture and the use of flavorings such as vanilla extract at about 0.75% results in stable whipped products. Certain difficulties may arise with the use of citrus fruit pastes some of which give quite appreciable losses in over-run and result in a product that collapses on storage unless a suitable stabilizer is added. Dairy cream products should be stored and sold under refrigeration.

Synthetic cream is of variable formulation but normally consists of an emulsified hydrogenated oil blended with sugar, flavor and a stabilizer.

Buttercream

There are three main types of filling which have this designation (a) A simple beaten mixture of butter and sugar. (b) A blend of gelatin marshmallow with either butter or a hydrogenated fat. (c) Fondant and nonfat milk powder blended with an emulsifier and water and whipped with butter or margarine.

In the case of the butter/sugar mixture, any flavoring in the form of a liquid or paste can be added prior to the beating. Due allowance should be made for some masking of the flavor by the fat. Because of this, there is a considerable advantage to the use of either of the above alternatives. The marshmallow used consists of a whipped aqueous solution of gelatine and sugar. Flavor, either in a liquid or paste form, is generally added to the mix just before whipping although there are a number of factors which affect the quality of the foam. The optimum conditions to achieve the desired over-run and consistency may have to be established experimentally. The flavored marshmallow is allowed to cool and set; the butter, margarine or other fat is softened and added first with slow-speed and then with high-speed beaters until a smooth cream results. A similar procedure is used in making the fondant-based product.

Although the above creams are quite satisfactory as a carrier of sweet or fruit flavors, they are not as acceptable if a savory cream is required. For such products, it is necessary to use a modified formulation using micromilled cracker meal or rusk in place of the sugar. Unless the milling and subsequent sifting of the powder is carried out, the resulting cream will have an unacceptable gritiness. The cream is prepared by blending the powder base with 10% of its weight of glycerine, allowing this to stand for several hours to ensure complete absorption and obtain the maximum softening effect. After this period, the base is whipped with butter, margarine or other fat and the flavoring is added; the mix is beaten to form a smooth cream.

Cheese is frequently used as a flavor for savory filled biscuits as well as in savory crackers. This material presents peculiar problems as its flavor changes during cooking. For best results, cheese should be used in a cream filling in which its individual flavor can be appreciated to the full. A wide range of natural spray-dried cheese powders is now available; these give very satisfactory results. If, however, a more positive flavor profile is required, this can be achieved by the use of natural or imitation flavor boosters. If spray-dried cheese powder is included in the formulation, the proportion of milled meal can be reduced—this will result in a softer cream having a more pleasing texture and mouth-feel.

Fondant

Fondant consists of dry fondant base admixed with water and corn syrup using a slow-speed mixer. It is usually applied warm at 38°–43°C (100°–110°F). This is the most useful of carrier bases and is suitable for use with almost all flavorings. Care should be taken that the hot mix is not beaten and that the minimum of air is incorporated during blending as this reduces its shelf-life.

Custard

Custard filling may be of two types.

(a) Traditional starch-based: The custard mix is boiled and thickened as necessary, allowed to cool to 60°–70°C (140°–158°F) and any flavoring added by gentle stirring. If a fruit paste is used, the custard should be allowed to cool to 50°C (122°F) before making the addition.

(b) Citrus pectin-based: The dry powdered ingredients consisting of milk powder, citrus pectin, cornstarch and sugar are uniformly blended, mixed with cold water and heated to thicken as necessary. The flavoring can be added by gentle stirring. This gives a rich base which is particularly suitable for strong flavorings (e.g., maple, caramel, pecan, etc.)

Glazes

Transparent and translucent glazes are widely used as top finishes for fruit-filled confectionery and may be appropriately flavored and colored. Any such flavorings are normally added to the prepared glaze after it has been allowed to cool to 80°C (176°). Care should be taken to avoid incorporating air.

Coatings and Covertures

Many baked goods carry a coating or coverture of chocolate which may be either milk or plain (dark) chocolate. A pleasing alternative to the simple chocolate theme may be achieved by the addition of flavorings such as coffee, peppermint, almond, orange or lemon. The chocolate itself may be modified by the use of imitation booster flavors. The principal criterion to the use of added flavorings is that they shall not over-ride the basic flavor of the chocolate itself but should merely supplement it.

FLAVORS FOR DRY CAKE-MIXES, ETC.

So far, only the addition of flavor to factory-produced items has been discussed. There is, in addition, a closely-related flour processing industry which is ultimately governed by the same technological considerations—the production of dry cake mixes, cookie mixes, etc. These units are packed so that the housewife prepares the batter as instructed (adding eggs, milk or water as necessary) and bakes the finished product herself. These dry powder mixes pose certain problems when it comes to incorporating either flavorings or colors. The product as purchased consists of a blended powder packed in a hermatically sealed, impervious pack or sachet so that any flavorings must be:

(a) In powder form.

(b) Readily dispersable in the aqueous medium which is used to prepare the batter from the dry mix.

(c) Those of the encapsulated type which ensure minimum deterioration due to volatile loss or oxidative change during the possibly long shelf-life of the product.

(d) They should be of the correct particle size so that no separation occurs during packaging, transportation or handling of the product.

(e) They should remain stable during all stages of preparation.

Similarly any added coloring materials must be:

(a) Permitted and stable within the pH range of the prepared product.

(b) Uniformly mixed so that unsightly specks are not present either in the premix or in the finished batter.

This list of requirements has resulted in the development of a very comprehensive range of suitable flavors encapsulated either in gum acacia (Arabic) or one of the modified starches. The advice of manufacturers should be sought on the most suitable flavorings available.

The baking industry is experiencing a constantly advancing technology in which continuous processing and automated production systems are replacing the traditional personal skills. Product developers in this field are well advised to consider the implication of flavorings very early in the project program so as to ensure the optimum results.

Chocolate chips and "flavor pods"

Throughout North America and in many other regions the chocolate chip cookie is very popular. The idea of imparting discrete bursts of flavor in baked goods by incorporating chocolate and/or other flavored chips of hydrogenated vegetable fat makes possible a wide range of interesting products. The prepared chips have a melting range of 35°–45°C (95°–110°F) and melt pleasantly in the mouth during eating. The method of making these flavored chips was patented in the United States in 1971 (Anon. 1971).

BIBLIOGRAPHY

Baked Goods and Bakery Technology

An annotated bibliography covering books issued between 1967 and 1974, is published by International Food Information Service, Commonwealth Agricultural Bureau, Farnham Royal, Slough, SL2 3BN, England. The following are the major texts listed covering the subject (excluding Bread Technology).

DeRENZO, D. J. 1975A. Bakery Products, Food Technology Review No. 20. Noyes Data Corp., Park Ridge, New Jersey.

DeRENZO, D. J. 1975B. Dough and Baked Goods, Food Technology Review No. 26. Noyes Data Corp., Park Ridge, New Jersey.

HERRINGSHAW, J. F. 1969. Baking. In Food Industries Manual, 20th Edition. A. H. Woolen (Editor). Leonard Hill Books, London.

MATZ, S. A. and MATZ, T. D. 1978. Cookie and Cracker Technology, 2nd Edition. AVI Publishing Co., Westport, Conn.

MATZ, S. A. 1972. Bakery Technology and Engineering, 2nd Edition. AVI Publishing Co., Westport, Conn.

MEISNER, D. F. 1974. Baking Industry—Cakes and Pies. *In* Encyclopedia of Food Technology. A. H. Johnson and M. S. Peterson (Editors). AVI Publishing Co., Westport, Conn.

SULTAN, W. J. 1976. Practical Baking, 3rd Edition. AVI Publishing Co., Westport, Conn.

TRESSLER, D. K. and SULTAN, W. J. 1975. Food Products Formulary, Vol. 2. Cereals, Baked Goods, Dairy and Egg Products. AVI Publishing Co., Westport, Conn.

WHITELEY P. R. 1971. Biscuit Manufacture—Fundamentals of In-line Production. Elsevier Publishing Co., Barking, Essex, England.

Flavors and Baked Goods

ANON. 1971. Flavor pods. Food Technol., *25*, No 11, 23.

ANON. 1972. Garlic flavoring from within opens new bakery product opportunities. Food Prod. Dev., *6*, Dec-Jan., 64–65.

BRADSHAW, R. C. A. 1969. Baking Ingredients—Dairy and Egg products. Food Prod. Dev., *3*, No. 6, 108–112.

HEATH, H. B. 1970A. Flavourings in baked goods. I. Correct use for best results. Baking Ind. J. Mar. 20, 24,26,30.

HEATH, H. B. 1970B. Flavourings in baked goods. II. Flavour influence on product acceptance. Baking Ind. J. Nov., 27–28, 30.

KAISER, V. A. 1974. Modeling and simulation of a multi-zone band oven. Food Technol., *28*, No. 12, 50–53.

MURRAY, D. G. and McCORMICK, R. D. 1970. Greater flakiness, better machinability demonstrated for improved pie crust formulation. Food Prod. Dev., *4*, No. 6, 69–70.

SWITZER, R. M. and HALUSKA, E. L. 1974. Bakery flavors. Baker's Dig. Febr., 57–59.

12

Snack Foods

The term "snack food," as it is understood within the food industry, can be applied to almost any manufactured commodity which is conveniently sized and packaged so that it can be eaten to provide sustenance between formal meals. Current usage would suggest that snack foods are: (a) savory (thus excluding candy and sugar confectionery); (b) conventionally packed to provide convenient and easily consumable portions for one person; (c) normally consumed between meals either for pleasure, as an accompaniment to drinks, for sustenance or even as a substitute for a main meal; (d) often eaten as a supplement to a main meal (e.g., potato chips) in which case the product may be purchased in larger family packs.

This is a highly competitive area of modern food marketing in which the flavor of the product has become one of the main selling points in gaining customer acceptance and keeping it. Convenience, size, shape, color, texture and cost are all important, but it is the flavor which makes the greatest impact, the most lasting impression and is the key to repeat sales.

TYPES OF FLAVORING

There are three types of flavoring materials used in snack foods:

(a) Herbs and spices—generally used for their visual appeal and to a more limited extent for the flavor they impart. Sesame seeds and ground paprika have a pleasing eye-appeal as do seasonings colored with tomato powder or turmeric. As a surface dusting for potato chips, blended seasonings convey an attractive flavor impression.

(b) Natural extracts—mostly essential oils and oleoresins derived from herbs and spices but also extracts of onion, garlic, mushroom, etc. Autolysates of yeast and hydrolyzed vegetable protein are also widely used in creating savory flavors. In most cases, these extracts are plated onto a salt carrier at a strength to enable direct use.

(c) Imitation and blended artificial flavorings—used to give specific flavor profiles such as "roast beef," "ham," "fried chicken," etc. There is an almost unlimited range of such flavorings designed to give a high level of customer appeal.

The only limitation to the use of these flavorings is that dictated by the product form and its method of manufacture (Chesson 1976).

PROCESSING TECHNOLOGY

The following are the main types of snack foods: (a) cookies, crackers and biscuits—the flavoring of which has already been discussed in Chap. 11; (b) potato chips (in the United Kingdom they are called "crisps"); (c) reconstituted potato, corn and other starch-based snacks; (d) extruded products; (e) popcorn; (f) pretzels; (g) nuts.

The processing of such a wide range of product types involves several quite different technologies, namely:

	Conditions	
	Temperature	Time
Baking	150°–260°C (300°–500°F)	10–30 min
Deep frying	162°–182°C (325°–360°F)	2–3 min
Extrusion:		
low pressure	below 100°C (212°F)	Followed by drying. Time depends on the machine compression ratios and moisture content.
high pressure	110°–150°C (230°–300°F	

All of these processes involve high temperatures and loss of water from the product as steam. In the case of extrusion techniques, there are the added problems of high pressure within the extruder associated with a high temperature followed by the rapid release of pressure as the product leaves the machine and the need to dry the end-product usually in a stream of warm air. These conditions are all conducive of volatile loss. So far as the use of imitation flavorings is concerned, high processing temperatures coupled with the loss of steam, restricts the choice in just the same way as flavorings for biscuits. Most of such flavorings have to be specially designed to withstand these conditions without significant change in profile and to suffer the minimal physical loss from the product. This being so, it is obviously advantageous to add any flavoring after the cooking stages are completed. Before considering the various ways in which flavoring may be added to snack foods it is necessary to review the processing technology involved so as to establish the optimum conditions.

Baking

This consists of the following stages: (a) blending of the dry ingredients in a suitable mixing machine; (b) preparation of the wet dough; (c) fermentation, in the case of yeast-raised products; (d) drop forming or sheeting and cutting as appropriate; (e) baking; (f) cooling and packaging.

The problems associated with this technology have already been discussed. Similar considerations apply to snacking cakes although the baking temperatures are generally lower at 150°–222°C (300°–430°F) and the baking times ranging from 10 min to 3 hr depending on the type of cake and its size.

For snack products, the flavoring may be incorporated into the biscuit dough (e.g., cheese straws, cheese and onion crackers) or dusted onto the surface.

Deep Frying

The techniques used in the production of potato chips may vary depending on the sophistication of the machinery used. The following stages are involved in traditional deep frying: (a) potatoes are peeled and washed well to remove surface starch grains; (b) sliced and fed onto a conveyor; (c) fried by immersion in fat at 188°–193°C (370°–380°F), the chips being carried through automatically in a constant stream and lifted onto a stainless steel mesh for draining; (d) flavor application.

The rate of throughput and, hence, cooking time is determined by the degree of cooking required; this may vary with the condition of the potatoes used.

The following modern techniques are now gaining wider acceptance in Europe, the United States and South Africa:

(a) Microwave cooking—chips are only partially fried and cooking is completed in a microwave cooker.

(b) Vaccum frying—chips are partially cooked in a traditional frier and are transferred automatically to a vacuum maintained at 100°C (212°F) for completion.

(c) Infrared drying—partially fried chips pass through an infrared heater to complete the cooking cycle.

The advantages and disadvantages of the above methods are discussed by Smith (1977) and Talburt and Smith (1975). Successful processing of potato chips by the long-established deep frying process imposes considerable constraints on the manufacturer to select potatoes of very special characteristics and to provide special storage conditions. These newer methods are much less severe and give a greater flexibility in the selection and handling of the raw materials.

A recent development in potato chip processing is based on the use of dehydrated potato powder which is rehydrated, sheeted and cut into chip-sized discs prior to frying. Such products are uniform in shape and have a more consistent textural and flavor character. Whereas with natural potato slices it is not feasible to incorporate flavoring, these reconstituted products enable seasonings and/or flavorings to be included in the initial dry mix so that the end-product may be flavored uniformly throughout. Presently, only one such product is being marketed in the United States by Proctor & Gamble. It is called "Pringles" and is made in thin chips and ridged chips, the latter a thicker product more suitable for chip-dipping.

The majority of flavored potato chips are made by the application of powder flavorings onto the surface of the hot chip as it emerges from the fryer or by passing the chips through a tumbler applicator (Smith 1974). Attempts have been made to incorporate flavor into the cooking oil and hence onto the surface of the fried chip. This is not practicable as almost no flavors are stable under such temperature conditions and there is the added danger that such materials may adversely affect the stability of the cooking oil itself.

Extrusion

There are two types of extrusion used in the production of snack and other food products:

(a) Cold forming or low-pressure extrusion—materials are blended dry, moistened to form a soft dough which is then passed through an extruder to give the desired shape. The resulting extruded strip or rod is usually cut into convenient sized pieces. These require further cooking by baking or frying.

(b) High-pressure cooking and forming—ingredients are fed into the extruder, either dry or as a premade dough, and continuously compressed by the helical screw in the extruder barrel. Sufficient heat is generated or applied to ensure that all the starch is cooked and gelatinized. When the extrudate emerges from the die, pressure is suddenly released, the product looses moisture which flashes off as steam and it expands to produce a long tube or ribbon having a cellular porous texture. The expanded product is usually shaped or cut into pieces automatically and passes on a conveyor to a suitable hot-air dryer.

The process is very versatile and modern extruders are capable of producing a wide spectrum of end-product shapes, sizes and texture by the simple adjustment of throughput rate, jacket temperature, compression and shear ratios, die shape and size, and cutter blade speed.

The extent to which the physical properties of the final product can be varied depends upon the composition of the feed stock, particularly the source of the starchy components. Texture can be modified by the incorporation of fat. Flavor can be imposed by the use of added heat-stable flavorings, although currently the range of these is very limited.

FLAVORING APPLICATION

As with baked goods, there are two basic ways in which flavor can be put into snack foods: (a) Internal to the product, implying that the flavoring must be able to withstand the often severe processing conditions. (b) Externally applied by spraying or dusting on or applied as a filler or coating.

These methods have already been discussed in relation to the flavoring of baked goods and will not be further considered except for the dusting-on process which is more applicable to the flavoring of snack foods, particularly potato chips.

Dusting-on Flavors

This technique is used for almost all types of snack product including potato chips, extruded corn curls and nuts. In some cases, the surface is already sufficiently sticky for the powder to adhere; in others, a predressing of oil or gum mucilage is necessary, otherwise the powder merely falls off the product during subsequent handling. Where the product is deep-fried, the cooked pieces pass directly from the hot cooking oil on a wire mesh conveyor under an automatic dispenser which may be either a vibratory tray feeder or a grooved roller and brush mechanism. The appropriate flavoring is filled into the hopper and the hot chips pass under the applicator set as a bridge over the conveyor belt. This results in a fixed weight of powder dropping onto the product, the motion of the individual pieces being sufficient to ensure distribution. Unfortunately, although the dosage can be controlled the distribution is far from uniform. By this method only one surface of the chips will receive a thin layer of the flavoring; any excess is collected and recycled.

Far better distribution is achieved by passing the cooled chips through a seasoner drum which consists of an inclided slowly rotating cylinder through which the product can be passed continuously and the flavoring metered accurately. The internal motion is sufficiently violent to ensure an even coating of the powdered flavor but demands close attention otherwise the product may be broken and become unacceptable for sale. Details of the equipment available are given in Matz (1976).

For the most part, blended flavorings and seasonings designed specifically for this type of application give little trouble; but three sources of difficulty may arise.

(a) Use of an excessive percentage of anti-caking agents (e.g. silicon oxide, sodium alumino-silicate, etc.) which, though giving a free-flowing powder, creates very dusty conditions. This makes it very unpleasant for the plant operatives and results in an unsightly layer of dust throughout the plant. The resulting product often has an undesirable dryness in the flavor profile.

(b) Presence of hygroscopic components (e.g., dehydrated onion powder, hydrolyzed vegetable protein, etc.) makes the blended flavoring sticky, causing it to "ball" and flow unevenly in the applicator. In the case of cheese powders, the oily stickiness results in a tendency to coat fluted rollers and eventually block the dispenser head.

(c) Separation in the hopper occurs when the blended flavoring contains materials of widely different particle size. The coarser particles tend to separate to the surface of the mix in the hopper under the vibratory conditions usually employed. To overcome this, many machines are fitted with two reciprocating plates and a pendulum action comb which effectively redistributes the flavoring components and prevents agglomeration.

The nature of flavoring mixes and their flavor characteristics are discussed in an article by Smith (1974). A wide range of flavorings is now available; most of these are reminiscent of savory flavors associated with main meals (e.g., salt and vinegar which reminds the eater of French fries (or in the United Kingdom, potato chips); cheese-and-onion or barbeque flavors, both of which have pleasing mental associations).

Salt.—The use of salt for the flavoring of potato chips is long standing; almost all such products carry a dusting of salt on the surface. This may pose a problem in the formulation of seasonings or flavorings due to the different production methods employed. In North America, it is normal for all chips to receive a coating of salt as they leave the fryer. If any additional flavoring is required, then the salted chips are diverted below a second applicator or pass to a drum applicator. Obviously, any added flavoring must take into account that the chips are already adequately salted and, hence, such flavorings must be salt-free. In Europe and many other countries where these products are made, batch rather than automatic in-line processing is used. Flavoring is added to the unsalted product and sufficient salt must be included in the total formulation to give the correct level on the finished product.

Seasoning and flavor blends are generally tailored to suit the needs of the marketplace. In this product area, consumers show wide and erratic flavor preferences depending on the region and age-group involved. By far the largest demand is for a salted chip, probably because it is so widely eaten not as a snack but as an accompaniment to a main meal in place of other forms of potato. The flavored products are frequently chosen at random and by impulse depending on the products displayed at the time of sale.

BIBLIOGRAPHY

ANON. 1972. Snack food seasoning line tailored for application. Food Process. *33*, No. 11, 28.

CONWAY, H. F. 1971. Extrusion cooking of cereals and soyabeans. Food Prod. Develop. *5*: Part I Apr., 27–31; Part II May, 14–22.

CHESSON, K. 1976. Putting the bite into savoury snacks. International Flavours Food Additives *7*, 261.

GUTCHO, M. 1973. Prepared Snack Food. Food Technol. Rev. No. 2. Noyes Data Corp., Park Ridge, New Jersey.

MATZ, S. A. 1976. Snack Food Technology. AVI Publishing Co., Westport, Conn.

SMITH, K. H. 1974. A survey of the snack food industry and its flavourings. Flavour Ind. *5*, 36–38.

SMITH, O. 1974. Potatoes. In Encylopedia of Food Technology. A. H. Johnson and M. S. Peterson (Editors). AVI Publishing Co., Westport, Conn.

SMITH, O. 1977. Potatoes: Production, Storing and Processing, 2nd Edition. AVI Publishing Co., Westport, Conn.

SCHOONYOUNG, F. 1974. Attacking snack food problems through flavor application. Snack Food *63*, No. 10, 40–41, 56.

TALBURT, W. F. and SMITH, O. 1975. Potato Processing, 3rd Edition. AVI Publishing Co., Westport, Conn.

13

Sugar and Chocolate Confectionery

Sugar confectionery consists of products characterized by a high sugar content which is responsible for their stabilization. They may be classified as follows, based on the major ingredient used in their manufacture:

Principal Ingredient	Product Type
Sugar	Hard candies, boilings
	Fondant
	Center fillings
Sugar and nonsugars	Gum
	Pastilles
	Marshmallow
	Nougat
	Caramels, toffees
	Jellies
Cocoa powder	Chocolate
Sorbitol and synthetic sweeteners	Nonsugar dietetic candies
Nuts	Products containing coconut, hazel and other hard nuts, almond (marzipan).

In general, most forms of sugar confectionery have little or no intrinsic flavor other than a sickly sweetness due to their high sugar content. This may be modified by the use of selected acidulants (e.g., citric acid), but the characteristic flavor is almost entirely conferred by the added flavorings. Some flavor results from change occurring during the cooking but this has relatively little effect in the final profile as compared with that of the powerful flavorings usually employed in these products.

The processing constraints which determine what flavorings may best be used are so different within this product group that it is necessary to consider each separately. The following techniques will be reviewed in this section:

(a) High-boiled confectionery: hard candies, sugar boilings.
(b) Low-boiled confectionery: toffees, caramels.
(c) Starch-deposited confectionery: pastilles, gums.
(d) Chewing gum.
(e) Chocolate and chocolate confectionery.

PROCESSING TECHNOLOGY FOR HIGH-BOILED CONFECTIONERY

The various forms of high-boiled sugar confectionery demand quite different processing conditions and have lead to development of sophisticated machinery to handle large volume throughput. Most operations were originally carried out by hand and may still be done in this manner in the laboratory for establishing the usage levels of added flavorings. These products represent a large market for flavoring materials and, although the following outlines may be adequate for the flavor technologist, those wishing for more precise information should refer to one of the many standard texts given in the bibliography. From a flavorist's point of view the following topics are of particular interest.

Sugar

Sugar is the mainspring of the confectionery industry and demands close attention by all involved in product development and production to ensure consistent and high quality goods. Details of the grades commercially available can be readily obtained from the principal suppliers who, generally, are able to give technical advice on usage. Terminology is, unfortunately, not universal; neither are the standard grades available so that usage must often be established experimentally. Even small differences in sugar quality can have a significant effect on the profile of the end-product.

Of the various sugars used, sucrose, either from cane or beet, is by far the most important. In some products, the sugar is added dry, directly into the product mix. In many, it is incorporated as a 70° Brix syrup, this being stored in bulk. In formulations, this material is often referred to simply as "syrup."

Sugar Boiling

In many confectionery products the main feature is the concentration of the syrup, either alone or with other ingredients, to give a mass having a solids content of about 99%. This may be carried out by one of the following methods:

(a) Open-pan cooking—gas, oil or solid fuel fired.
(b) Rapid-boiling steam pan with internal high-pressure steam coil.
(c) Vacuum batch-cooking.
(d) Nonvacuum microfilm continuous cooking.
(e) Vacuum continuous cooking.

The processing conditions encountered in each of these methods are very different. Most hard candy is currently made in machinery designed for continuous or semicontinuous handling of the product. Where batches are still made these are generally on a 100-lb basis.

The cooking state is critical to the nature of the end-product. The following conditions are of importance when considering new product development.

Batch Cooking. —The syrup is usually preheated to 104°–110°C (220°– 230°F), the mixture is then fed into a steam-heated kettle and cooked at 135°–138°C (274°–280°F). Full vacuum is then applied for a period of 10 min and the mass reduced to a water content of 0.5–1%.

Semicontinuous Vacuum Cooking. —The syrup is preheated to 110°– 115°C (230°–240°F) and the water partially removed at atmospheric pressure. The concentrated syrup is fed into a heat-exchange cooker in which it flows through a steam-heated coil and finally is discharged into a vacuum kettle which can be removed without breaking the vacuum or interrupting the flow.

Continuous Vacuum Cooking. —This is carried out in plants designed to expose the syrup to a minimum of heat and involves fast-flowing thin films. The preheated syrup is continuously pumped, concentrated under vacuum, and then continuously withdrawn. The degree of concentration depends on the temperature, the vacuum and the flow-rate. Conditions are adjusted to obtain maximum concentration with minimum heat to limit inversion and discoloration. The end-product contains 12–17% reducing sugars.

Graining and Doctoring

When syrup is concentrated to form a fluid or plastic mass of 0.5–1% moisture content, crystallization rapidly occurs; this is called "graining." Except for fondants and fudge where this condition is induced, this is normally an undesirable quality defect. Traditionally, part of the sucrose was inverted to dextrose and levulose by the addition of an inversion agent, or "doctor," which is either an acid or an acid salt. Alternatively, part of the sucrose could be replaced by invert sugar in the formulation. In both cases, this minimizes the tendency to graining.

Inversion, or the use of invert sugar, requires careful control in order to avoid unsightly discoloration of the product and, as levulose is hygroscopic, excess of this sugar may result in undesirable moisture pickup leading to reduced shelf-life and deterioration of the product on storage. The use of refined glucose syrup, made by acid or enzyme treatment of starch, is increasingly the modern practice. This minimizes the level of sucrose without the introduction of levulose and is economically advantageous.

Tempering and Kneading

The term "tempering" has a different meaning as applied to sugar boilings from what is implied in chocolate manufacture. In many kinds of high-boilings, the plastic mass after cooking requires conversion to a consistency suitable for forming into the final sweets; the method of manipulation depends on the nature of the end-product required. It is at this stage that any flavoring or coloring can be incorporated into the mass. In normal batch processing, the practice is to discharge the hot mass onto a mixing area coated with a release agent; the flavoring, color and acid, etc., are added and the mass mixed to ensure even distribution. The mixture is then transferred to a warmed tempering slab. As it cools, it is kneaded to produce the correct consistency. In semicontinuous or continuous processing the additives are incorporated in an in-stream mixing pan and kneaded mechanically. The kneading action, which takes about 5 min and during which the mass cools to about 100°C (212°F), is a critical stage as too low a temperature results in granulation and over-kneading gives an opaque product.

For soft opaque products, the mass is formed into a thick "rope" which is then specially pulled. This technique, which may be carried out by hand or mechanically, is designed to transform the transparent mass into a translucent one by the incorporation of minute air bubbles. The resulting product has a satin finish and an open texture. Any flavoring is added as the mass is being manipulated. The amount of air incorporated depends on the viscosity of the mass. A soft mix may enable as much as 30% of the mass weight of air to be added.

Forming and Finishing

The production of the finished sweets involves three sequential stages followed by their packaging for sale:

(a) *Batch rolling*—the automatic formulation of a rolled-out cylinder of the tempered mass.

(b) *Rope sizing*—to control the size of the roll to a diameter equal to the finished candy. This may be done either by hand or by a series of rollers which successively reduce the diameter of the rope.

(c) *Forming*—the rope is passed through a set of dies and cutters the shape and size of which determine the form of the end-product. It is at this stage that imperfections in the processing become apparent.

The final stage of manufacture involves cooling the formed candy. This is normally carried out on a wire-mesh conveyor mounted in a stream of dehumidified air. The final temperature should be below 32°C (90°F), but too low a temperature induces moisture pickup.

Automatic depositing

In the continuous vacuum boiling operation all of the above stages are in-line operations, the mass being fed from an electrically heated hopper into release-coated molds moving in a continuous band. All of the critical parameters can be set to obtain optimum processing conditions.

Flavorings in Sugar Boilings

Flavorings for use in sugar boilings and hard candies are added to material which is hot and remains hot for some considerable time after the addition. These conditions are, obviously, conducive to the loss of flavoring volatiles as well as to some degradation. The flavorings used range from essential oils through concentrated fruit juices to imitation flavorings. Because of the adverse processing conditions, the fixation of these materials is of considerable importance. In the creation of flavorings for this product group advantage is taken of the less volatile solvents such as propylene glycol, di- and triacetin.

Essential Oils.—The major class of flavorings used in candy manufacture is the essential oils including such popular flavors as peppermint, spearmint, lemon, orange and other citrus fruit oils, ginger, anise, eucalyptus, clove and certain others in minor quantities. Of these, peppermint oil is probably the most widely used in high boilings and in panned goods (e.g., mint imperials and coated chewing gum). Essential oils present certain problems for the manufacturer:

(a) *Contamination of environment.* Even with efficient extraction and air-conditioning systems, the high levels of volatiles in the atmosphere can make these departments uncomfortable for operatives and may indeed contaminate other products being made in the same area. Attempts to limit this by the use of encapsulated essential oils have not been successful as the resulting products have a poor appearance and a gritty mouth-feel.

(b) *Cloudiness in the product.* This is due to poor dispersion and by the differing properties for light refraction exhibited by the essential oil and the sugar mass. The first remedy is to ensure that the oil globules are of a minimum size by an efficient blending system. If this does not prove satisfactory, then the use of an approved solvent is indicated. The use of terpeneless or concentrated oils usually overcomes the problem but at an increased cost.

(c) *Deterioration on storage.* Essential oils stored in partly-filled drums on warm production floors rapidly deteriorate with the formation of off-notes. Balance stocks should be transferred to smaller containers and storage should always be in a cool place away from direct sunlight.

A real difficulty confronting the manufacturer of sugar confectionery is the wide divergence of food laws, particularly as these relate to flavoring and coloring materials. It is necessary to ensure that any additives are permitted in the country in which the finished confectionery is to be sold to the public. In the case of exportation of finished goods, it may be necessary to submit a list of ingredients the purity of which is established by a certificate from the manufacturer.

It will be appreciated that these are topics associated with one very small branch of sugar confectionery manufacture and cannot be taken in isolation as most manufacturers handle high boilings or hard candies as only one line of their total operation. Flavoring manufacturers generally have facilities for the production of laboratory or pilot-scale batches of high boilings and are thereby enabled to establish the optimum usage conditions, dosage, etc., and can make recommendations of the most appropriate flavoring for any given product concept.

PROCESSING TECHNOLOGY FOR LOW-BOILED CONFECTIONERY

Low-boiled confectionery includes caramels and toffees; the distinction between them has tended to disappear and the terms now are used synonymously. The manufacture of this class of goods involves the cooking of the ingredients to remove water and at the same time the development of flavor, color and texture. The main ingredients include sugar, which may be either white or brown; glucose syrup to inhibit crystallization or graining; milk solids in the form of unsweetened condensed milk, evaporated milk or spray-dried milk solids; milk whey; and fats. In some products, butter is used; this must be added near the end of the boiling process. Other ingredients include salt and emulsifiers as well as specific additives such as molasses, treacle, malt, licorice and other flavorings.

As browning and caramelization are essential to the nature of the finished product, the boiling process is usually carried out at atmospheric pressure in an open steam-jacketed pan boiling at 120°–132C (250°–270°F) depending on the characteristics required. Automated processing in a microfilm continuous cooker followed by exposure of the water-reduced sugar mass on a scraped-plate heater to induce caramelization is now widely used. The technique is not without its problems owing to the high viscosity of the product and the difficulty of achieving reproducible results. The cooked mass is fed continuously through a cooler unit and then to batch formers feeding the cutting and wrapping machine. As an alternative, the plant may be used to produce a toffee mass for feeding into a depositor, as used in the manufacture of high boilings, except that the molds must be made of plastic to enable the deposited caramel to be ejected without sticking. The formed caramels may then be wrapped or conveyed to an enrober.

Fudge

Whereas in caramels a smooth creamy texture is desirable and graining is a defect to be prevented, in fudge, which is a product intermediate between caramel and fondant, graining is essential to the texture of the end-product. This is induced either by replacing part of the glucose syrup and sugar solids with fondant or by rapidly cooling and beating the mass in order to produce a fine crystal matrix. Any flavorings should be added after the boiling stage.

Dutch Caramels

There is a marked demand for softer, chewy caramels having a light creamy texture. In these, the cooked sugar mass is pulled to achieve aeration which induces a very short texture. Flavor and color can be added to produce a wide range of confectionery having an attractive appearance and a high level of acceptability.

Flavorings in Toffees and Caramels

The dairy-type caramels and fudge, utilizing butter, whole milk and brown sugar, require little by way of added flavorings other than vanillin to accentuate a basic creaminess. Where formulations have to be modified for commercial reasons or because of a shortage of raw materials, then there are opportunities for the use of imitation flavorings (e.g., imitation butter, cream, caramel, etc.) as well as a host of traditional and characteristic flavors such as butterscotch, rum and butter, Devon cream, etc.

Essential oils, fortified fruit juices and imitation fruit flavors may be used in soft Dutch caramels. A range of fat-based powder flavors are also available for just such products. These flavors are easily handled and may be added to the cooking pan immediately before discharging. Being nonvolatile, they store well and are easy to handle by virtue of their free-flowing properties.

PROCESSING TECHNOLOGY FOR STARCH-DEPOSITED CONFECTIONERY

This group of products includes creams and fondants, jellies having a base of either agar-agar or pectin, gums with a base of gum acacia (Arabic), as well as various types of foam confectionery such as marshmallow. These products, in their many varieties, enjoy wide popularity. All are comparatively low-boiled and retain about 20% moisture. The hardness of the product, which is one of the major characteristics, is determined by the base used. Pectin, agar-agar and low concentrations of gelatine have a soft texture; as the gelatine is increased the product becomes more tough. A hard but chewable

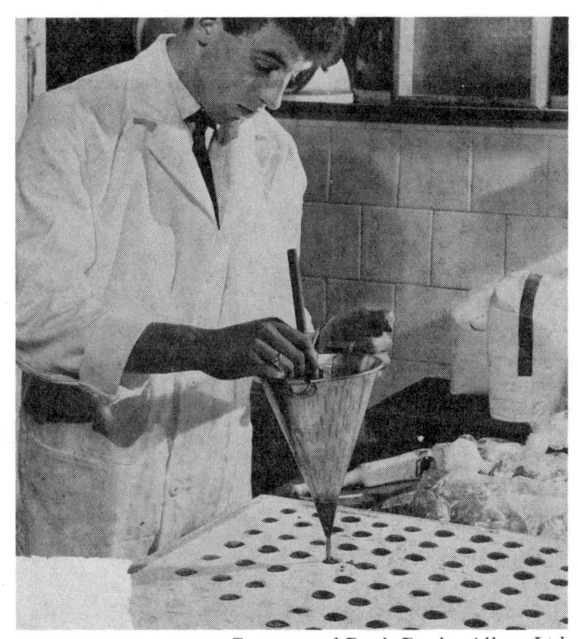

Courtesy of Bush Boake Allen, Ltd.

FIG. 13.1. PREPARATION OF A SAMPLE BATCH OF FLAVORED PECTIN JELLIES BY DEPOSITION IN A STARCH MOLD

"gum" is produced by using gum acacia which may be present at up to 50% of the mass weight.

The following are the main classes of deposited products.

Fondant Creams.—These are confections which consist of a solid phase of sugar crystals dispersed in a high sugar solids syrup which gives a noncrystallizable liquid matrix. The manufacturing technique is designed to convert a highly viscous mass into one which flows freely for depositing. The grain of the sugar crystals produced depends on (a) degree of inversion brought about by the cream of tartar (potassium acid tartrate); (b) amount of glucose syrup used in the formulation; (c) boiling time; (d) cooking temperature; (e) method of heating; and (f) use of added colloids such as gelatine.

Fondants make a very useful test medium for flavor evaluation so that their correct production is important in any laboratory involved in this work. It is a standard practice for the fondant base mix to be heated with a "bob batch" of sugar syrup; a "bob batch" is, in effect, a fondant syrup which has not been crystallized. Any flavorings and colorings are added and the mix cooled for depositing. The seeded cooling induces an even, uniform crystallization of the total mass. If

Courtesy of Cadbury-Schweppes, Ltd.

FIG. 13.2. FONDANTS, CREAMS AND JELLIES, PREFLAVORED AND COLORED AS NECESSARY, ARE INJECTED INTO A MOVING CHAIN OF MOLDS

the deposited articles are intended for sale as such, then it is usual to provide them with a hard outer coating of sugar crystals deposited from a supersaturated sucrose solution (72% sucrose).

Many fondants are intended as chocolate centers, in which case the forms are conveyed directly to a chocolate enrober. If a soft runny center is required, this may be produced by the use of added invertase which produces an inversion of the sucrose so that a fluid cream results after the hard fondant has been chocolate enrobed. Normally, 2 oz of invertase are added to 100 lb of fondant base and the product allowed to stand at room temperature for 2 weeks after enrobing. The temperature at the time of the addition of the enzyme must not exceed 38°C (100°F) as the invertase system is destroyed above 63°C (145°F).

Agar and Pectin Jellies.—These are soft, generally fruit flavored, jellies suitably colored and sanded with sugar. They are frequently deposited in fancy shapes corresponding to the fruit with which they are flavored. The deposited jellies are allowed to set at room temperature for 24 hr before they are cleaned of starch, steamed and sugar-dusted.

Starch Jellies.—Based on cooked starch with or without the addition of gelatin, they are firmer than pectin jellies and should be allowed to set for 48 hr.

Pastilles.—Pastilles are firmer than starch jellies but similarly based and manufactured. The difference is that pastilles have a higher solids (about 84%) and a lower moisture content achieved by depositing at a higher concentration with subsequent conditioning at 44°–49°C (115°–120°F) for 72 hr.

Gums.—These are the firmest of the deposited products being based on gum acacia (Arabic) and gelatine. Processing is similar but stoving is at 44°–49°C (115°–120°) for a period of 3 to 10 days.

Foam Products.—Of these, marshmallow is the best known and most widely manufactured. It consists of a sugar syrup, gelatin and egg albumin beaten into a stable foam. Batch and continuous processing are used. In each case, air is injected under pressure to produce a final foam having small air cells of uniform size with thin cell walls. The over-run is directly related to the texture of the finished product, the chewability being improved with increased over-run. Two basic methods are used in the production of marshmallow: either all the ingredients are mixed and beaten together; or the whipping agent and sugar are first beaten and a precooked batch of glucose syrup, sugar and water is added to the whipped foam. This later method gives greater flexibility for modifying the texture of the end-product.

The formulations for the above products show wide variations of gelling agents used and reference should be made to one of the standard texts for details (Pratt 1970; Lees and Jackson 1973). The general method of production falls into six well-defined stages: (1) preparation of the gelling agent, flavoring and color mix; (2) cooking and concentration as necessary; (3) depositing into preformed starch molds; (4) drying and conditioning; (5) de-starching and cleaning; and (6) steaming and crystallizing or glazing.

Large production volumes of these products can be achieved by the use of a Mogul starch machine in which the whole sequence of operations (3), (4) and (5) is carried out automatically.

Flavoring of Deposited Articles

The following factors are of importance in deciding the flavoring to be used:

Cooking.—All of these products were formerly made in an open steam-heated pan, but modern processing demands continuous cookers and depositors. These call for a modification in the basic formulation to give the correct texture in the end-product—a factor which must be allowed for in relating laboratory- and factory-made goods. The maximum cooking temperature is about 140°C (285°F). Prolonged cooking results in a loss of gel strength and discoloration. Processing temperatures are, therefore, not so demanding upon the stability of any added flavoring but this is offset by the limited amount of acid which can be accommodated to achieve a true fruit flavor.

pH and Acidity.—Most fruit flavors depend for their appeal on their sharply acidic character. As the pH of the end-product is a critical factor in gel setting and stability, it may not be possible to incorporate a sufficiently high level of acidity to produce the desired flavor balance. It is usually necessary to compromise within the following limits:

Gelling Agent	Added Acid (as Citric Acid) (%)	pH of Finished Product
Agar-agar	0.2–0.3	4.8–5.6
Pectin	0.5–0.7	3.2–3.5
Gelatin	0.2–0.3	4.5–5.0
Starch	0.2–0.3	4.2–5.0
Gum acacia (Arabic)	0.3–0.4	4.2–5.0

Buffering.—In the manufacture of jellies, it is essential to avoid over-inversion of the sugar which results in "sweating" and spoilage during storage. This can be controlled by the use of a buffering salt (e.g., sodium citrate) during the cooking stage and the incorporation of any acids at as late a stage as possible.

Gel Strength.—Availability of flavor to the palate is, to some extent, influenced by the nature and quality of the gelling agent used, which, in turn, dictates the texture and eating properties of the end-product. The most acceptable results are achieved with citrus pectin, followed by agar-agar and then gelatin. Fruit juice concentrates, fruit pastes and powders give superb flavoring results in pectin and agar jellies but with gelatin-based products the stronger, less subtle flavorings are required to cover the intrinsic background note of the gelatin itself. Loss of flavor on storage is often very pronounced and is due to drying out.

Drying, Stoving or Conditioning.—Stoving times vary according to the type of product and the efficiency of the ovens or drying rooms used. Loss of flavor during this period is substantial and allowance must be made by an increased initial usage level. Because of these inevitable losses, entirely natural fruit flavors in the soft-fruit range are not really practicable. Essential oils will stand up well to this processing and are generally used at 180–240 ml per 50-kilo batch.

PROCESSING TECHNOLOGY FOR CHEWING GUM

Chewing gum may be composed of a mixture of natural gums (e.g., chicle and Jelutong-pontianak), natural resins (e.g., balsam of tolu, pine resins), synthetic materials (e.g., polyvinyl acetate), starches, sugars and flavoring. The manufacturing process is by no means a simple operation and considerable experience is necessary to convert an apparently simple formulation into an acceptable product having the correct chewing characteristics.

The process may be considered in two stages:

(a) Preparation of the gum base, involving grinding of the constituents to a fine powder, followed by partial air drying. The dried powder mix is then heated to 90°C (194°F) and water is added. The mass is then kneaded for 2½ hr, bleached with a weak solution of sodium hydroxide and washed repeatedly in hot water for several hours. The washed gum is then dried and heated to 100°C (212°F), any other ingredients added and the mass cast into a block for the next stage.

(b) Blending and extrusion forms the second stage. The gum base is heated at 90°–95°C (195°–203°F). When the mix is soft, the requisite amount of icing sugar, glucose syrup, sorbitol and dextrose is added and the mass mixed for 90 min at a steady temperature of 80°C (170°F). At this stage any flavoring is added and mixing continued for a further 30 min after which the product is rolled into 2-in. thick sheets.

The prepared sheets may then be machine cut or extruded into ribbons. The final product is sugar-coated, if desired.

Flavoring of Chewing Gum

The nature of the raw materials used in the manufacture of strip and bubble gum, as well as the method of manufacture, demands special attention to the physical characters of any added flavoring. The type and amount of solvent present as well as the volume of flavoring required are critical. Too high a dosage level results in a sticky mass which will almost certainly give problems at the cutting stage. A further prerequisite is that the flavoring should be able to mask the background notes of the gum base itself, particularly where these are artificial, as may be the case with bubble gum. Finally, the flavor must have a good persistence and last throughout the extended chewing period.

For these reasons, the most satisfactory flavoring for chewing gum must be: (a) highly concentrated, (b) contain the minimum of solvent or none at all, or (c) be encapsulated.

The essential oils are widely used as chewing gum flavors. Both peppermint and spearmint are extremely popular in this medium. Low grade oils are generally a false economy as any adverse sensory attribute tends to be accentuated as the product is chewed.

SELECTION OF FLAVORS FOR SUGAR CONFECTIONERY

The technology discussed in outline above sets the constraints on flavor usage. The following general comments are made as a guide to the selection and evaluation of suitable flavor types. In most cases, small laboratory-scale batches of end-product can be made quite successfully in order to demonstrate the quality of any particular flavoring as compared with another so long as processing conditions are observed and the same raw materials are used as under factory conditions.

Probably sugar confectionery products offer the greatest opportunity for flavor diversification and exploration into new flavor experiences. Yet most manufacturers seem reluctant to change from the hardy, traditional fruit flavors such as lemon, orange, raspberry, strawberry, black currant, cherry and lime with a few "acid" varieties such as peppermint and menthol/eucalyptus. Looking through any flavor manufacturer's literature, it is at once obvious that the list of available flavors is comprehensive. Recommendations for use levels are normally given as fluid ounces per 112-lb batch of base or milliliters per 50 kilos of base; these being based on experience of acceptable flavor levels, although variations may be necessary to meet local conditions and tastes in different countries. In general, the amount of flavoring that can be added to a hot sugar mass is limited; it is not desirable to add more than 360 ml of flavoring per 50 kilos.

Continuous and automatic processing imposes special constraints on flavorings. In most plants, the flavorings are injected into the cooked sugar mass as it is moving through the collecting chamber at the discharge end of the microfilm cooker, or in the mixing cone, or into the discharge pipe from the vacuum chamber. Wherever this is done, the mixing time is limited; hence, ease of dispersion is vital. Many flavorings cannot be added at an earlier stage in the process owing to unacceptable high volatile losses and poor dispersion due to turbulence and aeration. Modern plants are designed to allow flavor addition directly into the evacuated section of the collecting chamber where a stirring mechanism efficiently mixes the flavoring without fear of graining or separation. Flavorings recommended for sugar confectionery are formulated to withstand 154°C (310°F); but the dwell-time in the hopper may be considerable and exposure at 140°C for periods of up to 10 min can induce considerable changes in profile. In the modified continuous plant, the flavoring is added under vacuum and this, too, must be taken into account. All of these considerations may mean that up to 25% more flavoring must be added in order to achieve the same flavoring effect as that in a batch-made product. For this reason, the usage recommendations made by the flavor manufacturer must be treated as a starting indication and adjustments made to suit particular processing conditions.

Deposited boilings, as opposed to formed boilings, have brought a new concept of quality to this type of candy. The end-product must be smooth, free from air bubbles and completely clear. The flavorings chosen must satisfy the following criteria: (a) must not produce bubbles in the hot sugar mass so that low boiling solvents are counter-indicated; (b) must be soluble and not produce a haze or cloud; and (c) must be readily dispersible.

An imitation of almost every well-known fruit is now available in a form suitable for use in sugar confectionery. The qualities offered are widely different and, to some extent, are dictated by price. Rarely can one flavor manufacturer's product be directly substituted for that of another. With very cheap candies the cheaper versions of an imitation flavor may be perfectly adequate to satisfy the consumer but in quality products it generally does not pay to economize on the quality of the flavoring which plays such an important part in successful sales.

CHOCOLATE AND CHOCOLATE PRODUCTS

Chocolate was introduced into Europe by the Spaniards after its discovery in Mexico in the late 15th century. At that time, it was a drink whipped into a stiff foam. Chocolate as we now know it was not developed until the 19th century. Chocolate processing is elaborate and for success demands purpose-designed machinery and considerable expertise. It is an industry which lends itself particularly well to mass production techniques.

Technology of Chocolate Production

The production of chocolate is very well described in the literature (Lees and Jackson 1973; Minifie 1970; Allerton 1974). It involves the following stages:

(a) Selection and blending of the cacao beans. The ultimate flavor of the chocolate depends to a large extent on the source of the beans used. These may be in one of three categories—basic or bulk beans, flavor beans or aroma beans. Basic beans have a strong very harsh flavor and are used in cheap covertures. Of the remainder, those from Sri Lanka, Java, New Hebrides and Samoa are most prized for their flavor; those from Ecuador, Trinidad, Caracas and Venezuela being appreciated for their fine aroma.

(b) Cleaning.

(c) Roasting. Before cacao is ready for further processing it must be stripped of its shell and have the characteristic chocolate flavor developed by roasting. Roasting time and temperature are important and are dictated by the particular plant used.

(d) Winnowing and dehusking. The resulting "nibs" represent about 78% of the original beans.

(e) Refining and melangeuring. Chocolate liquor is produced by grinding the nibs prior to mixing with sugar, milk solids and other ingredients.

(f) Conching. This is essential to the production of a high quality chocolate. It consists of subjecting the heated chocolate liquor to a rolling pressure for a long period.

(g) Tempering. Tempering is necessary to induce the cacao butter to crystallize in a uniform and stable form in a fluid chocolate mass.

Tempering involves heating the mass to 49°–55°C (120°–130°F) with stirring until it is uniformly smooth, then cooling to 31°–33°C (87.5°–90.5°F).

(h) Molding, depositing or enrobing. Although chocolate from different manufacturers may be of similar composition, there is often a considerable difference in flavor and texture. In most cases, this is due more to the selection of the beans than to any manufacturing techniques used. Large manufactures are able to make their chocolate liquor on a continuous basis as required, but smaller concerns may purchase this in molded 10-lb blocks. For further processing, either to make small molded goods or for the enrobing of flavored centers, etc., the chocolate must be carefully melted, tempered and correctly cooled otherwise "blooming" will result, making the end-product unsightly and unacceptable.

Flavorings in Chocolate and Chocolate Products

Added flavor can be of value in the manufacture of chocolate in three ways:

(a) As a modifier of the characteristic chocolate notes. This use of imitation flavorings enables beans of lower quality to be included in

Courtesy of Cadbury-Schweppes. Ltd.

FIG. 13.3. THE ENROBING AND CHARACTERIZING OF CHOCOLATE CENTERS

the blend. Other flavorings such as natural vanilla extract, vanillin, ethyl vanillin and imitation milk and cream flavors may also be of value. Certain spice flavors such as cinnamon and cassia are also used to give a distinctive flavor to the end-product.

(b) As an over-riding but compatible flavoring. This gives to the finished chocolate an entirely different profile (e.g., orange, peppermint, rum or coffee flavor).

(c) As a flavoring for centers.

The only criterion is that any added flavoring should be entirely compatible with the chocolate which itself has a high flavor level.

The development and application of flavors in aspects of chocolate manufacture demand close attention to the balance between intrinsic and added notes. In the majority of products, only mild flavorings such as vanilla are used in the chocolate itself and any specific flavoring effect being achieved in the various fillings used (e.g., fondant creams, jellies, praline, almond and other nut pastes, truffles, fudge, caramels, nougat, liqueurs, and compounded fillings of many kinds.)

BIBLIOGRAPHY

ALIKONIS, J. J. 1974. Confections. *In* Encyclopedia of Food Technology. A. H. Johnson and M. S. Peterson (Editors,). AVI Publishing Co., Westport, Conn.

ALLERTON, J. 1974. Chocolate and cocoa products. *In* Encyclopedia of Food Technology. A. H. Johnson and M. S. Peterson, Editors. AVI Publishing Co., Westport, Conn.

ANON. 1976. The use of gelatin in confectionery products. Confect. Prod. *42*, 359–360.

BUSH BOAKE ALLEN. 1957. Skuse's Complete Confectioner, 13th Edition. W. J. Bush & Co., London. (Now Bush Boake Allen, London.)

CAKEBREAD, S. H. 1972A. Confectionery ingredients—flavour. I. Confect. Prod. *38*, 190–192, 218.

CAKEBREAD, S. H. 1972B. Confectionery ingredients—flavour, II. Confect. Prod. *38*, 242–244,246.

CAKEBREAD, S. H. 1972C. Confectionery ingredients—flavour III. Confect. Prod. *38*, 314–316.

CAKEBREAD, S. H. 1972D. Confectionery ingredients—flavour, IV. Confect. Prod. *38*, 356–360,362.

CAKEBREAD, S. H. 1975A. Confectionery ingredients—acids, bases and salts. I. Confect. Prod. *41*, 24–25.

CAKEBREAD, S. H. 1975B. Confectionery ingredients—acids, bases and salts II. Confect. Prod. *41*, 68–70.

CAKEBREAD, S. H. 1975C. Confectionery ingredients—acids, bases and salts, III. Confect. Prod. *41*, 120–123,136.

CAKEBREAD, S. H. 1975D. Confectionery ingredients—acids, bases and salts, IV. Confect. Prod. *41*, 248–251.

CAKEBREAD, S. H. 1975E. Confectionery ingredients—acids, bases and salts, V. Confect. Prod. *41*, 293–295.

DEBSKI, M. S. 1973. Flavouring in confectionery. Flavour Ind. *4*, 431–432.

FELLOWS, G. 1972. Confectionery flavours, problems and factors affecting taste and choice of ingredients to protect or enhance flavors. Manuf. Confect. *52*, April, 70, 72, 74, 76, 78.

HEATH, H. B. 1969. Flavours for sugar confectionery and chocolate: Distillation and processing of essential oils. Confect. Prod. *35*, 790–798.

KENDALL, D. A. 1974. The release of methyl salicylate flavouring from chewing gum. Flavour Ind. *5*, 182–183.

KENNEDY, J. 1970. Flavors for sugar confectionery. Confect. Manuf. Market. *7*, 4, 6–7.

KNOCH, H. 1976. Processing of confectionery—Licorice. Confect. Prod. *42*, 461–467.

LEES, R. 1969. Confectionery. *In* Food Industries Manual, A. H. Woolen (Editor). Leonard Hill Books, London.

LEES, R. 1973. Flavouring sugar confectionery. Confect. Prod. *39*, 539–556.

LEES, R. 1974A. Citrus oils can achieve a 'different' flavour. Confect. Prod. *40*, 191–192, 194.

LEES, R. 1974B. Essential oils in confectionery production—recipes for mint caramels and creams and humbugs. Confect. Prod. *40*, 110–112.

LEES, R. 1974C. Essential oils in confectionery production—suggested trial usage levels. Confect. Prod. *40*, 57–58.

LEES, R. 1974D. Avoiding deterioration of stock and new flavourings. Confect. Prod. *40*, 403–404.

LEES, R. 1976A. Manufacture of caramels and toffee—recipes, etc. I. Confect. Prod. *42*, 123–124.

LEES, R. 1976B. Manufacture of caramels and toffee—recipes, etc. II. Confect. Prod. *42*, 222–223, 239.

LEES, R. 1976C. Manufacture of caramels and toffee—recipes, etc. III. Confect. Prod. *42*, 267–268, 280.

LEES, R. 1976D. Manufacture of caramels and toffee—recipes, etc. IV. Confect. Prod. *42*, 327–329.

LEES, R. 1976E. Manufacture of caramels and toffee—recipes, etc. V. Confect. Prod. *42*, 363–364.

LEES, R. and JACKSON, E. B. 1973. Sugar Confectionery and Chocolate Manufacture. Leonard Hill Books, London.

LENDERINK & CO. N. V. (Undated) Hyfoama—Manual for the Production of Aerated Confectionery. Lenderink & Co. N. V., Schiedam, The Netherlands.

MEINERS, A. and JOIKE, H. 1969. Handbook for the Sugar Confectionery Industry, No. 1. Silesia-Essenzenfabrik Gerhard Hanke K. G. Norf, W. Germany. (German)

MERORY, J. 1968. Food Flavorings: Composition, Manufacture and Use, 2nd Edition. AVI Publishing Co., Westport, Conn.

MINIFIE, B. W. 1970. Chocolate, Cocoa and Confectionery: Science and Technology. J. & A. Churchill, London; U.S.A. Edition, AVI Publishing Co., Westport, Conn.

PERRY, F. 1972. Essences, flavours and colours of confectionery. Confect. Prod. *38*, 412–414.

PRATT, C. D. 1970. Twenty Years of Confectionery and Chocolate Progress. AVI Publishing Co., Westport, Conn.

PRUDHOMME, P. 1970. The use of chocolate as a flavoring. Fr. Ses Parfums *13*, No. 69, 265–266. (French)

RIEDEL, H. R. 1976. Flavours and "perfumes" for confectionery industry. Confect. Prod. *40*, 63–64.

VERNON P. F. 1975. Sensory evaluation and its relationship to the chocolate and confectionery industry. Manuf. Confect. *55*, No. 6, 42, 44, 46–47.

WIELAND, H. 1972. Cocoa and Chocolate Processing 1972. Food Process. Rev. *27*. Noyes Data Corp., Park Ridge, N.J.

14

Pickles and Sauces

Pickles, sauces and relishes form a product group in which spices play a significant part in producing the desired flavor profile having the correct balance of piquancy, spiciness, acidity, saltiness, sweetness and fruitiness to blend pleasantly and give character to the dishes with which they are eaten. Most sauces contain acetic acid as the main ingredient. Traditionally, this is spiced and was formerly made by percolation of the ground spices, or by steeping a bag of the crushed spices in hot vinegar, or, even, by simmering the spices in vinegar—methods which are laborious, uncertain, time-consuming and wasteful of the aromatics in the spices.

MALT VINEGAR

In broad outline, the making of malt vinegar consists of:

(a) Grinding the malted barley.
(b) Preparation of the wort by digestion with water to enable diastatic conversion of starch. This is carried out at 63°–65°C (145°–150°F).
(c) Filtration of the wort.
(d) Fermentation with yeast to ethanol at 27°–32°C (80°–90°F).
(e) Filtration prior to storage for several weeks.
(f) Acetification with *Acetobacter*.
(g) Filtration prior to storage for several months.
(h) Standardization.

The color and flavor of the resulting vinegar can significantly alter the color and flavor of any pickle or sauce in which it is used. For this reason, many manufacturers prefer to use distilled vinegar which is malt vinegar from which the color has been removed by distillation. White distilled vinegar or spirit vinegar is also produced directly from ethanol (usually prepared from molasses); it is this product which is most preferred as it has a good full-bodied and consistent flavor. Artificial vinegar may be used in certain products; this is compounded from acetic acid and may or may not be colored with caramel. It lacks the flavor character of genuine vinegar. Most vinegars contain between 5 and 6% of acetic acid (Young 1974).

FLAVORED VINEGARS

The most widely used flavored vinegars include tarragon, garlic, mixed herbs and spiced. The traditional methods of manufacture, commented upon above, have now largely been replaced by the use of the appropriate herb or spice essential oil or oleoresin. This has resulted in considerable economies and, because of the complete utilization of the available flavor, it is generally necessary to modify any existing formulations based on the use of natural spices, the relative contribution of which to the profile is both variable and indeterminate. The correct equivalency may have to be determined experimentally but, once established, no further adjustment is likely to be required and continuity of flavor in the end-product can be assured. Not all manufacturers agree with this alternative technique claiming that the resulting vinegar is "raw" and lacks the mellow fullness of flavor associated with infused or percolated products. There is some justification for this as the aqueous vinegar does extract water-soluble fractions which are not present in either the essential oil or the oleoresin; but modern processing demands, economic considerations and the need for consistency in end-products for mass markets has tended to force the change.

There are few problems associated with the production of flavored vinegars once the formulation has been established. The essential oils/oleoresins may be more readily dispersed and solubilized by:

(a) making a concentrated solution in 80% acetic acid;
(b) admixing with a polysorbate (e.g., Tween 80);
(c) making an emulsion with gum acacia (Arabic);
(d) dispersion onto a soluble carrier (e.g., sucrose, dextrose or salt); or
(e) microencapsulation in a vegetable gum or modified starch.

The correct weight of these concentrates is added to the bulk of the vinegar with vigorous stirring. Where necessary the final product should be adjusted to contain not less than 3.6% of acetic acid. After a period of settling, the flavored vinegar is filtered.

Apart from their flavoring properties, it was formerly thought that spices enhanced the preservative action of vinegar-based products. Certain spices do have inhibitory effects on the growth of microorganisms due to the presence of phenols and other active bacteriostats in their essential oils (e.g., clove, allspice, cinnamon, chillies and mustard). These are generally present at such low levels of concentration that the inhibitory effects are negligible.

SAUCES AND PICKLES

These products may be classified as follows:—

(a) Thick Sauces

- Fruit sauces: cereal-thickened and finely comminuted sauces based on pureéd fruits and vegetables with added sugar (up to 25%), vinegar and spices.
- Mustard and other piquant sauces.
- Salad cream and mayonnaise: an egg-stabilized oil-in-vinegar emulsion.
- Tomato ketchup (catsup): a sauce consisting of pureéd tomatoes with vinegar, sugar, salt, onions and spices.

(b) Thin sauces

- Worcestershire: suspension of micromilled spices, vegetable extracts and other constituents in a highly seasoned vinegar.

(c) Thick Pickles

- Piccalilli: mixed chopped vegetables, mostly cauliflower, in a thickened mustard sauce.
- Relishes: finely chopped vegetables, mostly cucumbers, in a lightly spiced vinegar.
- Chutney: sweet pickles made from comminuted fruits and vegetables in a thickened, sweetly-spiced sauce.

(d) Clear Pickles

- Clear-pack vegetables: embracing a whole range of vegetables in lightly spiced vinegar, some products contain whole spices for their appearance (e.g., onions, cauliflower, gherkins, cucumbers, beets, red cabbage, walnuts, olives, mango, and mixed red peppers).
- Sweet pickles: vegetables in a sweet, lightly spiced vinegar (e.g., beets).

PROCESSING TECHNOLOGY OF FRUIT SAUCES

Fruit sauces are characterized by having a high viscosity, a sweetly acidic taste with a marked spicy fruity flavor and a level of piquancy specific to the end-product. The principal raw materials include: sugar, dates, tamarinds, mango, apples, raisins, sultanas and other fruits together with an appropriate thickening agent which is stable in acid media. The variety of thick sauces which can be produced from these materials is legion; some are internationally regarded, others are made to satisfy local demands.

The ingredients are cooked to the desired consistency; sieved to remove stones, pips, skin, etc.; and then seasoned and thickened prior to passing through a colloid mill or other homogenizer to produce a smooth-textured sauce. The product is deaerated and vacuum filled into capped bottles or jars. In the larger automated plants, use is made of precooked and sieved pulps so that the ensuing operation involves only mixing and milling.

Problems encountered in the manufacture of thick fruit sauces include: (a) fermentation, (b) separation, and (c) formation of air bubbles in the product on cooling. None of these can be attributed to the added flavorings which are often blamed for these processing defects.

Flavoring of Fruit Sauces

Traditionally, ground spices were added to the mixed fruit and vegetables base a short time before completion of the cooking period. This is a wasteful practice as many of the aromatic volatiles are lost in the steam and most of the flavoring components are but poorly soluble in either water or vinegar. The use of spice essential oils and/or oleoresins in either a solubilized or dispersed form is a far better way of adding seasoning. As these products are sterile, it is unnecessary to add these during the cooking stage; they may be incorporated immediately prior to the transfer of the bulk product to the holding tank for bottling. Under such circumstances, the finished sauce generally has a brighter color and, as the aromatic losses are minimal, a fuller flavor profile the consistency of which can be more readily controlled. There is also likely to be an economic advantage to this method of adding seasoning.

Formulations for seasonings appropriate to this group of products are given in the literature (Binsted *et al.* 1971).

PROCESSING TECHNOLOGY OF PREPARED MUSTARD

Prepared mustard is available in many varieties of flavor and pungency and is a universally popular product. The technology of its manufacture is not complicated. It involves the fine milling together of the blended mustard seeds with additional spices, followed by admixture with vinegar to form a smooth paste and then filling into suitable containers such as jars or tubes. Unfortunately, the full flavor and piquancy of the mustard tends to fade on storage and it is not uncommon to incorporate a small percentage of mustard oil (allyl isothiocyanate) to the final mix. In many countries the formulation of prepared mustard is controlled by legislation, particularly the level of use of added cereal flour and color (e.g., tartrazine or turmeric).

The flavor profile of the end-product is dictated by local tastes, many products having a distinctive flavor as well as varying levels of piquancy ranging from hot to mild. For example, French mustard often contains garlic, clove and a little horseradish; German mustard may be flavored with clove, cinnamon and allspice and is usually made with wine vinegar; the famous Dijon mustard is flavored with a subtle blend of herbs and spices in which tarragon predominates. The diversity of formulations reflects the importance of this condiment throughout the world.

Mustard sauce, the fluid component in piccalilli, is basically similar to prepared mustard except that the milled mustard flours and seasoning are blended with starch and gently cooked with vinegar to thicken prior to pouring over the prepared vegetables.

PROCESSING TECHNOLOGY OF MAYONNAISE AND SALAD CREAM

These products are essentially oil-in-water emulsions with egg yolk as the emulsifying agent together with edible vegetable oils, salt, vinegar, spices, flavorings and colorings as appropriate to the formulation. In many countries, the designation "mayonnaise," "salad cream" or "salad dressing" imply specific formulations and standards.

The method of manufacture consists of:

(a) Preparation of a gum tragacanth or other vegetable gum mucilage in part of the vinegar and water.

(b) Preparation of a premix of sugar, lactic acid, salt, flavorings and coloring (if permitted) with the remainder of the vinegar, into which the egg yolk is stirred and the mix sieved into a stainless steel steam-jacketed mixing pan.

(c) Temperature of the mix is raised to 82°C (180°F) with gentle stirring.

(d) Gum mucilage is added and mixing continued until the bulk is uniform.

(e) Vegetable oil is then slowly incorporated with vigorous stirring—a stage which can cause problems unless careful attention is paid to details.

(f) Product is cooled, homogenized and filled out into prepared bottles.

Flavoring of Salad Dressing

The most commonly used flavoring in this type of product is mustard together with tarragon and onion, preferably added in the form of the essential oil so as to avoid unsightly speckiness or an unacceptable discoloration of the product. The flavor profile of these products is generally delicate so that any defects in the quality of the raw materials are readily apparent, particularly any off-notes due to rancidity. Because of this, care should be taken to ensure that the vegetable oil used is fresh and free from any taints. Color changes may occur if the product is displayed or left exposed to direct sunlight—a common fault in products containing mustard or turmeric as a source of natural color. Where permitted, this may be overcome by the use of a permitted food color such as tartrazine, which is light-stable.

PROCESSING TECHNOLOGY OF TOMATO KETCHUP

The composition of this universally popular product is defined by legislation in many countries. Although the formulation may differ in details, it consists of tomato paste, sugar, vinegar, salt, spices and other flavorings blended with a suitable thickener. The resulting product is smooth, flows easily but is not too liquid, and is free from any discrete solid particles such as tomato seeds, skin, etc.

The method of manufacture comprises the following stages:

(a) Preparation of a gum and/or starch mucilage in part of the vinegar and water.

(b) Cooking together of the tomato paste, minced onions or onion powder, sugar and salt in the remainder of the vinegar, in a stainless steel steam-jacketed mixing pan, at 40°C (120°F).

(c) Addition of the gum mucilage with gentle mixing until uniform (about 10 min).

(d) The cooked mass is then sieved and homogenized prior to transfer to a vacuum concentrator and deaeration unit.

(e) On completion of this stage, the product is transferred to a holding tank at 88°C (190°F) and vacuum filled hot into capped bottles. It is usual to cool the filled bottles rapidly.

Flavoring of Tomato Ketchup

Finely ground spices, particularly clove, cinnamon, cassia, pimento (allspice) and nutmeg were originally used in making this tomato-based product but all of these produce dark-colored powders on milling and their incorporation into a smooth creamy product often resulted in unsightly speckiness and an overall dull and unattractive appearance. Seasonings are best added as dispersions of the essential oil or oleoresin on salt or dextrose. These products are pale-colored and in this form the seasoning does not detract from the bright color of the tomato. Being sterile, they can be added to the product mix after the cooking stage and so avoid undue exposure to heat. Solubilized spices in the form of emulsions may be added to the mix as it enters the holding tank prior to filling; but care must be taken to ensure that it is uniformly dispersed.

PROCESSING TECHNOLOGY OF WORCESTERSHIRE SAUCE

Worcestershire sauce is by far the most widely used of the thin sauces in which micromilled spices, vegetable extracts and various other flavorings are macerated in vinegar and allowed to mature in wooden casks to develop a full smooth flavor profile. The product is finally passed through a coarse-meshed sieve to remove any larger particles while allowing the passage of fine particles which contribute to the characteristic sediment of this product.

Flavoring of Worcestershire Sauce

The basic ingredients in thin sauces include many having a high intrinsic flavor level, e.g., anchovies, tamarinds, walnuts, mushrooms, onions and garlic as well as a whole range of spices such as pepper, pimento (allspice), coriander, nutmeg, mace and clove. The level of pungency is controlled by the use of cayenne pepper or a dispersion of oleoresin capsicum on salt. The use of the equivalent spice essential oils and/or oleoresins is not so successful in thin sauces as the ultimate appearance depends upon the presence of the fine deposit of milled spices which has to be dispersed, each time the product is used, by shaking the bottle.

PROCESSING TECHNOLOGY OF PICKLED VEGETABLES

Vegetables, preserved either in brine or vinegar, are very popular throughout the world and most products represent the traditional preservation of locally grown produce. The principal vegetables used are beets, cauliflower, cucumbers, brocolli, gherkins, green peppers, mushrooms, red cabbage, red peppers, runner beans, and walnuts.

Before bottling in vinegar (which is usually spiced) the vegetables are soaked in brine (15% sodium chloride) in order to preserve and tenderize them. Whereas in the case of sauces where the ingredients are heated together before being cooled and sieved, the vegetables in these conventional clear pickles are not heated at any stage. Consequently, enzymic deterioration is not inhibited and the pickled vegetables are, therefore, subject to the gradual development of enzymic effects such as softening (due to the degradation of pectin by pectinase), darkening, flavor deterioration, liquor clouding, sediment formation and the development of yellow quercetin on the surface of onions. All of these are detrimental to the quality of the end-product.

Products such as the sauces and sweet pickles contain enough sugar to offset the acid flavor; the plain pickles lack this modifying effect and if the acetic acid is increased to improve preservation the resulting flavor is generally considered to be unacceptably sharp. Ways are now being sought to reduce this acidity by lowering the level of acetic acid in the product and ensuring adequate preservation by pasteurization. This has been successful with dill pickles but poses a danger when the product is opened by the consumer and then only partially used. The remaining pickles are inadequately preserved and soon display a cloudy liquor and a strong off-odor. In spite of this, the method is widely used. Pasteurization achieves:

(a) In conjunction with the action of the acetic acid present, it preserves the sealed pack against microbiological spoilage during its shelf-life and for a very limited period after opening.

(b) It inactivates the enzymes responsible for various degradative changes in the product.

(c) By suitable application in conjunction with an hermatic seal, it helps to remove air from the tissues of the component vegetables.

(d) By establishing a headspace vacuum, it minimizes oxidative changes in the product during storage.

It is possible to arrive at a very satisfactory compromise in providing the necessary safeguard against microbiological spoilage and a too-acid flavor; but the optimum formulation and processing conditions generally have to be determined experimentally for each type of vegetable.

PROCESSING TECHNOLOGY OF DILL PICKLES

This is a very popular and is the most important class of pickles. Dill pickles consist mainly of cucumbers preserved in a distinctly spiced liquor containing a high level of dill weed. The characteristic flavor and crunchy texture of the cucumbers is achieved by controlled fermentation. Genuine dill pickles were traditionally prepared from freshly harvested cucumbers; but as these are now cultivated in huge quantities and collected over a very short period, it is impossible to process them immediately. This necessitates that the cucumbers be preserved by immersion in brine by one of two methods depending on the locality:

(a) Low-salt brine. Cucumbers are covered with 8% (30° salinometer) brine, further salt being added during the ensuing storage under carefully controlled conditions.

(b) High-salt brine. A similar process as that above but starting with 10.6% (40° salinometer) brine, the rate of salt addition is higher and dependent on the ambient temperature.

The ultimate processing of salt-preserved cucumbers is a matter of choice between three methods of fermentation:

(a) True lactic fermentation by *Lactobacillus* organisms.
(b) Yeast fermentation which results in gas formation.
(c) *Acetobacter* fermentation.

In order to produce a really first-class product, this fermentation stage must be controlled; this involves considerable expertise. The fermented cucumbers are washed, packed into jars and covered with the prepared seasoned brine. Formerly, ground spices and fresh dill herb were used to produce the characteristic flavor profile and in many regions this method persists and is still preferred. Most large-scale manufacturers have now changed over to the use of solubilized seasonings based on dill weed oil and suitable spice oleoresins. These give an acceptable and consistent flavor in the finished pickles.

In the United States the term "Kosher" dill pickles is often used. This does not necessarily imply Rabbinic approval but rather that the flavor contains onion and garlic in the Jewish tradition. Generally, they are also a little more highly spiced than those from mid-European origins.

Two aspects of flavor quality are of importance:

(a) Off-Flavor. Some packs develop a hay-like or musty odor associated with a strong rank taste. This is thought to be enzymatic in origin and may be prevented by pasteurization. The use of insecticides on the cucumbers during development may also be a source of unpleasant off-notes.

(b) Bitterness. Some types of cucumbers and gherkins are initially very bitter, a character which is most obvious in fresh packs. This defect tends to fade on storage although the reasons are not well established.

Seasonings for Pickles and Sauces

Although the prime ingredients of pickles and sauces make a significant contribution to the overall flavor of the end-product, it is the judicious use of herbs and spices coupled with the correct degree of acidity which impose the major characteristic aroma and flavor attributes. In developing an appropriate seasoning, two quite different factors have to be taken into account: (a) to what extent must the seasoning over-ride or contrast with the predominant flavor of the base; and (b) whether or not it is necessary for the seasoning to be completely soluble in the liquid phase of the end-product.

In spite of the fact that herbs and spices are used at relatively low concentrations, their effects directly influence the quality of the end-product. Too little seasoning results in a product that is flat and lacking in impact; too much, and it tires the palate, swamps the flavor of any food with which it is eaten and eventually becomes objectionable. Fortunately, consumer spice tolerance is very wide, particularly with regard to piquancy; most seasonings are aimed at providing a sufficient level of pungency and a balanced aromatic impact so that the product is compatible with a wide range of dishes and is pleasingly stimulating to the appetite.

For this group of products, spices may be used in one of several ways: (a) as whole or broken spices to contribute a distinctive appearance to the product; (b) as finely ground spices to give flavor and texture; (c) as essential oils or oleoresins either dispersed on a soluble carrier or solubilized with an appropriate solvent or emulsified to give flavor and color; (d) as an acetic acid or vinegar infusion of the herbs and spices to give a clear end-product.

Appearance certainly contributes to the acceptability of any pickle or sauce and, traditionally, many clear pickles contain a few whole spices such as coriander, black pepper, mustard seeds and pieces of ginger or sprigs of fresh herbs, such as tarragon or dill, in order to give the impression that the product is "home made" or "naturally" flavored. In fact, the flavor of the product is generally determined by the use of a premade seasoned vinegar, the whole spices being added for purely aesthetic reasons. It is important that spices used in this way should be judged on their appearance rather than on any other attribute as so often clean bold spices are less flavorful than those of poorer appearance.

The use of ground spices in the making of a seasoned vinegar has already been commented upon. In deciding the most appropriate form of seasoning to use, the following disadvantages should be noted:

(a) Ground spices are variable in flavor strength and quality depending on source, age and grade.

(b) The method of making seasoned vinegar by boiling or otherwise extracting the ground spices with vinegar is poorly incomplete, wasteful of available flavor and likely to yield a product having an inconsistent profile.

(c) Unless micromilled, the ground spices contribute unsightly dark brown or black specks to the product which makes a smooth creamy product like mayonnaise appear dull and "dirty."

(d) Spices are more or less contaminated with microorganisms which, when conditions are right, may give rise to off-odors, a cloudy liquor and other less obvious product defects.

Automated, continuous processing is now commonplace in this branch of the food industry and many long-established methods of preparing and adding seasonings are no longer considered practicable. There has been an increasing acceptance of processed spices the many advantages of which enable the manufacturer to achieve a product having a consistent profile and appearance. Being sterile, they are consistent with good hygienic manufacturing techniques and can be added to the pickle or sauce liquor at the end of or even after any cooking stage so reducing losses to a minimum without detriment to product stability.

The herbs and spices have been reviewed in an earlier section of this book but for convenience certain widely used herbs and spices are worthy of special note in this context.

HERBS

Mint

There are many varieties of mint used in the production of mint sauce and jelly. Culinary mint is derived from varieties of spearmint (*Mentha viridis*, L.) and, as harvested, consists of both stalk and leaves. Only the latter are of value and must be stripped off the stem as soon as possible and chopped into small particle size. Delay in doing this results in the cut herb overheating with resultant blackening of the leaves and the development of an unpleasant rank odor. The cut leaves may be preserved in acetic acid until required.

Other species of mint are available and may be used to give products having a distinctive profile. These include: apple mint (*Mentha rotundifolia*, L.), American apple mint (*M. spicata*, Huds.), curly mint (*M. spicata*, var. *crispata*, Schrad.), and Russian mint (*M. verticillata*, L., var. *strabala*, Briq.).

For the production of mint sauce and mint jelly, the herb is usually harvested before the flower heads have formed. At this stage, the essential oil has not yet achieved the same balanced composition that it has at maturity (i.e., in full bloom). For this reason, the flavor of the herb tends to change becoming smoother and more spearmint-like on storage and it is difficult to obtain the preferred profile by substituting spearmint oil blends. The best results are still achieved by macerating the freshly cut leaves in acetic acid.

Dill Weed

The umbelliferous plant, *Anethum graveolens*, L., is normally cultivated for its seed which is distilled to yield oil of dill. If, however, the plant is harvested when the oldest seeds are just starting to turn brown and the whole plant is distilled, then the resulting oil is known as oil of dill weed. It is this oil which is used to give the characteristic flavor to dill pickles as it most nearly and much more conveniently replaces the use of a vinegar infusion of the freshly cut herb. The profile of this oil compared with that of dill seed has already been described. The difference depends on the balance between terpenes such as β-phellandrene and d-limonene and carvone at the time of distillation. As the herb ripens the carvone content of the herb oil increases and approaches that of the seed oil (i.e., 40–60%) with the consequent development of a marked caraway-like character. Weed oils containing as little as 20% carvone are preferred for use in dill pickles. The usage level varies between manufacturers but is in the order of 1 oz of oil per 20 gal. of pickling liquor.

Tarragon (Estragon)

The fresh herb is now largely replaced by oil of tarragon which is obtained by distillation of the freshly harvested, tender leaves and tops of *Artemisia dracunculus*, L. The oil has a pleasantly sweet, green, anise-like profile which makes it of particular value in the flavoring of mayonnaise and salad dressing.

PUNGENT SPICES

Many sauces depend for their efficacy on a marked level of pungency derived from the use of such spices as capsicum, red pepper or cayenne, ginger, black and white pepper, mustard and horseradish. The relative pungency of these materials is not easy to determine as the effect in the mouth is so very different. For this reason, the desired "heat" effect can only be determined by trial and error, remembering that pungency tends to be cumulative and that the overall effect in the sauce can only be satisfactorily assessed when the product is eaten at normal levels.

The pungency in these spices is variable and particularly so in the case of capsicums which are available from many sources in a wide range of forms, colors and capsaicin content. Because of this, most manufacturers prefer to use the equivalent oleoresins, the pungency of which can be controlled and standardized to within narrow limits. The following products are of interest to the pickle and sauce manufacturer.

Oleoresin Capsicum

The pungency of this product is often standardized in terms of Scoville units, a sensory method based on threshold pungency detection at different levels of dilution. It has been demonstrated that this method of assessment is open to wide interpretation depending upon the sensitivity of the panel used. The pungent principle of capsicum is capsaicin (8-methyl-N-vanillylnon-6-enamide) and the direct determination of this chemical gives a more accurate measure of the available pungency. Many methods have been proposed in the literature; these are included in the bibliography. Opinions differ on the exact equivalent between Scoville units and capsaicin content but generally:

1 million Scoville units = 6.38% capsaicin

or

1% capsaicin = 157,000 Scoville units.

Oleoresins containing 500,000 and 1 million Scoville units are readily available. These products are, of course, intensely pungent and present some hazard in handling; consequently dilutions in a vegetable oil or dispersions on salt or dextrose are preferred.

Oleoresin Red Pepper

This product is obtained by the extraction of redder and less pungent varieties of capsicums and is generally standardized for both capsaicin content (or pungency) and color. The pungency may vary between 100,000 and 250,000 Scoville units. The strong red color is measured spectrophotometrically, the absorbance being multiplied by an empirical factor of 61,000 to give arbitrary "color units." The specification of the Essential Oil Association of U.S.A. (EOA No. 245) requires that the product shall have a pungency of 240,000 Scoville units and a color value of 20,000 units. By comparison, oleoresin paprika, which lacks pungency, has a color value between 40,000 and 100,000 units. Where oleoresin red pepper is not available or where different pungency and color levels are required, these can be achieved by the use of blends of oleoresin capsicum (high test) and oleoresin paprika.

Oleoresin Black Pepper

This oleoresin is obtained by the solvent extraction of black pepper derived from various commercial sources. The pungent principle is mainly, but not entirely, piperine which forms the basis for standardization of the pungency; the essential oil content is a measure of its flavoring strength. It has long been the practice to quote the piperine content in terms of Kjeldahl nitrogen, but this method has been shown to give results which are variably 10–15% higher than those obtained by the direct spectrophotometric determination of piperine (Genest *et al.* 1963). When comparing products, therefore, it is necessary to know by what method the figures were obtained. Various methods for the determination of piperine are quoted in the bibliography.

Oleoresin black pepper is a more or less pastey solid-to-viscous fluid consisting of a mass of piperine crystals in an oily matrix. In some products, this may separate into two phases during storage and where this has happened the product must be gently warmed and remixed well before use. Many stable fluid products are now available and these, though having a lower piperine content, are preferred as they are so much easier to handle.

Chili Powder

This is a product often encountered in formulations for pickles and sauces. Chili powder is not just ground capsicums, but consists of a blend of dried sweet and pungent varieties of Mexican chillies together with cumin seeds and oregano herb. The flavor of this product depends on the precise formulation used.

AROMATIC SPICES

The aromatic profiles of all the spices used in pickle and sauce seasonings have already been discussed in an earlier section. The following additional notes are germain to their use in this group of products.

Cinnamon/Cassia

Genuine cinnamon is derived from the inner bark of *Cinnamomum zeylanicum*, Blume., cultivated in Sri Lanka (Ceylon). This has a fine aromatic profile which blends perfectly with tomato-based products but it is relatively expensive. Alternative sources of other *cinnamomum* species yield much less expensive material but at some loss of profile. The major commercial sources of cinnamon or cassia are China, Laos, Cambodia and Vietnam (exported via Saigon), Indonesia (Batavia) and the Phillipines. The following are the principal species involved:

Saigon cinnamon (cassia)	*C. loureirii*, Nees.
Chinese cinnamon (cassia)	*C. cassia* (L), Blume.
Korintje cinnamon Java cinnamon	*C. burmanii*, Blume.

In commerce these are usually ground and blended without an indication of source; but, of these, Saigon cinnamon is regarded as having superior aroma and flavor.

It is not easy to prepare a stable oleoresin of cinnamon as the product tends to polymerize on standing and forms a brittle plastic-like mass which is insoluble in any of the normal solvents. Commercially available oleoresin cinnamon generally contains an added solvent such as propylene glycol or glycerin which is incorporated during the final stages of solvent removal.

An alternative consists of using the resin obtained from predistilled cinnamon; this product is used with the appropriate quantity of the distilled essential oil. The main flavor of cinnamon is carried by the essential oil which may be substituted quite successfully in seasonings. This, of course, lacks the color which characterizes cinnamon bark. It should be remembered that cinnamon bark and cinnamon leaf oils have quite different profiles—the former is largely that of cinnamic aldehyde and the latter that of eugenol.

Clove

Ground cloves are very oily and are difficult to reduce to a fine powder. In consequence, their use may result in an unacceptable level of speckiness. Cloves contain at least 12% of quercitannic acid and other tannins which are soluble in dilute acetic acid and give to the solution a very distinct astringency. These tannins are also present in oleoresin clove but are not carried over into the distilled essential oil. This may or may not be considered an advantage but will most certainly make a difference in balancing a spice formulation, depending on which form is used.

The presence of tannins can pose other problems particularly associated with color; the product may darken on exposure to air and even give rise to black discoloration in contact with iron. This is often the cause of black ringing round the necks and caps of sauce bottles.

Other parts of the clove plant also contain essential oil and it is common practice to blend the expensive clove bud oil with much less valuable oils obtained from the stem and leaves. To improve their character, these oils are frequently rectified; but their profile still falls short of that of the genuine bud oil.

COLORED SPICES

Reference has already been made to paprika, but by far the major source of natural color in many pickles and sauces is turmeric.

Ground Turmeric.—Turmeric is the dried rhizome of *Curcuma longa*, L., cultivated in India, Sri Lanka (Ceylon), Indonesia, China, Taiwan, Peru, Brazil, Mexico and Jamaica. The quality and coloring power of the ground spice varies widely with source, the best quality having a rich orange-yellow color, a characteristic and highly aromatic odor and a bitter taste. Madras and Alleppy turmeric are the best and most widely used varieties.

Oleoresin Turmeric.—An extract of ground turmeric is a highly aromatic pastey oleoresin having a brilliant orange-yellow color due to curcumin (*bis* 4-hydroxy-3-methoxy-cinnamoyl). The tinctorial power may be quoted either in terms of curcumin content or as arbitrary color units determined by multiplying the absorbance at 422 mμ by an empirical factor of 2000.

The color of turmeric is light sensitive so that pickles containing it (e.g., piccalilli) may fade on storage, particularly if exposed to sunlight. It is also sensitive to pH, being a full yellow in acid media changing to a dull reddish-brown as the pH increases.

This oleoresin may be offered premixed with a solubilizing agent, such as polysorbate 80, where this is permitted. In the United States, it may be used in pickles and sauces so long as the maximum present does not exceed 0.05% in the finished product.

STORAGE CHANGES IN FLAVOR PROFILE

Seasonings tend to mature on standing so that a freshly made batch of pickle liquor or sauce may be different from the same product after, perhaps, a month's storage. This is an aspect of seasoning that must be considered during product development. Appropriate shelf tests should be set up to establish the nature and extent of any maturation changes in the flavor profile.

BIBLIOGRAPHY

ALLGEIER, R. J., NICKOL, G. B. and CONNER, H. A. 1974A. Vinegar; history and development. I. Food Prod. Dev. *8*, No. 5, 69–71.

ALLGEIER, R. J., NICKOL, G. B. and CONNER, H. A. 1974B. Vinegar; history and development. II. Food Prod. Dev. *8*, No. 6, 50,52,56.

ANON. 1974. Freeze/thaw/heat stable salad dressing from mix. Food Process. *35*, No. 11, 37.

ANON. 1975. Worcestershire sauces vary in quality, solids content—affect formulation, usage, Food Process. *36*, No. 3, 58.

BINSTED, R., DEVEY, J. D. and DAKIN, J. C. 1971. Pickle and Sauce Making, 3rd Edition. Food Trade Press, London.

BLANCHFIELD, J. R. 1969. Pickles and sauces. *In* Food Industries Manual, 20th Edition. A. H. Woollen (Editor). Leonard Hill Books, London.

CRIPPS, H. D. 1966. Oleoresin turmeric application in pickle production. Glass Packer/Process. *46*, Oct., 24–25.

CUMMING, D. 1966. What makes a dill pickle taste just right? Glass Packer/Process. *45*, Oct., 33, 36, 38.

DEAK, T. 1974. Isolation and identification of lactobacilli associated with pickle fermentation. Conf. Proc. 4th Int. Congr. Food Sci. Technol. *4a*, 26–28.

DIECCO, J. J. 1976. Gas-liquid chromatographic determination of capsaicin. J. Assoc. Off. Anal. Chem. *59*, 1–4.

EISERLE, R. J. 1966. The role of oleoresin turmeric in the pickling process. Glass Packer/Process. *45*, Oct., 48–49.

FERNANDEZ, D. *et al.* 1974. Pectin substances in pickling products related to softening problems. Conf. Proc. 4th Int. Congr. Food Sci. Technol. *1a*, 19–21.

GENEST, G. *et al.* 1963. A critical study of two procedures for the determination of piperine in black and white pepper. J. Agric. Food Chem. *11*, 509–512.

GOVINDARAJAN, V. S. and ANANTHAKRISHNA, S. M. 1974. Paper chromatographic determination of capsaicin. Flavour Ind. *5*, 176–178.

HAMBOLD, J. 1974. Pickling liquor: Basis for the automation of concentration control. Ind. Obst Gemueseverwert. *59*, No. 13, 351–353. (German)

KARASZ, A. B. *et al.* 1973. Detection of turmeric in foods by rapid fluorometric method and by improved spot test. J. Assoc. Off. Anal. Chem. *56*, 626–628.

KUHN, M. 1974. Technology, properties and analysis of mayonnaise. Dtsch Lebensm. Rundsch. *70*, No. 10, 352–356. (German)

LAL, G. *et al.* 1967. Preservation of Fruits and Vegetables. Central Food Technological Research Institute, Mysore, India.

MAGA, J. A. 1975. Capsicum. C.R.C. Crit. Rev. Food Sci. Nutr. *6*, No. 2, 177–199.

MATHEW, A. G. *et al.* 1971. Capsaicin. Flavour Ind. *2*, 691, 693–695.

PETERSON, M. S. 1974. Pickles. *In* Encyclopedia of Food Technology, A. H. Johnson and M. S. Peterson (Editors). AVI Publishing Co., Westport, Conn.

YODA, H. 1974. Mustard Paste. Assigned to S. & B. Shokuhin Co. U.S. Pat. 3,852,488, Dec. 3.

YODA, H. 1975. Mustard Paste. Assigned to S. & B. Shokuhin Co. Br. Pat. 1,396,950, June 11.

YOUNG, H. 1974. Vinegar. *In* Encyclopedia of Food Technology, A. H. Johnson and M. S. Peterson (Editors). AVI Publishing Co., Westport, Conn.

WALFORD, J. 1976. Solubilizers for essential oils in flavour formulations. Food Manuf. *51*, No. 2, 35,37.

WREN, J. 1974. Food emulsifiers. Nutr. Food Sci. *35*, 10–14.

15

Soups

Virtually anything that can be eaten can be used in the making of a soup. It is one of the traditional forms of presenting food and may be consumed as an appetizer or as the main dish. Soups can be manufactured as dry products to be reconstituted as required by the consumer; as a single-strength product which, when heated, is ready to serve; as a canned concentrate which requires dilution with water before cooking; or, more recently, as a frozen product which is of particular value in institutional feeding. Each of these types of end-product presents its own problems of manufacture but, so far as the blending of suitable seasonings is concerned, they are all similar.

PROCESSING TECHNOLOGY

From the above it will be appreciated that the production of soups falls into three categories, namely: canned, dehydrated and quick-frozen. Details of the methods employed are discussed in standard texts such as Binsted and Devey (1970). The following outlines are intended only as a guide to those interested in the flavoring of these products.

Canned Soups

Products include thick (creamed) and thin soups with or without the presence of particulate ingredients such as meat and/or vegetables. In many countries, there is either legislation or a code of practice covering the formulation and designation of this class of products as well as regulations governing the declaration of ingredients.

Production is carried out in steam-jacketed stainless steel vessels fitted with automatic stirrers. The ingredients are first cleaned, as necessary, or otherwise prepared and gently heated with the liquid

Courtesy of The Nestle Company

FIG. 15.1. HAND PREPARATION OF ONIONS FOR ONION SOUP

Courtesy of The Nestle Company

FIG. 15.2. CHARGING THE PREPARED VEGETABLE INTO A COOKER/BLENDER

base with slow stirring until the mix is boiling. If a thickener is used it is prepared as a cold slurry in part of the recipe water and added at this point. Particulate solids (e.g., meat, diced vegetables, etc.) may be incorporated into the batch but more generally are filled out separately into the final containers to ensure even distribution. Batches are pumped, in turn, to the filler where the soup is automatically metered into cans at usually not less than 80°C (175°F). The filled cans are seamed and autoclaved at 115°–121°C (240°–250°F) for a time dependent upon the can size and the nature of the contents. Times usually range between 1 to 1½ hr. After retorting the cans are water-cooled.

In the case of cream soups, the same process is carried out except that the final soup mix is passed through an homogenizer at 1500–2000 psi before filling. Any particulate solids must, of course, be added separately to the cans.

Dry Soup Mixes

The dry ingredients are blended together in a suitable ribbon blender. In some instances a special order of mixing is called for but as the formulations are so diverse one cannot generalize. The particle size of the several ingredients should be reasonably uniform, otherwise separation may occur during handling and packaging operations. After mixing, the homogenous powder is machine-filled into plastic/aluminum foil laminated envelopes or sachets and heat sealed. In some packets the residual air is removed before final sealing.

Mixing and filling operations should be carried out in an air-conditioned atmosphere with a relative humidity controlled within the range 40–45%. Modern techniques involving computerized batching of ingredients from bulk storage hoppers are now employed by the larger manufacturers but true automation is very difficult owing to the variable nature of the ingredients required to give an end-product having the desired textural and visual characteristics.

Inevitably, the method of mixing and blending will involve considerable air movement through the product. This can lead to volatile losses for any seasoning or flavoring used, particularly if this is merely dispersed on a carrier, although once in the final pack such losses are minimal. Because of this, it is an advantage to use encapsulated herbs and spices; the flavoring is then protected from volatile loss or oxidative change until the mix is ultimately reconstituted. Hence, the product is assured of a satisfactory shelf-life even under adverse storage conditions. In the case of onion and garlic, which are generally added as dehydrated powder, there is an added advantage in that the encapsulated products are virtually odorless, are more pleasant to handle and there is less risk of crossover of flavors.

Quick-Frozen Soups

Soups for quick freezing are prepared in a manner similar to that used if the soup were to be immediately consumed (i.e., it is fully cooked). The bulk is vacuum deaerated under carefully controlled temperature conditions and is then cooled prior to unit packaging and deep freezing or bulk freezing in blocks. The techniques are described by Tressler *et al.* (1968).

FLAVORS IN SOUPS

It is from the point of view of flavor application that the main differences in formulation occur. The dehydrated soups require the use of dry powder flavors and seasonings, added as finely ground, dispersed or encapsulated herbs and spices blended with other powdered flavoring materials as necessary. In the manufacture of canned and quick-frozen soups it is advantageous to add liquid flavorings, although the use of dispersed spice products yield totally satisfactory results. Certain soups are traditionally clear and free from particulate matter (e.g., consommé) and in these it is essential that any flavoring be completely soluble.

Most manufacturers have developed distinctive flavor profiles in their products by which they hope to capture the public imagination and be so appealing that consumers will pick their product first. Local tastes tend to dictate product formulations and, usually, a compromise has to be accepted to achieve mass-market coverage. Apart from the necessity of making a soup to satisfy consumer demand, there is also the need to maintain consistency of flavor. In an end-product which is consumed without further cooking except heating, this implies close attention to processing techniques, the use of good quality raw materials and the incorporation of standardized flavorings wherever possible, consistent with the commercial viability of the product.

The following flavoring materials are of particular interest to the soup manufacturer:

Basic Taste Modifiers (Salt, Sugar, Vinegar and Wine)

Unless a reasonable proportion of salt is present in a soup, the main proteinaceous and starchy ingredients will be uninteresting or even unpalatable. In most formulations, salt should be present at about 1.0–1.25% in the soup as consumed although in some regions somewhat higher percentages are tolerated.

Sugar should be used only to enhance any naturally sweet flavors present and not impose a marked sweetness per se. The exception to this is in the formulation of Norwegian fruit soups.

In many liquid soups a little vinegar or diluted acetic acid adds a pleasantly sharp pungency. Malt vinegar may be a little too powerful for this purpose and wine or cider vinegars are generally preferred. Spiced vinegar based on tarragon or blended herbs may be a convenient way of introducing subtle flavor notes. In dry soup mixes the use of low levels of citric acid, tartaric acid or sodium diacetate may achieve the same result.

Wine added to a soup formulation can considerably enhance the flavor profile producing a full-bodied, rounded effect and, with certain wines, a pleasing astringent mouth-feel. It is necessary to establish the best type of wine and the usage level under exact processing conditions as the ultimate profile may alter considerably, particularly if the product is retorted.

Herbs and Spices

The domestic practice of hanging a muslin bag of broken herbs and spices in the stockpot when making a soup is far too imprecise and variable to be applied under factory conditions. Although finely ground spices are used, best results are achieved by employing processed spices in one of the following forms:

(1) standardized spice extractives plated onto an edible carrier (e.g., salt, dextrose, MSG);

(2) encapsulated spice essential oils and/or extractives;

(3) liquid flavorings which may be in an acceptable solvent or emulsified with gum acacia (Arabic) or a modified starch; or

(4) spice essential oils and/or oleoresins solubilized by admixture with a polysorbate (e.g., "Tween" 80).

The question of seasoning blends and the actual flavors to be achieved in any given product is a matter of opinion and will depend to a large extent on the nature of the prime ingredients, consumer demands and price. Each product calls for individual development and market appraisal but the basic seasonings given in Table 15.1 may be used as a start for further adjustments to suit local tastes or product concept. The usage rate for the seasoning may be determined in the first instance experimentally in a simple neutral soup medium.

STOCK SEASONINGS

The formulation of seasonings demands considerable skill and attention to subtle changes in flavor profile during the development

TABLE 15.1
SEASONINGS FOR SOUPS

Ground Spice or Equivalent Spice Extractives on a Salt/Dextrose Carrier (% w/w)	Cream of Celery	Chicken & Noodles	Clam Chowder	Lobster Bisque	Mexican Bean	Cream of Mushroom	Noodles & Beef	Onion (French Style)	Cream of Tomato	Spring Vegetable
Allspice									0.5	
Basil			3							
Bay laurel	5			0.5		0.7				
Cayenne (capsicum)					5.5		0.1	0.1	0.8	
Celery	75	45	8	50			15	16.5	26	58
Cinnamon/cassia									9	
Clove						0.3		0.1	1	
Coriander		7								
Cumin					0.1					
Garlic powder		0.5	8	20	4		1.5	1.5	16	5
Mace							1			
Marjoram									3	
Mushroom (*Boletus*)						70				3
Onion powder		5.5	18	3.5	50.4	20	80.4	80	35.7	12
Oregano					25				5	
Paprika					10					
Parsley herb		2	50							
Black/White Pepper	10	10	8	15	5	9	1.5	1.2		17
Rosemary	10			1			0.5	0.6		5
Tarragon			2							
Thyme		1	3						3	
Turmeric		30		10						

and ultimate storage of any product; but once established, the spice blend can be manufactured and handled in bulk or in unit packs as may be convenient for any given processing program. In this respect, it should be borne in mind that holding large stocks of premade seasonings, whatever their actual nature, not only ties up capital but may result in stock deterioration through loss and change during the storage period. Ground spices lose strength rapidly over weeks rather than months so that a speedy and regular turnover is essential. In all cases, ground and plated spices should be stored in air-tight containers in a cool dry place away from direct heat or sunlight. Containers should always be re-lidded after use and any small residual stocks should be transferred to smaller containers.

Curry Powder

Curry powder is often listed as an ingredient in soup formulations. This is not a single spice but a blend of spices, there being almost as many formulations as there are compounders of this useful seasoning product. Curry is essentially an Indian flavor based on the grinding together of locally-produced spices as required. This implies that the blending of curry is best carried out immediately before use; but this is just not possible on the manufacturing scale. However, if possible, it is preferable to obtain supplies fresh from a local producer rather than import the material with all the attendant delays. Curry does change in character on storage and unless the best quality spices are used it may rapidly lose its initial freshness.

The flavor of curry has now spread and is almost universally popular. It must be recognized that different blends are required for different types of food to obtain the most satisfying results. Both usage level and intrinsic pungency may be varied to suit local palates. A selection of typical curry powder formulations is given in Table 15.2 which indicates the type of profiles available. Manufacturers of this product will almost certainly have a wide spectrum of blends suitable for many different basic products.

Although curry powder may be used to give a predominant flavor to a soup or gravy (e.g., Mulligatawny soup) it is often used at much lower levels in various other soups such as oxtail, beef with noodles, etc., to give an interesting profile.

Flavor Enhancers (MSG, the 5′-Ribonucleotides)

The use of flavor enhancers in soups is widespread and, unless care is taken, may result in an unpleasant lingering after-taste. The level of use must be balanced to achieve the desired effect as increased usage does not imply greater enhancement once a certain level has been reached.

Monosodium Glutamate (MSG).—The optimum use level for MSG in soups is 0.1–0.5%, although as little as 0.05% will probably be sufficient, particularly in the presence of salt. The quality of MSG has improved considerably over recent years and that most widely available is almost odorless and flavorless, although cheaper grades may have a slightly meaty note and a metallic after-taste. The popularity of MSG depends on its main effects which are:

(a) It intensifies and enhances any natural flavors present in the product.

(b) It prevents flavor fading.

TABLE 15.2
TYPICAL FORMULATIONS FOR CURRY POWDER

Spice	General Purpose I (%)	General Purpose II (%)	General Purpose III (%)	General Purpose IV (%)	Mild (%)	Sweet (%)	Hot (%)
Coriander	35	37	48	38	37	40	35
Turmeric, Madras[1]	25	19	16	19	10	10	25
Cardamom, Alleppy	4.75	2.5	3	3	4	5	
Black pepper	9.5	7	4	1.5	5	15	5
Mustard seed	9.5						3
Cayenne (red capsicum		8	6	2	1	1	5
Ginger, Cochin[2]	9.5	14	4	2	2	5	5
Allspice	4.75		2	1.5		3	
Clove			1	1.5	2	3	
Cumin	2	10		6	2		15
Fenugreek			12	9			7
Cinnamon/cassia		2.5	2	12	2	10	
Nutmeg			2	1.5			
Bay laurel				2		5	
Poppy seed				1	35		
Lemon peel, dried						3	
Garlic powder[3] or garlic oil							

[1]Madras turmeric is preferred as this has a good color and a pleasant flavor. The overall color should be orange-brown rather than bright yellow. Turmeric should be limited to a maximum of 25% of the formulation.

[2]Cochin ginger is preferred as this has a marked lemony note which adds a pleasing nuance to the curry blend. Australian ginger may be substituted as this, too, has a similar character.

[3]Garlic in the form of powder or garlic oil may be added to suit particular tastes.

Note

Salt is sometimes added but the inclusion may be restricted by legislation.

Spices should preferably be bulked whole or broken, coarsely ground through a mill and then repassed and finely ground to the desired particle size.

(c) It produces a pleasant mouth feel and a sense of satisfaction.

(d) It reduces sharp unpleasant notes such as onion, the earthiness of potato and the bitterness associated with many vegetables.

(e) It gives to the overall flavor an acceptable richness and a more rounded fullness.

Ribonucleotides

5′-nucleotides.—Disodium 5′-inosinate and 5′-guanylate (5′-IMP and 5′-GMP) are white, odorless powders with a very slight meaty flavor. Their effect as flavor potentiators is complex and foods containing very low levels have a lingering feeling of satisfaction after the food is eaten. Their usage rate is very much less than that of MSG being in the range of 0.015–0.02%. When used with MSG, they have a synergistic effect and enable the usage rate of the latter to be significantly reduced. In liquid preparations such as soups, the 5′-nucleotides create a sense of viscosity, the product having a richer, more full-bodied profile.

Added Flavorings

These may be either natural or synthetic and include: (a) meat extracts, (b) protein hydrolysates, (c) yeast extracts, (d) imitation flavorings (where permitted), and (e) vegetable extracts (e.g., onion, mushroom).

Flavors to stimulate or boost natural constituents in soups are not widely used, although replacement of meat by textured vegetable proteins may alter this. There would appear to be a considerable potential for the use of imitation meat flavors (e.g., roast beef, stewed beef, pork, ham, kidney, etc.), poultry flavors (e.g., roast chicken, fried chicken, etc.), fish flavors (e.g., lobster, crab, etc.) and tomato flavor. The dehydration of most prime components used in soup manufacture invariably leads to a significant loss of characteristic top-notes which give to the original materials their pleasing freshness. The use of suitable imitation flavorings could replace these missing notes, although care is necessary to ensure that the overall profile does not become too "chemical" as a consequence, particularly after retorting. There is no doubt that dehydrated tomato powder cannot produce a tomato soup having anything like the same profile as one made from the fresh vegetable or even from tomato purée or paste. The addition of a low level of a topping-flavoring could result in the differences being less obvious. The use of any such flavorings must, of course, be within the regulations governing these products.

Onion Extract.—Onions, either as a fresh vegetable or as a dehydrated powder, are widely used in soup manufacture. The fresh vegetable gives not only flavor but texture to the product and in many specialist soups (e.g., French onion soup) this is an essential character. There are no problems of achieving this in a canned soup, apart from the difficulties of cleaning, peeling, and cutting fresh onions in bulk; but it is necessary to compromise in the case of dry powdered versions of such soups. Kibbled, dehydrated pieces may be used but these do not rehydrate readily and remain hard in the reconstituted soup instead of giving a smooth soft mouth feel. There are many soups in which only the flavor of onion is required. This is best achieved by the use of:

(a) Onion oil—a powerful essential oil obtained by distillation of freshly crushed onions. As it is so powerful, it is preferable for this to be incorporated as a dispersion on salt or in the form of an encapsulated flavor.

(b) Onion extract (or oleoresin)—a viscous, dark-brown concentrate of onion juice which is readily miscible with water. A dry dispersion on salt or dextrose can be used in dry soup mixes.

These products are available commercially and their usage rates and relative strengths should be established with the respective manufacturer.

Mushroom Extract.—There are many varieties of edible mushrooms, the common field mushroom or champignon (*Agaricus campestris*, L.) being the most widely known. These are used in soups but rather for their texture than for the available flavor which is very weak. Commercially, the more flavorful Boletus mushroom (*Boletus edulis*, L.) is preferred, this being available in dehydrated pieces or powder. This species, which grows in Poland and parts of South Africa, gives to soups a full, rich meaty flavor with a characteristic mushroom-like back-note; quite different, however, from that of the field mushroom. Extracts of *Boletus* mushroom are available commercially. These are dark-brown, viscous pastes miscible with water. Because of their variable concentration, the usage level should be ascertained from the manufacturer.

BIBLIOGRAPHY

BINSTED, R. and DEVEY, J. D. 1970. Soup Manufacture: Canning, Dehydration and Quick-freezing, 3rd Edition. Food Trade Press, London.

LYALL, N. 1965. Some Savoury Food Products. Food Trade Press, London.

MERORY, J. 1968. Food Flavorings: Composition, Manufacture and Use, 2nd Edition. AVI Publishing Co., Westport, Conn.

TRESSLER, D. K., VAN ARSDEL, W. B. and COPLEY, M. J. 1968. The Freezing Preservation of Foods, 4th Edition, Vol. 3 and 4. AVI Publishing Co., Westport, Conn.

16

Ice Cream and Frozen Goods

Ice cream is a frozen dessert food the formulation of which and the manufacturing conditions under which it is made are controlled by legislation in most countries. It is prepared from milk (full cream or skimmed), fat (cream, butter or hydrogenated vegetable oil) and milk solids (skimmed milk powder or condensed milk) in various proportions together with sugar, an emulsifier which is usually a monoglyceride, a stabilizer such as gelatine or sodium alginate, flavoring and coloring as appropriate and, possibly, sodium citrate or phosphate to prevent syneresis. Air is incorporated into the mix during freezing. The texture of ice cream is an all-important quality attribute dependent on the ice crystal size which is controlled by whipping and stirring during the freezing stage. There are almost limitless variations in the formulation of ice cream products within the regulated limits, as well as methods of handling, manufacture and presentation to the consumer.

The following are the main types of frozen desserts loosely regarded as "ice cream."

Frozen custard—ingredients are cooked to a custard before freezing; usually high in egg-yolk solids and contains not less than 10% of milkfat. This was the original type of ice cream and is richly creamy. It is generally made as required and may or may not be whipped during freezing.

Plain ice cream—the most popular form of this product in its many varieties. Standards are generally imposed on milkfat and total milk solids content. The flavoring may be either completely dissolved in the mix (e.g., vanilla extract, imitation strawberry flavor, etc.) or present as particulate solids (e.g., cocoa powder, fruit and nuts) in a range of specialty products.

Low-fat ice cream—a product in which the fat content is usually between 2 and 5%.

Sherbet—made from fruit juices, sugar, stabilizer, and some milk solids with added flavoring and color as necessary. The flavor is generally sharp and refreshing.

Water ice (Popsicles, ice lollies)—made from fruit juices, sugar and stabilizer with added flavoring and color as necessary. Fruit acids may be added to improve the fresh fruit flavor impact. The sugar content may be regulated but is usually about 30%. No dairy products are used in the formulation.

The composition and designation of these products may be precisely defined in many countries and reference should be made to the appropriate regulations to ensure compliance, particularly with regard to the nature and quantity of the raw materials which may be used in the manufacture of any given product.

PROCESSING TECHNOLOGY OF PLAIN ICE CREAM

Detailed formulations and precise methods of manufacture differ between countries and indeed between manufacturers and the following outline process is open to considerable interpretation of detail depending on plant facilities available.

The milk, stored at about 2°C (36°F) is raised to 50°C (122°F) in a stainless steel mixer unit. The temperature is slowly raised to 71°C (160°F) as the sugar, milk powder, stabilizer, etc., are added. When the bulk has attained the correct temperature, the fat and any other condensed products are mixed in. The mixing is continued until the emulsion is homogeneous, the bulk being maintained at this temperature for at least 30 min in order to effect pasteurization. The mix is then passed through a coarse screen into the homogenizer at 3500 psi at 65°–71°C (145°–160°F). This stabilizes the fat-in-water emulsion giving globules of about 2 μ diameter; at the same time, it gives the product a smoother texture and a better whipping characteristic. The mix is then pumped into a continuous freezer, of which there are many types (Arbuckle 1972), the freezing time and temperature being determined by the plant used. Filtered air is incorporated at the same time and the whipping process takes place with scraped-surface action on surfaces refrigerated at –20° to –33°C (–4° to –28°F). The finished ice cream emerges at –5°C (22°F), passes to a pressure relief valve and is then either formed into blocks or filled into retail containers before passing to an air-blast tunnel. This latter stage may take 40 to 60 min at temperatures as low as –46°C (–52°F). The blocks or tubs are finally hardened for 2–3 days in storage at –29°C (–20°F).

Over-Run

The degree of air which is incorporated into the product is expressed in terms of "over-run." This is defined as the volume of ice cream obtained in excess of the volume of the starting mix expressed as a percentage. The amount of over-run is determined by the composition of the product mix and the manner in which it is processed. The levels generally found in different products are as follows:

	%
ice cream	70–80
ice cream in bulk	80–100
sherbet	30–40
soft ice cream	30–50
water ice	25–30

Flavorings for Ice Cream

The selection of flavorings for ice cream and related products is of utmost importance. No matter how smooth and creamy the base product may be, it is the added flavoring which ultimately characterizes and establishes the success or otherwise of the product. Chilled products, such as ice cream, are eaten cold and this has a marked effect upon the flavor impact. It is necessary to ensure that any flavoring materials are used at an adequate level and are not subject to "freeze-out." It is well recognized that the palate becomes numbed and partially anesthetized when eating anything as cold as ice cream; consequently, there is a lowering of flavor perception. It is this phenomenon which causes flavor fade or "freeze-out," making it essential to evaluate any flavorings in end-products under normal processing and consumption conditions. Ice cream is not palatable if served too hard and cold. The flavor and texture are best appreciated when the product is served at a reasonable temperature which for most people lies between –12° and –9°C (10° and 16°F), the normal temperature range of a conservator cabinet.

There is an enormous range of flavorings available for ice cream that include natural products (e.g., cocoa powder, citrus fruit pastes, etc.), natural extracts and essences (e.g., vanilla extract, lemon essence, etc.), imitation flavorings (e.g., imitation strawberry, raspberry, banana flavors) and synthetic chemicals (e.g., vanillin, ethyl vanillin, etc.) as well as fruit pieces, nuts, crystallized fruits, etc.; generalizations on their application are valueless. The ingredients used in the basic mix play an important part in determining the nature and quantity of the flavoring to be added. For example, a mix having a high butterfat content requires less flavoring than one of low butterfat. Similarly, variations in the milk solids content and the amount of over-run also affect the level of added flavorings. When checking the usage level of any flavoring, it is essential to establish the basis for its calculation in relation to over-run. Most manufacturers make recommendations based on a 60% over-run and should this figure be exceeded then the quantity of flavoring may have to be increased; and vice versa, the exact usage level being determined experimentally.

Flavor Application

Before considering a few of the more important flavorings, the following general aspects of flavor application should be noted.

Accuracy of Dosage.—Good manufacturing practice demands consistency of the flavor in the end-product and this can only be achieved if hand measurement or metering devices for flavorings are accurate. Addition may be by weight or by volume, the former generally being considered the more accurate. But it should be remembered that the gravity of a liquid flavor depends on the solvent present.

Economics.—The cost per pound (kilo) of flavoring is ultimately judged in terms of cost per gallon (liter) of the finished ice cream. Although the flavoring ingredients constitute one of the smallest individual items in the cost of manufacture, the contribution to the quality of the product is disproportionately large. In terms of the total cost, it is false economy to use cheap flavorings.

Addition to the Product Mix.—This depends on the processing technique. In batch processing, flavorings are added just prior to freezing and with both horizontal and vertical freezers this poses no problems. In continuous processing, the mix is held chilled in an aging tank and then passes directly to the freezer so that the addition is not so simple. A flavor tank with automatic metering can be interposed between the holding tank and the freezer but care must be taken to ensure that the mix is uniform. The flavoring may also be added to the bulk mix in the aging tank and subjected to a very slow continuous agitation to ensure dispersion. The technique of adding the flavor prior to pasteurization is not recommended as this exposes delicate flavor constituents to unnecessary heat.

Hygiene.—Ice cream must be produced to strict microbiological standards and, in consequence, hygiene requirements are strict. As flavorings are added after pasteurization, it is essential that they, too, be microbiologically acceptable. Generally, liquid flavorings containing permitted solvents such as ethanol or propylene glycol are commercially sterile although no special precautions are taken in this respect. When using nonpasteurized flavor products such as flavoring emulsions and natural fruits, special handling precautions are necessary. Pasteurized products (e.g., citrus pastes) are normally supplied in sealed cans and stored under chilled conditions; once opened, the contents of the cans should be used and not left in a partly filled container.

CHOICE OF FLAVORING

The choice of available flavorings is wide and in their selection it is necessary to consider not only the quality of the end-product but the regional and age-group preferences of consumers. In some areas, fine delicate flavors are not appreciated and the demand is for a strong high-impact flavor. In other markets, the reverse applies. For the bulk of ice cream, the flavor choice is limited although some attempt is made to introduce speciality lines based on less well-known flavors. The following flavorings are of particular importance owing to their undisputed popularity in this medium.

Vanilla

This is without doubt the most popular and widely used flavor for ice cream; about 75% of all such products contain one or another of the following vanilla preparations: vanilla beans, vanilla bean extract, vanilla paste, vanilla sugar, compounded vanilla flavor, imitation vanilla flavor, vanillin, ethyl vanillin.

Vanilla Beans.—As a natural source of flavor, vanilla beans have already been discussed. One would not normally consider that vanilla beans could be used per se in the flavoring of ice cream, but there is a manufacturing technique which converts the beans into a smooth purée by passage through a colloid mill. The inside of the vanilla bean is filled with numerous very small black seeds and it is essential that these be reduced to avoid any grittiness in the end-product. The chief aromatic constituent of vanilla beans is vanillin but the purée contains other aromatics, many of which are very heat sensitive. It is claimed that a vanilla bean purée produces a fuller and more highly esteemed flavor then can be achieved with either extracts or flavoring; but, as the beans are nonsterile, the purée should be added to the mix before pasteurization. Hence, the flavor components are endangered. Also, beans contain about 10% fixed oil which is unstable and could lead to the development of off-odors and off-flavors in the end-product, even when stored under chilled conditions. Although it may have its advocates, the use of vanilla beans in this manner is not widely used.

Natural Vanilla Extracts.—The following are available commercially:

Vanilla extract, single-strength—prepared by the extraction of finely chopped vanilla beans with ethanol (50–55%). In the United States, this product is defined as the soluble matter from 10 g of vanilla beans having a moisture content of 25% in 100 ml of diluted alcohol with or without the addition of sugar, glycerin and color. In practice, this is equivalent to 13.35 oz of bean per U.S. gallon (128 fl oz) or 16.69 oz per Imperial gallon (160 fl oz).

Ethanol-free extracts of an equivalent strength using propylene glycol and/or glycerin as a solvent are also readily available and are, of course, much cheaper than the alcohol-based extracts.

Vanilla extract, 5-fold, 10-fold, etc.—concentrated extracts made by distilling off part of the alcoholic solvent, usually under vacuum at the lowest possible temperature, or by countercurrent methods of extraction, or by dissolving resinoid vanilla in an appropriate strength alcohol. The concentration achieved is expressed by the "-fold" in terms of the single strength product. That made by countercurrent techniques is preferred as having the finer top-notes unaffected by any heat treatment.

The usage of these products is directly related to strength in terms of the original beans. In practice, an average of 5 fl oz of single-strength extract is sufficient to flavor 5 U.S. gal. of mix (equivalent to about 10 U.S. gal. of finished ice cream) depending on the percentage over-run.

Vanilla Paste.—This is prepared from so-called vanilla resinoid, an extract from which all of the solvent has been removed under vacuum. This concentrated extract is mixed with a suitable, edible carrier such as liquid glucose and may contain added vanillin. The flavor strength of the product is quoted in terms of the single-strength extract.

Vanilla Sugar.—Made by finely grinding vanilla beans with sugar or by blending vanillin and finely ground sugar.

Compounded Vanilla Flavors.—In order to reduce costs, part of the natural vanilla extract may be replaced by synthetic vanillin. The following relative strengths are important in calculating formulations:

(a) 1.125 oz of vanillin is equivalent in flavoring strength to 16 oz of vanilla beans.

(b) 0.94 oz vanillin is equivalent to 1 U.S. gal. of single-strength extract.

(c) 1.17 oz vanillin is equivalent to 1 Imperial gal. of single-strength extract.

Of course, these equivalents are only very approximate as there is really no comparison between the profiles of a natural extract and of vanillin. In the United States, there is a limit of 1 oz of vanillin to be used with the equivalent of 13.35 oz of vanilla beans, not less than 1/2 of the available flavor being obtained from the bean extract.

Imitation Vanilla Flavors.—These are very popular because of the considerable cost savings in their use. Most imitation vanilla flavors are composed of a relatively small number of synthetic chemicals (Merory 1968) including vanillin, ethyl vanillin, piperonal and low levels of a few esters. The solvent is usually a mixture of ethanol, glycerin and water but alternatives are available. Caramel is added as a coloring to match that of the natural product. These products are marketed under the name of the bean they replace (i.e., Bourbon, Tahiti, Mexican, etc.). The flavoring strength is quoted in terms of the single-strength extract (e.g., Imitation Bourbon Vanilla flavor, 10-fold).

Such flavorings, apart from being economical in use, enable the ice cream manufacturer to achieve distinctive notes in the end-product. The flavor manufacturer's recommended usage rate should be followed initially and adjustments made as necessary to suit local tastes.

The following approximate flavoring equivalents are of value in establishing formulations:

1 part vanilla beans is equivalent to 0.07 parts vanillin
1 part ethyl vanillin is equivalent to 3–4 parts vanillin
1 part piperonal is equivalent to 2 parts vanillin

Chocolate

The most widely used flavoring ingredient for the production of chocolate flavored ice cream is cocoa powder. A good grade should be chosen as this not only contributes a full flavor but spreads well. Cocoa powder gives an even color and does not cause undue speckiness—a common fault when using poor grades of cocoa powder. The quantity may vary, but generally 6–8 oz per gal. of mix will give an adequate level of flavor depending on the percentage over-run.

The best method of incorporating cocoa powder into the mix is as a syrup made by stirring the powder with an equal weight of sugar and gradually adding boiling water equivalent to the combined weight of the solids. The mix is then brought to the boil, strained to ensure freedom from any aggregates and cooled rapidly. It is then ready for use. If it is necessary to store the syrup, it should be refrigerated and used within one week.

Special grades of chocolate liquor are available for flavoring ice cream. These may be prepared to suit the requirements of the ice cream manufacturer. Since they do not have the same flavoring power as cocoa powder, it is necessary to seek guidance for the appropriate usage level. These products have a much more interesting profile than can be obtained from cocoa powder, the profile of which is not particularly full and giving little room for individual character.

A more convincing chocolate profile may be achieved by the judicious use of imitation chocolate flavors and boosters as well as by the addition of vanilla extract to improve the rich creaminess of the product.

Fruit Flavors

Ice cream is a very good carrier for fruit flavors and such products make a pleasant change from standard vanilla. The fruits used may be fresh, frozen or canned and are normally added in the form of a purée or paste, as concentrated juice or even as an alcoholic extract. In some products, pieces of fruit may also be added to improve the attractiveness and appeal of the product.

The flavoring power of natural fruits is very low. To make a good quality ice cream it may be necessary to add as much as 15 to 30% of good quality fruit. This implies the use of a base mix sufficiently high in fat to compensate for the presence of so much fruit so that the final mix contains at least 10% fat. In some instances, it may be considered preferable to reduce the quantity of fruit and to supplement the flavor by the use of concentrated juice or by the addition of imitation booster flavors.

Of the many fruit flavors, strawberry is the most popular and is universally acceptable over all others. This fruit has a delicate, evasive profile which changes considerably as the fruit ripens and even more so when it is cooked. Consequently, there is a wide spectrum of "strawberry flavor," each manufacturer having products representing several aspects of the fruit profile. Generally, in ice creams it is the just-ripe character which is preferred. This is best achieved by the use of a sweetened, fresh fruit pulp; but there are several very good imitations on the market.

When using natural fruits, it may be necessary to increase the quantity of sugar in the formulation and to add citric acid to bring out the full fruit flavor. It is usual to add 2–3 oz of 50% citric acid solution for each 30-lb pail of fresh strawberries used. Strawberry ice cream is susceptible to oxidation and the development of off-notes due to the presence of this acid—an aspect of quality control requiring some attention. In using any of the natural fruit products care should be taken to ensure that they are hygienically acceptable.

Butter Flavor

The profile of a good quality ice cream includes a rich creaminess imparted by the butterfat present. In products made with butter or high-fat cream there is little need to accentuate this attribute; but in products made with substitute fats the use of an imitation butter flavor is very desirable, otherwise the attractive creamy back-notes will be missing. Several such flavorings are available. They are usually very strong and as little as ⅛ fl oz per 10 gal. of mix (about 0.08%) should be quite sufficient to give the desired effect without introducing any unnatural "chemical" nuances.

Butter flavor is very compatible with vanilla flavor and a combination of the two in varying proportions opens the door to a further range of attractively-flavored products.

Flavors in Ripple Ice Cream

Ripple ice cream is an attractive marketing idea and has gained considerable popularity internationally. The basic formulation is that of plain ice cream which, while passing through the freezer outlet, receives an injection of flavored and colored syrup which may be based on natural fruits with or without added flavorings, or imitation flavors. The addition is achieved by the attachment of a small machine comprising a syrup tank, a variable-speed motor and a pump; the rate of flow of the syrup is controlled by the speed of the motor. The resulting product has a uniform flavor and an appealing "marbled" appearance.

The syrup for use in this method of application must be freshly prepared. A range of flavors is available in a convenient powder form in heat-sealed packs sufficient to make 1 Imperial gal. of syrup with the following formula:

Ripple powder flavor	170 g	6 oz
Sugar	3.175 kg	7 lb
Citric acid	35.4 g	1¼ oz
Boiling water	2.556 liter	90 fl oz

The syrup is allowed to cool before use. It is claimed that 1 gal. (4.546 liters) of flavored syrup is sufficient to "ripple" 25 to 40 gal. (112–182 liters) of ice cream depending on the over-run.

Flavors in Topping Syrups

The use of a flavored syrup to pour over a serving of ice cream or as a constituent in a prepared dessert is widespread in restaurants and cafes. Such a syrup can be made to the above formulation but as the product is likely to be in use for some extended time it is necessary to incorporate a permitted preservative. For this purpose ½ fl oz (14 ml) of a 20% solution of sodium benzoate is satisfactory.

Suitable liquid flavorings may be used in topping syrups, these being incorporated directly into the prepared base with sufficient gentle stirring to ensure uniform dispersion.

PROCESSING TECHNOLOGY OF WATER ICES

Basically, water ices consist of an acidulated, flavored and colored sugar syrup, suitably stabilized with gelatine or sodium alginate, and frozen. The mix may be frozen in bulk with a low over-run or, more usually, is formed around a holder (e.g., a wooden stick) which acts as a handle to the individually packed item. The following procedure is followed in their manufacture.

The base syrup is prepared in accordance with the formulation; the stabilizer is mixed in accordance with the manufacturer's instructions and allowed to stand for 15 min before use. The mix is then made in the following order: To the stabilizer solution add the exact quantity of base syrup and some of the remaining water; add the recommended dose of flavoring, a solution of the coloring and, lastly, the calculated quantity of citric acid solution. When the acid is added the mixture will thicken and must be stirred continuously to maintain a smooth consistency. The remainder of the water is then added to produce the required volume. The product is filled out into molds and quick-frozen by immersion of the molds in brine at about −29°C (−20°F). When the mix has commenced to freeze, the sticks may be inserted either by hand or automatically. When frozen hard, the lollies are removed from the molds after a short period to allow release, then are air-dried for a few minutes and packaged.

Flavors in Ice Lollies

The most popular flavors in what has been called "a drink on a stick" are orange, lemon, lime, pineapple, raspberry and strawberry with other fruits such as black currant, blackberry, apple and pear having only a limited appeal. In most cases, the main flavoring is based on concentrated fruit juice or fruit paste with or without the use of imitation booster flavors. A cheaper range of these products can be made with imitation flavorings. Most flavor manufacturers offer a full range of fruit flavorings suitable for water ices. Many of these products are in the form of compounds requiring only the addition of stabilizer and sugar or sweetener to make the mix ready for freezing.

Sweetness, texture, flavor and color are all important quality attributes. Sweetening agents in the form of sucrose, liquid glucose or dextrose, as well as saccharine in the nonsugar products, normally provide more than 50% of the solids in the base mix. In order to ensure a full retention of the flavor and prevent separation of ingredients, it is necessary to create a very fine crystal structure and this can be done by quick-freezing.

YOGURT

This is a chilled product which is gaining in public acceptance and popularity and it is now available in several varieties and styles with or without the addition of flavors. Yogurt is produced as a result of the fermentation of milk by *Streptococcus thermophilus* and *Lactobacillus bulgaricus* under carefully controlled conditions (Davis 1969). The flavor of natural yogurt is strong and quite characteristic. Having a marked acidity, it forms a very good medium for added fruit flavors.

The choice of flavors is somewhat limited owing to the high intrinsic flavor of the natural yogurt base. In some countries, the nature of the flavorings permitted in this group of products may be defined by regulations; but, generally, most products currently available are flavored with natural fruit juice concentrates or pastes, used with or without imitation booster flavors at a level to overcome the flavor of the natural yogurt. To achieve the correct sweetness/acidity balance, it may be necessary to add sugar to fruit-flavored yogurts. Most of the flavored varieties contain pieces of the appropriate fruit; strawberry, raspberry, blueberry, peach and pineapple are the most commonly available (Christie, 1974).

The shelf-life of yogurts is limited to about 20 days. Distribution is effected under refrigerated conditions.

Frozen Yogurt

Compared with the European countries, where yogurt has for long been popular, the consumption of yogurt in North America over the past decade has shown remarkable growth. More significant, however, has been the development and market acceptance of frozen yogurt. Chandan (1977) reports that world business in this commodity is predicted at $1.5 billion with ice cream manufacturers contributing about 1/3 of their capacity to yogurt-based confections.

It is considered that frozen yogurt could account for 80% of these confections and soft-serve yogurt the remaining 20%. Food manufacturers are becoming interested in this market success and are developing other yogurt-based products including mayonnaise, savory salad dressings and even yogurt drinks.

Currently yogurt is marketed in three distinct forms, namely: (a) the conventional product having a soft smooth runny texture, (b) "soft-serve" having a consistency comparable to that of soft-serve ice cream and (c) "hard frozen" having a texture similar to normal ice cream.

Conventional yogurt is not amenable to freezing as it develops an unacceptable granual crystalline texture. In consequence, these frozen products have been specifically developed embodying various systems for improving their stability, texture, appearance and shelf-life (Andres and Hagan 1977). The difference between soft-serve and hard-frozen yogurt lies in the formulation and the way in which the product is manufactured and ultimately distributed.

Process Technology.—Frozen yogurt may be manufactured in plants designed for the production of ice cream. The ingredients of the base mix will normally fall between the following limits (Chandan 1977):

	(%)
Milk fat	1.5–2
Nonmilk solids	13–15
Gelatine (250 Bloom)	0.15–0.2
Sucrose	7–10 (although initially, figures as high as 16 are quoted)
Corn syrup solids (36 DE)	3–5

This results in a product having a total solids content of about 24 to 32%.

The manufacture of frozen yogurt involves several stages. The process starts with the blending of the dry ingredients and admixture with milk to form a smooth cream; the mix is then pasteurized at 88°C (190°F) for 40 min or at 90.5°C (195°F) for 20 min. After pressurized homogenization, the mix is cooled to 44°C (111°F) and transferred to a suitable vat in which it is innoculated with active yogurt culture. The mix is then allowed to stand until a pH of 3.9 is achieved, the time and temperature being dictated by whether the short-set or long-set method is employed. When ready, the mix is cooled to between 22°–24°C (70°–75°F) and further textured by non-pressurized homogenization. Fruit and other flavorings may then be added and the mix frozen as appropriate to the end-product.

Flavoring of Frozen Yogurt.—Although natural or unflavored yogurt is popular, the greater demand is for variously flavored products and a similar pattern is likely to develop in the growing demand for frozen yogurt. However, the unflavored product is less attractive in the hard-frozen form but as soft-serve may be dispensed as a suitable carrier for flavored topping syrups and/or fruit conserves to satisfy local demands without proliferation of flavored base.

Processing and textural considerations may limit just what may conveniently be added as flavorings but it is likely that those currently found acceptable in traditional yogurt will continue to be employed in this new range of products with a strong preference for natural products (Anon. 1977). These, subject to legislation in many countries, may consist of fruit pieces, fruit pulp, fruit juice or juice concentrate, jams and other fruit conserves, nuts, chocolate and natural flavoring extracts. Of course, imitation flavorings are technologically acceptable and their use is only limited by marketing considerations. The nature of the added flavoring material will depend to a large extent on the desired end-product and its method of packaging, distribution and presentation. Consumers now accept a wide range of flavorings in ice cream and frozen mousse and it is likely that these will also become the principal flavors of the frozen yogurt range with a strong emphasis on fruity flavors. With the exception of vanilla, which is so widely used in ice cream, the range of fruit flavors found to be satisfactory in that medium should be even more acceptable in the characteristically acidic medium of yogurt; but problems may arise in achieving an acceptable flavor balance with the predominantly sweet profiles associated with chocolate, nuts, tutti-fruitti and peppermint.

Legislation.—At present, legislation governing the sale of frozen yogurt products and, hence, of labeling and flavor incorporation, is far from precise. The question of whether these products should be classified under regulations for ice cream, yogurt or dairy products has yet to be satisfactorily resolved. To date (1977) only two states in the United States (California and New York) have defined standards for frozen yogurt but others are likely to follow suit in the near future.

BIBLIOGRAPHY

ANDRES, C. and HAGAN, E. 1977. Stability systems are designed for new yogurt/dairy-based products. Food Process. *38*, No. 8, 48–49.

ANON. 1977. Yogurt forms and flavors boom forecast $600 million sales by 1978. Food Process. *38*, No. 4, 66–69.

ARBUCKLE, W. S. 1972. Ice Cream, 2nd Edition. AVI Publishing Co., Westport, Conn.

ARBUCKLE, W. S. 1976. Ice Cream Service Handbook. AVI Publishing Co., Westport, Conn.

CHANDAN, R. C. 1977. Considerations in the manufacture of frozen and soft serve yogurt. Food Prod. Dev. 11, No. 9, 118–119,121.

CHRISTIE, R. G. 1974. Yoghurt. *In* Encyclopedia of Food Technology. A. H. Johnson and M. S. Peterson (Editors). AVI Publishing Co., Westport, Conn.

DAVIS, J. G. 1969. The Dairy Industry. *In* Food Industries Manual, 20th Edition. A. H. Woollen (Editor). Leonard Hill Books, London.

EOPECHINO, A. A. and LEEDER, J. G. 1970. Flavor modification produced in ice-cream mix made with corn syrup. 2. CO_2 production associated with the browning reaction. J. Food Sci. *35*, 398–402.

GORDON, J. and KIPLING, N. 1973. Flavour application in the dairy industry. Flavour Ind. *4*, 485–487,490.

LEWIS, J. A. 1973. The flavouring of dairy products. Dairy Ind. *38*, 274–277.

McCORMICK, R. D. 1976. New chocolate flavor enhancers. Food Prod. Develop. *10*, No. 1, 14.

PANNELL, R. J. H. 1973. Overcoming problems in flavouring ice cream. Dairy Ind. *38*, 268–270.

PEARSON, A. M. (Undated). Ice cream manufacture. Dep. Dairy Sci., Ontario Agr. Coll., Guelph, Canada.

TRESSLER, D. K. and SULTAN, W. J. 1975. Yogurt. *In* Food Products Formulary, Vol 2. Cereals, Baked Goods, Dairy and Egg Products. AVI Publishing Co. Westport, Conn.

17

Soft Drinks and Beverages

The terms used to describe this group of products differ considerably in meaning between countries. The term "soft drink" is universally recognized and widely used; "mineral water" is now a generic term applying, not as used originally referring to naturally-occurring mineral-containing waters, but to all the aerated products of the beverage industry; and "nonalcoholic beverage" which is at least an accurate description of these products but could also include such drinks as coffee and tea. However, the products themselves are so universally well known that the designation is of small importance except as defined in any legislation. The products which will be considered in this section include ready-to-drink beverages (both carbonated and noncarbonated or still); concentrates which require dilution by the consumer; and, for convenience, powdered "crystal" beverages which must be mixed with water to make the drink.

The various types of product on the market include: (a) carbonated beverages—clear and cloudy; (b) noncarbonated beverages—squashes, cordials, etc.; (c) specialist products—ginger, grape, cream soda, and stimulating beverages such as cola and root beer; and (d) "crystal" beverage mixes.

The typical soft drink, whether it be concentrated or not, is based on the following ingredients: fruit juices, natural essences, flavorings usually in the form of emulsions, colors, preservatives, heading and/or clouding agents, acidulants, sugar and/or artificial sweetening agents and water. In many countries, the composition and description of this group of products is controlled by legislation to which reference should be made, particularly with regard to the nature and quality of any permitted additives.

Reference should be made to the standard texts for a full treatment of the raw materials used in the manufacture of nonalcoholic beverages (Woodroof and Phillips 1974).

MATERIALS AFFECTING THE FLAVOR OF THE END-PRODUCT

Water

Water naturally plays a most important part in any beverage formulation and the flavor profile of the end-product is often considerably affected by the quality and character of the water used in its manufacture. Water does have a flavor and this is generally due to the presence of dissolved gases, mineral salts, traces of organic vegetable matter and, on occasions, of algae. Untreated water differs widely in hardness, an excess of which may affect added flavoring ingredients and in a clear product can result in the formation of an unsightly deposit on standing. Although distilled water is free from any dissolved matter it is really not satisfactory for beverage manufacture. It lacks dissolved air and in consequence is "flat" and "insipid." Water softened by passage through activated ion exchange resins is preferable and widely used; but some processors prefer a moderate degree of residual hardness as it is claimed that this improves the overall taste of the finished drink.

Water suitable for the manufacture of beverages must be clear, colorless, free from any objectionable taste or odor, free from metals such as iron, free from organic matter, have a low alkalinity and be sterile. The presence of excessive chlorine in most public water supplies renders these quite unsuitable for mineral water production.

Acidulants

Acidulants are necessary in a beverage formulation for three reasons: (a) They significantly modify the flavor profile. (b) By providing the correct pH, they enable sodium benzoate to be an effective preservative and create an adverse environment for microbiological growth. (c) They invert and modify the sweetness of sucrose.

Citric acid is by far the most widely used acidulant but may be partially or entirely replaced by malic or lactic acid in some products. Phosphoric acid is used in cola drinks and tartaric acid in grape products. The choice of the correct acidulant and the balance between it and sugars is most important and depends on the nature of the flavor. In carbonated beverages, the pH of the end-product generally falls between 3.0 and 3.5, or 2.5 for cola beverages. The amount of acidulant necessary is determined by individual manufacturers who must take into account local variations in the pH of the water and other raw materials used. In many formulations, citric acid is used as 0.3 to 1.5% w/v and the following balances are generally achieved:

pH	2.4 (cola) to 4.7 (club soda)
sugar	9.0% (ginger ale) to 13.5% (orange)
CO_2	1.5 vol (orange) to 3.5 vol (club soda)

This gives an enormous range of individual products to suit local tastes (Anon. 1966; Merory 1968).

Sweeteners

Sucrose is used in nearly all soft drinks at 9–12% by weight or at 5–8% when used with an artificial sweetener such as saccharin which may contribute 30–50% of the sweetness. Glucose is used in a few products where a lower level of sweetness is desirable. Saccharin is used to sweeten dietetic drinks, although its use is now forbidden in Canada and is under legislative review in the United States and other countries.

Emulsifiers

Although not contributing flavor as such, the emulsifiers have two functions related to the acceptability of the product:

(a) as a means of solubilizing and enabling a uniform dispersion of flavoring oils;

(b) as a clouding agent to give the product an appropriate appearance.

A number of emulsifying agents are available, the most widely used being gum acacia (Arabic), tragacanth, karaya and guar. Other hydrocolloids such as algin, sodium alginate, methyl cellulose, carboxy methyl cellulose are also used as dispersion stabilizers in cloudy drinks and as body agents in clear low-sugar beverages. The level of use of these additives is limited by the amount that can be dissolved in the solution and by the type and amounts of solvents and emulsifiers present.

Flavorings

In most soft drinks the added flavoring confers the characteristic note or augments any less powerful flavors present. The basic requirements for a flavoring for soft drink manufacture are:

(a) The flavor must impart the characteristic profile which by name it represents. For fantasy flavors (e.g., cola, root beer) it must have a recognized profile having consumer acceptance.

(b) It must be compatible with the physical system in which it is to be used.

(c) It must be stable to heat, light, acids and preservatives.
(d) It must impart the correct physical appearance to the product (i.e., cloudy, orange color, etc.).
(e) The flavor compound must be free from spoilage organisms as well as being resistant to spoilage.
(f) It must comply with all the current regulations governing the products in which it is used.

The application of flavors is dictated by the required appearance of the end-product as well as any governing legislation. This latter is complicated by an almost complete lack of uniformity between countries. The use of an imitation flavor may be desirable from a technological and consumer point of view but precluded by the labeling regulations in force for the product. For a clear beverage (e.g., lemonade, ginger ale, cream soda, etc.) an entirely soluble essence or flavoring is essential; whereas for cloudy beverages, the added flavoring may be in the form of an emulsified essential oil or flavoring blend with or without the addition of fruit juice. Flavor manufacturers offer a wide spectrum of such products designed for specific end uses and reference should be made to their literature for advice on applications (Anon. 1968).

Colors

Permitted artificial colors do not directly affect the flavor profile but do have a marked influence on flavor appreciation. They are widely used in carbonated beverages in order to impart an attractive and appropriate appearance to the product. Caramel is used to color cola drinks at about 0.25% in the bottling syrup. Natural colors such as β-carotene have a limited use for improving the color of citrus beverages.

PROCESSING TECHNOLOGY FOR CARBONATED BEVERAGES

Carbonated soft drinks may be classified into two main groups: (a) essence-based products giving a clear drink; and (b) fruit- or essence-based products having a cloudy appearance.

The basic process of manufacture consists of the following stages: (a) preparation of a syrup from sugars and water, filtration and storage until required; (b) addition of acid, flavor and color as appropriate; (c) blending the mix and adding an aliquot either to a bottle or a container; (d) filling with carbonated water; and (e) capping and finishing.

Preparation of the Syrup

The formulation of the syrup will depend on the nature and required sugar content of the end-product. The ratio of sugar to water is that necessary to give the desired sugar content in the diluted beverage. Simple syrups usually contain 55–65% sucrose, the strength often being quoted in terms of specific gravity expressed as degrees Twaddell or Baumé (Table 17.1).

TABLE 17.1
SYRUP TABLE

Sugar to Produce 10 liters (kilos)	Deg Twaddell	Deg Baume′ (approx)	Deg Brix (approx)	Specific Gravity 20°/20°C
2.613	20	13¼	23.8	1.100
2.742	21	13¾	24.9	1.105
2.875	22	14½	26.0	1.110
3.005	23	15	27.0	1.115
3.140	24	15½	28.1	1.120
3.270	25	16	29.2	1.125
3.406	26	16¾	30.2	1.130
3.540	27	17¼	31.2	1.135
3.670	28	17¾	32.3	1.140
3.800	29	18¼	33.3	1.145
3.937	30	19	34.3	1.150
4.070	31	19½	35.3	1.155
4.203	32	20	36.3	1.160
4.338	33	20½	37.3	1.165
4.473	34	21	38.3	1.170
4.600	35	21½	39.3	1.175
4.740	36	22	40.2	1.180
4.873	37	22½	41.2	1.185
5.010	38	23¼	42.2	1.190
5.140	39	23¾	43.1	1.195
5.280	40	24¼	44.1	1.200
5.955	45	26¾	48.7	1.225
6.637	50	29	53.2	1.250
7.329	55	31¼	57.6	1.275
8.022	60	33½	61.8	1.300
8.720	65	35½	66.0	1.325

The preparation of the syrup is not difficult but does call for accuracy and strict attention to hygiene. In practice, the syrup may be prepared in one of five ways: cold-process, hot-process, acidified cold-process, acidified hot-process and high density syrup. The method employed is that most suited to the individual manufacturer. The considerations involved are discussed by Jacobs (1959). The cold process is the most simple and involves the direct solution of sugar in cold water. The hot method employs a steam-jacketed pan in which the solution is brought to the boil, skimmed as necessary, and then rapidly filtered and cooled. The elimination of air during this process is advantageous from the point of view of flavor stability.

FLAVORING INGREDIENTS

Fruit Juices and Concentrates

Citrus fruit accounts for about 90% of all fruit juice-based drinks. The fruits are handled in a variety of ways depending on the fruit itself and the region in which it is grown and processed. The following are the major citrus products.

(a) Juice—the raw juice obtained mechanically, screened to remove solid matter and flash pasteurized. This single-strength product can be further processed to give a *concentrated juice* or may be dehydrated to give the citrus *fruit powder*. Particular care is taken to minimize the effect of heat on the quality and profile of the resulting product. The juice or concentrate is generally shipped in casks or drums preserved with either sulfur dioxide or benzoic acid; or it may be canned, in which case it is pasteurized.

(b) Comminutes—obtained by finely shredding the whole fruit or by recombining the separated components of the fruit.

(c) Peel—not used directly as such but may be macerated or distilled with alcohol to give a soluble essence.

(d) Essential oil—obtained by rasping the outer peel and mechanically separating the oil, with or without the use of water. The dried, clarified oil may be marketed as such or may be processed to give *concentrated oils* or *terpeneless oils*. These latter products are of particular value in the production of soluble essences.

(e) Pectin—a by-product of the residual fruit tissues.

(f) Washed cells—the separated pulp is processed to give whole cellular matter which may be used as a natural cloud in the finished beverage.

The operations are usually carried out in the country of origin immediately adjacent to the growing area although the manufacture of the concentrated and terpeneless oils is done in specialist factories in Europe, North America and other parts of the world. In some regions, the processing of citrus fruit is a large and highly sophisticated agricultural operation.

The processing of soft fruits such as strawberry and raspberry calls for a somewhat different technology. The juice is extracted by pulping and pressing the fruit; the extraction is carried out in plants designed for this purpose. The resultant juice is then treated with pectinase (which increases the yield and makes the resulting product easier to handle), depulped, filtered and flash pasteurized.

FLAVORINGS

Almost all materials used in the production of flavorings for use in soft drinks are insoluble or only sparingly soluble in water. There are two ways in which these substances can be incorporated into beverages: they can be dissolved in a permitted solvent so that when added to the bottling syrup and ultimately diluted they remain in solution; or, they can be emulsified with water using a suitable emulsifying agent. These latter products may be used to produce a cloud in a drink which would otherwise be clear.

Soluble Essences and Flavorings

It is generally accepted that only wholly natural products shall be classed as "essences" and that all other forms shall be called either "flavors" or "flavorings," even where these contain a considerable proportion of natural ingredients. The formulation of these products follows that employed in the creation and development of a flavoring for any other type of product except that the flavoring as offered to the beverage manufacturer must have an acceptable solubility when used at the recommended rate in the bottling syrup. Any solvents used must, of course, comply with government regulations. Although ethanol is without doubt the best solvent for this purpose, the high rate of duty which it normally attracts may make the products too costly and, hence, of limited application. Solvents that are duty-free (e.g., isopropanol, propylene glycol, glycerine, etc.) are generally used in the making of soluble flavorings.

In view of their popularity, citrus oils form the basis of many soluble essences. Their preparation consists of removing the insoluble terpenes either by distillation or by dissolving the oxygenated flavoring components in diluted ethanol. In practice, the concentrated and terpeneless citrus oils are used as constituents which offer many advantages, but lack some of the top-notes which characterize the straight citrus oil. An alternative process, which gives a better profile, consists of mixing the natural oil with ethanol and then diluting it with water to the desired strength. After thorough mixing and settling, the coarse emulsion breaks and separates; the terpenes form a supernatant layer and the flavoring constituents remain dissolved in the lower alcoholic layer. It is this phase which can be separated and used as the basis for the soluble essence.

Every flavor manufacturer has a range of soluble essences and flavorings suitable for beverage manufacture and their recommended usage rates must be followed for success in their application. The rates may vary between 2 and 10 fl oz per 10 Imperial gal. (0.625 and 3.125 liters per 50 liters) of bottling syrup, the formulation of which is usually suggested.

Flavoring Emulsions

Juice-based drinks and many nonjuice beverages are required to have a cloudy appearance. This may be produced either by particulate matter from the juice or by the scattering of light on the interfaces of minute oil globules suspended in the drink. If the globules are too small, light will pass straight through and a thin cloud will result.

The formulation and production of flavoring emulsions is theoretically simple; the aim is to break the oils down into tiny globules with the aid of an emulsifying agent so that when added to a bottling syrup and diluted out the oil will remain in stable suspension and produce a uniform cloud. In practice, there are several problems to be overcome. If an emulsion of essential oil is added to water, is well shaken and allowed to stand, the product will eventually display either a sludge, if the original disperse phase is heavier than water, or "ringing" if it is lighter. The same considerations apply in the dilution of a bottling syrup containing a flavoring emulsion except that the gravity of the finished drink is slightly higher due to the dissolved sugars. Ideally, the emulsified oils should have the same specific gravity as that of the finished bottled beverage. In the past, this was achieved by the use of "weighting agents" such as brominated vegetable oil (BVO), sucrose acetate iosbutyrate (SAIB), glyceryl benzoate or rosin esters, the gravity of the essential oil being raised from about 0.9 to between 1.03 and 1.15 depending on the ultimate application. Recent legislation in many countries now forbids the use of brominated vegetable oils and limits the use of other weighting agents. An alternative technology has been developed, based on the use of surfacants and lipophilic emulsifiers, to keep the globules small and hence produce a dense stable cloud. Again, it is recommended that the manufacturer's guidance be sought if problems of separation and unsightly product spoilage are to be avoided.

Good, stable, intense clouds in many products are associated with juice content and, in general, the minimal juice content required by regulations is barely sufficient to impart either an adequate flavor or an attractive stable cloud. To overcome these deficiencies, it is usual to employ fruit-based compounds containing an adequate level of natural juice to satisfy the standards together with emulsified natural citrus oil using various stabilizing agents to give an adequate flavor impact.

STORAGE OF FLAVORINGS

There are one or two points in connection with the storage of flavorings which merit attention. Soluble flavors are best kept at an even temperature of about 15°C (60°F). At lower temperatures the essence is likely to become cloudy due to a throw-out of the oil. Although this can generally be rectified by warming and shaking, it is not always easy to redissolve the last traces of oil and the product may then give trouble when added to the bottling syrup. In order to preserve the original freshness of flavor, essences should be stored away from direct sunlight.

Emulsions can be satisfactorily stored under similar conditions but the storage time is much less than that for soluble essences. The turnover program should be based on a six-month maximum storage. On any account, emulsions should not be allowed to freeze. Although they can be thawed out and then appear quite normal, the changes in their physical condition may easily upset the balance of the emulsion and its stability in the end-product. It is important that partly-filled bottles be kept securely corked or capped to reduce both volatile loss and oxidative changes in the flavor profile. It is a good practice to transfer small residual quantities to another appropriate-sized bottle.

PREPARATION OF THE FLAVORED BOTTLING SYRUP

The final bottling syrup will normally contain some or all of the following ingredients: (a) the desired level of sugars and/or sweetening agents; (b) fruit juice as necessary; (c) acidulants; (d) flavoring; (e) coloring; and (f) permitted preservative.

It is made up as required and is then ready for metering into the end container. The method of syruping may be one of two types:

(a) gravity filling in which the compound and syrup are measured into the container (an operation known as "throw") and the carbonated water added; or

(b) positive or premix where the carbonated water is mixed with the syrup prior to filling.

The general method of compounding a bottling syrup is to add the flavor slowly with constant stirring of the syrup in a suitable tank, then to add the preservative followed by any acid, coloring or heading required. Each item must be individually added to the syrup and mixed until homogeneous before making any further addition. When sodium benzoate is used as a preservative it is essential that it be added and well mixed before the addition of any acid. If this procedure is reversed it will be found that the sodium benzoate will be thrown out of solution as fine crystals which can only be reincorporated with considerable difficulty.

PRESERVATIVES

The preservatives permitted for use in soft drinks include:

(a) Sulfur dioxide—frequently employed in the form of a 6% solution in water. The maximum quantity allowed is defined by legislation and is often 100 ppm which with a syrup throw of 1 in 6 is equivalent to approximately 1½ fl oz of the solution per gallon of syrup.

(b) Sodium benzoate—benzoic acid is the most commonly used preservative but, as it is only soluble to a very low level in water, it is generally employed in the form of the sodium salt which is readily soluble. Since it is the acid itself which is the effective preservative, it is essential that when sodium benzoate is used there should be an adequate level of acidity in the syrup. The limits for benzoic acid vary but at 0.1%, which is often the permitted maximum, benzoic acid has a marked effect upon the palate. This dosage is reached by the use of 1 oz of sodium benzoate per Imperial gallon of syrup. For convenience, many manufacturers use a 10% solution of sodium benzoate.

CARBONATION

Correct carbonation is a very important factor upon which the success of a drink very largely depends. The degree of carbonation required varies according to the particular drink and to some extent to local consumer preference. Orange juice drinks, cream soda, etc., are carbonated at low pressures; cola, at higher pressure; while ginger ale and lemonades are carbonated at high pressure. The degree of carbonation is usually expressed in "volumes." At normal atmospheric pressure at 15°C (60°F), a given amount of water will dissolve an equal volume of carbon dioxide. This is called one volume of carbonation. As the pressure is increased, a greater volume of CO_2 may be dissolved; similarly, the lower the temperature the more readily the gas is dissolved. From Tables available for this purpose, the volume of dissolved carbon dioxide can be determined from a knowledge of these two variables. Most soft drinks are carbonated at between 2 and 4 volumes.

Specialty Products

Specialty carbonated beverages account for the large proportion of the soft drink market throughout the world. The following are of particular importance; also, a very informative series of articles by Beatie (1970–1971) should be consulted for specific information on their history and characteristics.

Cola or Kola.—The majority of cola drinks are franchise bottling of proprietory flavoring syrups of secret composition (e.g., Coca-Cola, Pepsi-Cola). Most flavor houses offer a version of the cola flavor containing cola nut extract in one base and a well-balanced blend of flavoring oils in another. These bases are designed to be used in equal parts together with the correct level of either phosphoric acid (1.750) or tartaric acid to make the finished bottling syrup.

The flavoring element has many variants but is usually based on essential oil of lemon, lime, orange, cinnamon/cassia, nutmeg, neroli and coriander with a fairly high level of vanillin. These blend well together and no one attribute should be outstanding. They have a tendency to mature and acquire a smoother profile on standing so that a regular order pattern is necessary for consistency of flavor in the end-product. Beatie (1970–1971) recommends several formulations suitable for this type of product.

Colas usually have the following characteristics:

sugar content	9–12% w/w
pH	2.8–3.0 (although much lower pH figures have been reported)
carbonation	3–3.5 volumes

Root Beer.—This product is no longer a fermentation of bruised sassafras bark, wintergreen bark and sarsaparilla root but is now prepared from a flavoring emulsion. These are based on methyl salicylate with a blend of spice essential oils including clove, anise and coriander.

The proportion of the various components may be varied depending on whether a sarsaparilla type or birch-beer type of product is required.

The carbonation is usually at 3 volumes.

Ginger Ale.—This is a very popular carbonated beverage for drinking as such or as a "mixer." The main flavor attribute is that of ginger but rounded and modified by citrus and/or spicy notes (Beatie 1970–1971).

There are two main types of ginger ale—"pale dry" and "golden," or "sweet," but within these groups there are very many variants.

The production of a stable ginger ale essence poses many problems and techniques have now been evolved which result in a totally soluble concentrate based on the extraction of ginger oleoresin and blended essential oils. Caramel is usually added to give the desired level of color in the final product.

Ginger ale has the following characteristics:

sugar content	8% (dry), 10% (golden)
pH	2.7–3.1
carbonation	3.5–4 volumes.

Ginger Beer.—Originally, this was a product brewed from crushed ginger rhizome, lemon peel and spices; when bottled, the carbon dioxide produced by secondary fermentation was retained in the bottles under considerable pressure—a constant source of danger to the consumer. The method was eventually modified. Currently, brewed concentrates, ready for dilution and carbonation, are available from specialist flavor houses. These products usually have a slight deposit which must be well mixed before use, but on dilution give to the drink the correct level of cloudiness, a good head and the traditional flavor which can only be achieved by the initial brewing process.

Ginger beer is usually bottled with a low level of carbonation of about 2.5 volumes.

Cream Soda.—This is a specialty flavor, existing in innumerable variations, which is particularly popular in the United States. The main flavor profile is that of vanilla with low levels of citrus and rosy notes.

The composition of cream soda is open to wide interpretation but the average characteristics fall within the following limits:

sugar content	9–12%	(although lower values are preferred in the U.S.A.)
pH	about 4.9	
carbonation	2–3.5 volumes	

NONCARBONATED PRODUCTS

Beverages based on fruit juices are widely used throughout the world, mostly in the form of concentrates which require dilution (1:4) with water before drinking. In many countries, these products are governed by regulations which define the description and composition, in particular the fruit juice and sugar contents.

As applied to orange products, Beatie (1970–1971) classified them into 11 categories of which the following are applicable to most other fruit types:

(a) Pure fruit juice syrups—may contain added sugars but no added water.
(b) Concentrated cloudy or pulpy squashes—may contain natural and/or artificial cloud.
(c) Concentrated clear cordials—may contain sugars with or without other juices and/or flavoring.
(d) Comminuted squashes—made from whole fruit.
(e) Drinks designed for blending with milk—flavor may be natural, blended or imitation and is usually buffered to prevent curdling of the milk.
(f) Concentrates used as "mixers."
(g) Balanced concentrates of two or more flavors.
(h) Heavy concentrates for domestic dilution to make the equivalent single-strength juice.

The addition of flavorings, whether natural and/or imitation poses no problems and follows the general rules.

CRYSTAL BEVERAGES

These are dry products designed for reconstitution with water as required by the consumer. When diluted, the product has a strength similar to that of the single-strength fruit juice. These products have gained considerable popularity over the past decade, particularly in North America. The major fruits represented include orange, lemon, lime, grapefruit and grape although other flavors will almost certainly be added to the range in the near future.

The following is a typical formulation for an orange crystal beverage:

	(%)
Sugar	62.75
Dextrose	23.60
Citric acid, anhydrous	7.50
Sodium citrate	1.00
Ascorbic acid	0.20

Dry powder flavor and color as recommended by the manufacturer

Where a cloudy product is required, a spray-dried vegetable oil may be added to the above formulation at 4%

The powders are blended until uniform and packed into 100-g sachets for reconstitution by mixing with 24 fl oz of water.

BIBLIOGRAPHY

ANON. 1966. Beverages Tech. Bull. *TS-26*, Allied Chemical Corp., New York.
ANON. 1968. Products for Soft Drinks Industry. Bush Boake Allen, London.
ANON. 1970. Correct type of flavour to use and its correct usage. The soft drink industry. Flavour Ind. *1*, 309–312.
ANON. 1974. New sweetener for low calorie drinks/foods (aspartame). Food Process. *35*, No. 10, 68–69.
ARNOLD, M. H. M. 1975. Acidulants for Food and Beverages. Food Trade Press, London.
BEATIE, G. B. 1970–1971. Soft drinks flavours: Their history and characteristics. Flavour Ind. *1*: Part I. Cola or "Kola" flavours, 390–394; Part II. Lemonade, 395–399; Part III. Gingerale, 454–458; Part IV. Orange drinks, 530–540; Part V. Concentrated orange drinks, 599–604; Part VI. Ginger beer, 702–706; Part VII. Lime, 772–776; Part VIII. Grapefruit, 836–840. Flavour Ind. *2*: Part IX. Cream soda, 28–32; Part X. Non-alcoholic grape beverages, 93–97.
BIRCH, G. G. 1974. Sweetness and sweeteners. Soft Drinks Trade J. *28*, 442–444.
CLARKE, K. J. 1970. Modern trends in flavouring and flavour creation for the soft drink industry. Flavour Ind. *1*, 388-389.
DOWNER, A. W. E. 1973. The application of flavours in the soft drinks industry. Flavour Ind. *4*, 488–490.
ESMOND, A. W. G. 1973. "Pop drinks." Flavour Ind. *4*, 120–121.
GUADAGNI, D. G. *et al.* 1970. Storage stability of frozen orange juice concentrate made with aroma solution or cutback juice. Food Technol. *24*, No. 9, 72–76.
HALL, J. R. 1971. Guidelines on the formation of a new or improved carbonated beverage. Am. Soft Drinks J. *126*, No. 3, 90–99.
HALL, J. R. and SWAINE, R. L. 1972. Trends in the carbonated beverage industry. C.R.C. Crit. Rev. Food Technol. *2*, No. 4, 517–536.
JACOBS, M. B. 1959. Manufacture and Analysis of Carbonated Beverages. Chemical Publishing Co., New York.
KNIGHTS, J. 1973. What is a kola drink? Flavour Ind. *4*, 118,120.
LIME, B. J. and CRUSE, R. R. 1972. Beverages from whole citrus fruit purée. J. Food Sci. *37*, 250–252.
MERORY, J. 1968. Food Flavorings: Composition, Manufacture and Use, 2nd Edition. AVI Publishing Co., Westport, Conn.
MERVIN, E. J. 1971. Vanilla extracts and flavors in carbonated beverages. Am. Soft Drinks J. *126*, No. 4, 26–30.
McCORMICK, R. D. 1973. Younger consumer is target for beverage marketeers. Food Prod. Dev. *7*, No. 3, 17,19,23.
MORGAN, R. H. 1938. Beverage Manufacture (Non-Alcoholic). Atwood & Co., London.
PRACTICUS 1971. Soft drink flavours and the E.E.C. Soft Drinks Trade J. *25*, 522–524.
PRACTICUS 1973. Looking at flavour and aroma. Soft Drinks Trade J. *27*, 356–358.
ROGERS, J. A. and EISERLE, R. J. 1971. Production, quality control, standards, test procedures and usage of essential oils in carbonated beverages. Am. Soft Drinks J. *126*, No. 3, 86–92,94.
ROTHSCHILD, G. and KARSENTY, A. 1974. Cloud loss during storage of pasteurized citrus juices. J. Food Sci. *39*, 1037–1041.
WOODROOF, J. G. and PHILLIPS, G. F. 1974. Beverages; Carbonated and Noncarbonated. AVI Publishing Co., Westport, Conn.
YOUNG, R. E. 1970. Correlation between gas chromatographic patterns and flavor evaluation of chemical mixtures of cola beverages. J. Food Sci. *35*, 219–223.

18

Quality Assurance of Highly-Flavored Products

ENVIRONMENT FOR SENSORY ASSESSMENT

One of the most important activities in any flavor laboratory is the sensory assessment of samples, be they of aromatic chemicals, spices, natural extracts, flavorings or products containing these materials. The flavorist is called upon to pass judgment on such materials based on his reaction to their smell and flavor. The methods used are frequently taken for granted as being part of normal experience but there are certain well-based precepts that should be followed if the evaluation is to be effective. Perhaps the most important and yet the least observed of these is the need for complete concentration and an absence of distractions. The environment in which sensory judgment is reached is most important. The essential requirements and optimum layouts of facilities suitable for the sensory assessment of a wide variety of raw materials and end-products is well described in the literature. A selection of references to this aspect of sensory assessment is given in the bibliography.

Ultimately, the facilities provided depend on the budgetary importance that any company or organization places on this aspect of its technical program. Even if the facilities fall short of the optimum, it is usually possible to ensure that the test conditions are as natural as possible, that the assessor is at ease but not overcomfortable, and is quiet and free from external distractions. Other provisions, such as air-conditioning and controllable lighting, though desirable, are not strictly essential. All too frequently, even these minimum conditions are not available and much evaluation is carried out at the laboratory bench with other normal routines progressing in the immediate vicinity. This sort of environment is far from satisfactory for the making of a considered judgment on sensory attributes, particularly where highly flavorful products are involved. Regrettably, this is the one most frequently encountered.

TEST METHODOLOGY

The technique adopted for any particular evaluation will depend to a large extent on the nature of the sample and the complexity of the information sought from the test. Obviously, one cannot evalute in the same way the sensory attributes of such widely divergent materials as whole spices, oleoresins, essential oils, concentrated flavors, aromatic chemicals, and end-products such as soft drinks, ice cream, sugar confectionery and meat pies. Each calls for a very specific methodology so as to reduce to a minimum the variables which may influence the judgment. A detailed treatment of such a complex subject falls outside the scope of this work and reference should be made to the standard texts (Amerine *et al.* 1965; Stahl and Einstein 1973: Goodall and Colquhoun 1967). We are here concerned more with establishing the basic principles upon which the methodology is founded, particularly as these apply to the assessment of highly flavorful raw materials used in food processing.

When drawing a representative sample of any aromatic preparation or material (whatever its actual nature), there are several fairly clearly defined aims in view. One may wish the evaluation to:

(a) Distinguish between one or more samples in some defined way.
(b) Establish and possibly characterize both qualitatively and quantitatively any difference that exists.
(c) Ascertain what changes have taken place (e.g., after processing, during storage, etc.).
(d) Establish a standard of acceptance.
(e) Establish compliance with an existing standard or reference material.
(f) Ascertain whether the relative quality of the samples can be expressed as a numerical index.
(g) Grade the samples according to a prespecified classification system.
(h) Establish a relationship between instrumental and sensory data relative to the samples.
(i) Establish a degree of acceptability between the samples (i.e., establish an hedonic value).

In all these cases the assessor is, in fact, being used as an "instrument," but whereas most mechanical instruments have a specific data output for any given input, the human "instrument" is capable

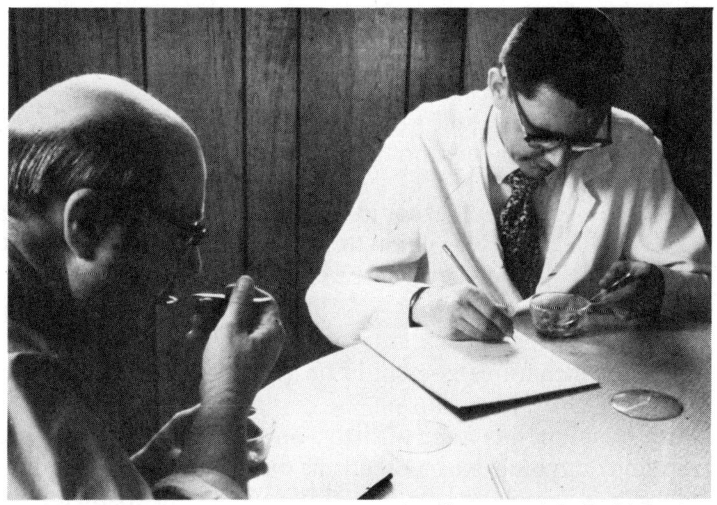

Courtesy of A. D. Little, Inc.

FIG. 18.1. FLAVOR PROFILE PANELISTS EXAMINING FLAVOR-BY-MOUTH ATTRIBUTES OF A PRODUCT

of providing a variety of information depending upon the questions asked. In all cases, personal judgment is involved and, hence, personal bias and hedonic considerations may condition the response. The design of the evaluation methodology and the use of panels to achieve statistically significant answers is well documented. Basically the methodology falls into two categories: (a) analytical—applicable to differences, ranking and grading of quality; and (b) hedonic—applicable to preference and acceptability tests.

From this it can be stated that there are three prime aims to sensory evaluation, each having its own problems when it comes to precise methodology:

(a) Difference testing—usually carried out by a single expert of considerable experience or a small panel of trained but not necessarily very experienced assessors.

(b) Rating into agreed categories—usually carried out by a panel of experienced assessors, the size of the panel often being determined by availability but preferably not less than ten assessors.

(c) Acceptability appraisal—often carried out by panels ranging from small local groups within a company, to large consumer panels. It is generally agreed that trained assessors should not be included in such panels as their views are often biased and not truly representative.

Preparation of Sample

The degree of preparation necessary will be determined by the nature of the material under examination:

Unmilled spices and rubbed herbs—should be reduced to a fine powder having a particle size of not greater than 250 μ.

Spice oleoresins, Essential Oils — should be dispersed on a suitable carrier prior to tasting but may be evaluated directly for odor.

Concentrated flavorings—should be diluted to an acceptable level prior to evaluation.

Aromatic chemicals—should be dissolved in ethanol and tasted at a suitable dilution.

In spite of what has been written on the subject, flavorings are still frequently evaluated by smelling directly from the sample bottle; a technique which at best can only give a very distorted impression of the flavoring capabilities of the product.

In the case of the concentrated preparations, it is necessary first to dilute these to an acceptable level to permit direct evaluation by tasting. This may be achieved in one of three ways, namely: admixture with a simple chemical substance (e.g., dextrose, sucrose, lactose or salt); dilution with an acceptable solvent (e.g., alcohol); or incorporation into a neutral food product or carrier at the correct dosage rate.

Carriers and Diluents

The carriers recommended for this purpose include:

(a) A *neutral soup* comprising 60% corn starch, 22% caster sugar and 18% salt, used at 4.5% with boiling water and thickened by cooking for one minute. The sample for evaluation is added to the prepared soup. This base is particularly suitable for the evaluation of spices, culinary herbs and blended seasonings.

(b) A *sugar syrup*—10% sucrose in potable water. This medium is the standard used for the evaluation of most flavorings, essential oils and the sweet spices such as ginger and cinnamon.

(c) *Reconstituted dehydrated potato*—this is a particularly useful base for the evaluation of the alliaceous vegetables (e.g., onion and garlic), paprika and the blended seasonings.

(d) *Fondant* using an instant fondant base reconstituted as required, the flavored base being formed in cornstarch molds.

(e) *Sugar boilings* using the following base:

	(g)
Sucrose	120
Water	40
Glucose syrup	40

Boil the glucose and water and mix until dissolved; add the sugar and boil rapidly at 152°C (300°F); remove from heat and stir in the flavor, any color and citric acid, if desirable; pour the mix onto a greased slab and allow to cool. Pass the mass through drop rollers to form the product.

(f) *Pectin jellies* using the following base:

Sucrose	100 g
Water	112.5 ml
Glucose (43 DE)	100 g
Citrus pectin (slow set)	6.25 g
Citric acid solution 50%	2 ml

Mix the pectin and sugar thoroughly; heat the water to 77°C (170°F) and add the pectin/sugar mix very slowly stirring all the time; bring the mix to the boil. Add the glucose with continuous stirring and raise the temperature to 106°C (220°F). Remove from heat; stir in the flavoring, any color and the acid solution; and cast quickly into cornstarch molds; allow to stand for 12 hr before removing from the molds.

(g) *Carbonated soft drink base*—there are many formulations but the following has been found to be of general application. Citric acid is used as an acidulant for most flavors but an equivalent of phosphoric acid is used for cola beverages:

Sodium benzoate solution, 10%	0.25 ml
Citric acid solution, 50%	1.0 ml
8 lb sugar syrup to make to	160.0 ml

Prepare the base and stir in the flavor and color at the appropriate dosage level recommended. Dilute the concentrate 1 plus 5 with potable water, bottle and carbonate.

(h) *Milk* forms a suitable carrier for the preliminary assessment of flavors for ice cream and other dairy products, although the reduced acid level possible poses a restriction. The concentrated flavor is diluted directly to the correct flavor level using whole milk sweetened with 8% sucrose.

In many cases the appropriate base to use will be that of the finished product into which the flavor is to be used. This should be made up according to instructions following exactly the recommended cooking or serving conditions for both the sample and any reference sample. Comparisons can then be made, using any flavor combination depending on the nature of the test.

The American Society for Testing and Materials (ASTM 1968B) in its Special Technical Publication *434* lays down the following conditions for the preparation of samples for flavor evaluation:

(a) The method of preparation shall impart no foreign tastes or odors to the sample.
(b) All samples shall be prepared in a similar manner.
(c) For difference testing the method used shall be that most likely to permit the detection of a difference keeping the preparation as simple as possible.
(d) Avoid preparations which may add flavor to the sample (e.g., frying).
(e) For preference testing, use a method suitable to the normal use of the product.
(f) Use any supplementary food carriers as may be judged necessary (e.g., ice cream for a topping syrup evaluation).

In many cases, the method of sample preparation used will be determined by common sense. The aim is to present the assessor with the sample in the best form and under the best conditions to enable a reliable judgment to be made.

Odor Evaluation

The conditions for odor evaluation are most important and particular care is necessary to ensure that no extraneous odors interfere with the judgment of the assessor. One would assume that this condition is obvious; but experience proves that many judgments are made with scant regard for this simple precaution. As with all other sensory techniques, odor evaluation depends upon complete concentration on the part of the assessor and even minor distractions (such as someone waiting for the answer) considerably reduce the efficacy of the judgment. This is particularly so when one is attempting to evaluate highly odoriferous materials or determine successive stimuli or attributes in an aroma profile.

The primary requirement for all smelling tests is the presentation of the vapor from the sample to the nasal receptors in a consistent and preferably standardized manner. Environmental conditions are important, particularly temperature and relative humidity, and, if possible, these should be maintained steadily throughout the testing period. The odor samples may be evaluated by: (a) natural breathing—using both nostrils; (b) forced sniffing—possibly using only one nostril; (c) "blasting" from an automatic dispenser.

The two former methods are simple to apply and, in consequence, are those most favored. For all general routing work, it is found that these natural techniques are quite adequate and give satisfactory results. The use of the "blast" technique (in the artificial conditions associated with an olfactometer designed to control all aspects of the presentation) is unduly complex and more appropriate to a research project than for routine odor assessment.

For most routine evaluations, two techniques are widely used:

(a) For *liquid materials* such as the essential oils, fragrance compounds, concentrated flavorings, etc., a standard smelling strip (usually 140–150 mm long by 5–6 mm wide) is used cut from suitable absorbent paper free from any extraneous odor. These should be dipped into the sample to a fixed level (1–2 cm) and the vapor inhaled through the nostrils from a distance of about 1 in. from the nose for a period of 2–3 sec at a time. After this interval, the smelling strip should be completely removed from the nose. The sequence of subsequent smelling should be rhythmic. It is bad practice to compare two odors, the one with the left hand nostril and the other with the right hand nostril. This may sound obvious, but it is quite usual to see this happening in practice. Similarly, it is undesirable to smell dipped strips which are mounted en masse on a spring clip, a common enough practice in almost all laboratories.

(b) For *odorants in sprays*, use standard smelling cards which are cut from heavier absorbent paper; otherwise used as described above. The spray is applied from a standard distance for a standard time.

Odor evaluation is most accurate when one compares one sample directly with a standard of similar identity and strength. It is least reliable when no comparison is possible and the assessor is called upon to compare his impression with a mental concept of what the odor should be. This can be achieved by the experienced expert whose odor and flavor memory have been trained; but, the results can be misleading and this type of assessment is to be discouraged.

The odor pattern over a period of time is generally very informative as the "air-off" characteristics and the residual "dry-out" notes are frequently very good indicators of differences between samples. For most samples, the odor assessment should be carried out and a judgment made at the following time intervals: immediately after dipping; after 1 hr, 2 hr and 6 hr; after standing overnight or for a period not less than 18 hr. During the intervals, the strips should be mounted in suitable clips at room temperature and away from any high concentration of extraneous odors.

An alternative method of presentation, which may be used for all types of aromatic materials, is the use of a closed glass vessel into which the sample is placed and allowed to reach equilibrium vapor pressure in the head space prior to smelling. In the case of dry materials, a representative sample is placed in the testing vessel so that the level is about ¼ of the total height. In the case of oleoresins, essential oils and compounded flavors, a 5-g sample is usually adequate in a 100-ml stoppered container. In each case, the closed vessel is allowed to stand at a steady temperature for at least 1 hr before evaluation is carried out.

Sniffing of the vapor from such a vessel must be controlled so that it is only the vapor phase which is inhaled. For this reason, care must be taken to ensure that dust from the sample does not swirl upwards which results in a particular cloud reaching the assessor's nostrils; nor must the nose be allowed to touch the rim of the container. Care must be exercised to ensure that equilibrium is achieved inside the closed vessel between repeated smelling tests and for this purpose an interval of 5 min is usually quite satisfactory. Obviously, gentle heating of the sample will release volatile components faster than can be achieved at room temperature; but, the aroma profile may be considerably altered in consequence. For this reason, it is preferable to adopt a standard temperature of say 20° or 25°C, depending on the normal ambient temperature.

The odor of certain products cannot be assessed directly as the aromatic components are encapsulated either naturally within cellular matter (e.g., herbs and spices) or within a gum or starch shell (e.g., spray-dried flavorings). In such cases, the evaluation can be carried out by taking 5 g of a sample and adding 25 ml of boiling water, swirling the contents in the loosely-topped vessel so as to mix well; then being allowed to stand to attain equilibrium. After 15–20 min, the temperature will have fallen to about 50°C and the odor can be then assessed. If time allows, it is preferable to let the contents cool to room temperature before making the judgment as it is then easier to make reassessments. Comparative evaluations must, of course, be carried out under identical test conditions.

Odor evaluation should always precede flavor evaluation, either as a separate exercise or as the first stage of a combined assessment.

Flavor Evaluation

The conditions for carrying out tasting tests are, if anything, more critical than those applicable to odor evaluation. Most commentators agree that a separate room should be provided for tasting as concentration and an absence of distraction are all-important to the achievement of meaningful results. It has been found that the sensitivity of tasters is appreciably lowered when other tasters are present, as in a panel session, and for this reason the provision of individual booths is highly desirable—even if these are made from interlocking hardboard screens mounted temporarily on a table or bench as required (Jellinek 1964).

One condition which is frequently overlooked is the necessity for assessors themselves not to be a source of foreign odors from their use of soaps, perfumes, toiletries, etc. Prior to tasting sessions, all assessors should be instructed to wash their hands with a nonperfumed soap or detergent and refrain from the use of odorous cosmetics, etc. Although opinions differ as to the effect of habitual smoking on the sensitivity of assessors, smokers can be a source of odors objectionable to the nonsmoker and it is strongly recommended that assessors should refrain from smoking for at least 30 min before taking part in a taste-test session so as to minimize the contribution of extraneous odors to the environment of the test.

When it comes to assessing end-products, the lighting conditions can make a considerable difference to the reliability of the judgments, although the use of abnormal colored lights may have an adverse and even an inhibiting effect upon assessors. The use of colored lights to eliminate differences in color or shade between samples or components of products is often advocated, but the efficacy or need for this refinement has not really been substantiated in the literature. It is appreciated that assessors tend to be influenced by irrelevant characteristics in samples and because of this every effort should be made to make the samples physically identical leaving only the flavor to be judged. This is, however, not always possible or desirable. End-products are best presented in their normal state, even if extraneous characters are present—these, after all, can be identified to the assessor before the test and their relevance, or otherwise, to the judgment made quite clear.

The following factors are of particular importance to the conduct of any flavor evaluation:

(a) Utensils: All utensils used in preparing the samples and throughout the test must be scrupulously clean and free from contaminating odors or tastes. Preferably, identical containers should be used for sample presentation. If disposable cups, spoons, plates, etc., are used, it must be confirmed that they do not contribute any odor or flavor and that they are rejected at the completion of the test.

(b) Water: The water used in making up samples must be potable and as odor- and taste-free as possible. If only local municipal water supply is available, it must be used in the preparation of comparative samples, otherwise an additional taste factor may be introduced. Distilled water should generally be avoided as it may introduce an unnaturally flat character to samples.

(c) Test conditions: Samples for direct comparison should be compared at the same time under similar test conditions. The assessors should receive sufficient of any sample, at the right temperature, to allow for several taste trials until a judgment is reached. Whether or not the sample is actually consumed is a matter for the test controller to define. Whatever the test methodology, the techniques employed must be as consistent as possible.

(d) Palate clearing: Not everyone is in agreement with the desirability of palate clearing between samples or taste trials; but, whereas this may not be necessary when evaluating end-products, it is very desirable when evaluating powerful flavors such as the spices, concentrated flavors and essential oils. Each assessor may develop his own technique and, again, consistency of procedure is all-important. The following materials have been recommended for this purpose.

(1) For most fruit flavors, herbs and mild seasonings:
- water (potable or carbonated)
- salt-free crackers
- puffed rice
- fresh bread (not crust)
- skimmed milk
- buttermilk

(2) For flavorings having a bitter, strong or oily flavor, or after-taste:
- diluted lime juice
- apple slices
- apple juice (slightly sweetened)

(3) For pungent spices:
- natural yogurt
- diluted syrup (10% sucrose)
- mashed potato

(e) Sample coding: The literature abounds with data on the effects of the order of sample presentations and a study of this subject is essential to anyone concerned with organizing taste-test procedures. All tests should be designed to ensure that the presentation is random and controlled to eliminate bias as far as possible. All samples should be identified with a three-digit random number code and simple letters such as A, B, or C should be avoided as these may introduce an unwarranted judgment factor.

(f) Test results: There is one rule which should always be followed—never discuss any findings or judgments until after the completion of the test by all assessors. Preferably, results should be written on predistributed proforms leaving the assessor the minimum of writing to convey his findings.

TASTE PANELS

Without any doubt, there are many occasions when the opinion of an expert assessor is sufficient judgment, but there are many more cases where the combined opinion of numerous judges is preferable and more reliable. Who then shall be the judge? This is a perennial and occasionally an emotive question, particularly when a person's status in a company may be affected, for panel members must be selected on proven ability to do the job, not on any hierarchical position. There is no simple answer, and many factors must be taken into account when deciding both the quality and the number of assessors that shall be called to form any panel. Of course, in the case of the larger consumer panels, the selection is random and not subject to the same considerations.

In the flavor industry, in particular, and also in many aspects of the food industry, the roll of the expert is traditional and of great importance. In the creative stages of flavor and food product development, the expert's judgment is paramount and there is a real need for what might be called the opinion of the connoisseur. But, in making this claim, one is referring to the real expert and not to the many *soi-disant* experts one finds around. The flavorist has a part to play as an expert but there are limitations to the acceptability of his unsupported judgments. It must be emphasized that whatever the opinion of the expert flavorist or the expert food technologist (and the same applies equally to perfumers in the field of cosmetics), one should not lose sight of the fact that the finished products are eventually judged by randomly selected, often irrational, non-experts—the consumers who alone make or break the success of any product.

Special Precautions

The quality control and assessment of certain spices and products made from them require special consideration owing to unusual organoleptic qualities. The following pose particular problems:

Pungency.—There are several clearly-defined pungent effects which may characterize certain spices. Pungency is detected as an irritation in the oral cavity (i.e., by the tongue, the inner cheeks and the throat). The location of the stimulus is usually an indication of the spice present. At normal dosage levels, pungency is detected as follows:

Pepper, along the front edge of the tongue.
Ginger, along the roots of the tongue.
Capsicum, deep in the throat.
Mustard, throughout the whole mouth.
Horseradish, similar to mustard.

Pungency also creates a difficulty in strength and quality assessment due to a rapid buildup of the stimulus with swamping and tiring of the sensory area. This being so, the evaluation of such materials as curry powder and other highly piquant seasonings containing capsicum must be carried out so as to reduce this effect to a minimum by:

(1) evaluating at low concentrations near to the threshold,
(2) allowing time for the effect to wear off,
(3) repeat testings of the samples in the reverse order of presentation,
(4) limitation of the test to a single attribute assessment (e.g., comparative pungency).

Scoville Test.—The pungency of oleoresin capsicum is frequently quoted in terms of Scoville heat units. The sensory test method is described in the ASTA (1968) Official Analytical Methods No. 21.0 and is generally used as follows:

Accurately weigh 0.200 g of sample and transfer it to a 50-ml calibrated flask with the aid of 95% ethanol; adjust to volume and mix thoroughly; allow any insoluble matter to settle.

Prepare a test solution by diluting 0.15 ml of the above solution with 140 ml of a 5% sucrose solution in water. If 5 ml of this test solution, swallowed all at once, produces a distinct sense of pungency in the throats of at least 3 of 5 assessors, then the sample is equivalent to 240,000 Scoville units. The test may be repeated after at least 30 min.

The dilution of the test solution can be further adjusted as necessary in accordance with the following to accommodate samples having a higher initial pungency:

Test Solution (ml)	5% Sucrose Solution (ml)	Scoville Units
20	—	240,000
20	20	480,000
20	40	720,000
20	60	960,000
20	80	1,200,000

Caution.—Oleoresin capsicum is a powerful irritant and even small quantities produce an intense burning sensation in contact with the eyes and tender parts of the skin. The use of a dilute solution of potassium permanganate on the skin or cocaine eye drops for the eyes is most likely to be effective in allaying the irritation (Martindale's Extra Pharmacopoeia 1967).

In the case of ground capsicum, it is necessary first to prepare an alcoholic extract by shaking 0.10 g of the sample with 50 ml of 95% ethanol and allowing to macerate for 24 hr; the extract is then filtered and adjusted to 50 ml with 95% ethanol.

Bitterness.—Spices such as nutmeg and mace are extremely bitter and this attribute may affect their evaluation. The palate is capable of detecting bitterness at very low concentrations and an assessment of relative bitterness must be carried out in dilute solution so that the palate is not saturated. A suitable level of concentration should be established experimentally before the presentation of a sample to a taste panel.

Alliaceous Vegetables.—Garlic has a particularly strong and, to most, an objectionable odor. The same applies to onions under certain conditions. The assessment of garlic and onion oils and seasonings containing high levels of dehydrated powders poses considerable problems due to a carryover of flavor in the mouth. Both are very persistent. Garlic, in particular, may even taint the breath for hours after consumption. In addition, the test rooms may carry a residual odor of garlic unless the air-conditioning is good.

Samples for assessment, involving the weighing of the highly odoriferous essential oils, must always be prepared well away from the test environment and preferably in a well-ventilated, fume-cupboard. Containers should be kept closed so as to minimize atmospheric contamination.

Onion and garlic powders, or products of a similar strength, are best evaluated in reconstituted mashed potato, using diluted lime juice as a mouth cleaner. It is preferable to taste these materials in a cold rather than a hot medium as differences in aromatic profile are then more obvious.

Anesthetic Effects.—Certain spices (e.g., clove and allspice) and flavoring essential oils (e.g., oil of peppermint) have a marked physiological effect on the nose and palate. The former are very astringent and also anesthetic; peppermint, on the other hand, has a strong cooling effect in the mouth and also is somewhat anesthetizing in the nasal cavity. These effects make direct evaluation uncertain but they can be minimized by careful dilution and by allowing an adequate recovery period between presentation of the samples. The odor evaluation of peppermint oil should always be made over a period so that the overpowering initial cooling effect is allowed to air off.

Appraisal.—One can readily appraise highly aromatic materials or very flavorful products so long as three prime rules are obeyed:

(a) that the material is diluted to a suitable level so that the nose and palate are not swamped,
(b) that due time is allowed between samples and between tests for the senses to recover fully,
(c) that the tests are presented in reverse order on a second sequence to eliminate any carryover effects.

The following initial dilutions are recommended, adjustments being made as necessary depending on the strength of the initial sample and the test medium chosen:

Aromatic chemicals: Dissolve 0.1 g (0.1 ml for liquids) in 5 ml of 95% ethanol; add 0.02 ml of the solution to 100 ml/g of the test medium.

Essential oils: Mix 0.25 ml with 10 ml of 95% ethanol; add 0.05 ml of the solution to 100 ml/g of the test medium.

Oleoresins: The usage depends on the strength. Guidance of the manufacturer is necessary. For general evaluation of all except the pungent spices, dissolve 0.25 g in 10 ml of 90% ethanol; shake well and allow any insolubles to settle; add 0.5 ml of the clear supernatent solution to 100 ml/g of the test medium. An alternative is to take the same quantity of sample but to disperse this on salt and taking 0.5 g of the salt dispersion for tasting in the test medium.

Flavorings: Dissolve 0.02 ml (0.02 g for dry flavorings) in 50 ml/g of the test medium.

TEST METHODS

The various methods used for the sensory evaluation of samples is well documented. Reference should be made to the several standard texts quoted in the following bibliography.

BIBLIOGRAPHY

AMERINE, M.A., PANGBORN, R. M. and ROESSLER, E. B. 1965. Principles of Sensory Evaluation of Food. Academic Press, New York.

ARTHUR D. LITTLE, INC. 1958. Flavor Research and Food Acceptance. Reinhold Publishing Corp., New York. (Chapman & Hall, London.)

ASTA 1968. Official Analytical Methods, 2nd Edition. American Spice Trade Association, New York.

ASTM 1968A. Basic principles of sensory evaluation. Spec. Tech. Publ. *433*. American Society for Testing and Materials, Philadelphia.

ASTM 1968B. Manual on sensory testing methods. Spec. Tech. Publ. *434*. American Society for Testing and Materials, Philadelphia.

ASTM 1968C. Correlation of subjective and objective methods in the study of odors and tastes. Spec. Tech. Publ. *440*. American Society for Testing and Materials, Philadelphia.

BEDNACZYK, A. A. 1973. Spices. In Quality Control for the Food Industry. Vol 2. Applications. A. Kramer and B. A. Twigg (Editors). AVI Publishing Co., Westport, Conn.

BUCHER, J. A. (Undated) Flavour Thoughts. I. Flavour Evaluation. N.V. Chemische Fabrick "Naarden", Naarden-Busson, The Netherlands.

CLONINGER, M. R. and BALDWIN, R. E. 1976. Analysis of sensory rating scales. J. Food Sci. *41*, 1225–1228.

DOVING, K. B. and SCHIELDROP, B. 1975. An apparatus based on turbulent mixing for delivery of odourous stimuli. Chem. Senses Flavor *1*, 371–374.

DRAKE, B. (Editor) 1968. Proceedings of International Symposium on Sensory Evaluation of Food—Principles and Methods. Kungalv, Sweden.

FERRIS, G. E. 1960. Sensory testing. Food Technol. Aust. *12*, 313–317, 385–393.

GOODALL, H. and COLQUHOUN, J. M. 1967. Sensory testing of flavour and aroma. Sci. Tech. Survey *49*, Br. Food Manuf. Ind. Res. Assoc., Leatherhead, Surrey.

HARRIES, J. M. 1973. Complex sensory assessment. J. Sci. Food Agric. *24*, 1571–1581. 1581.

HENDERSON, D. and VAISEY, M. 1970. Some personality traits related to performance in a repetitive sensory task. J. Food Sci. *35*, 407–411.

HIRSH, N. L. 1974A. Getting fullest value from sensory testing. I. Use and mis-use of test methods. Food Prod. Dev. *8*, No. 10, 33–46.

HIRSCH, N. L. 1974B. Getting fullest value from sensory testing. II. Considering the test objectives. Food Prod. Dev. *9*, No. 1, 10,13.

JELLINEK, G. 1964. Introduction to and critical review of modern methods of sensory analysis. J. Nutr. Diet. *1*, 219–260.

KRAMER, A. and TWIGG, B. A. 1970. Quality Control for the Food Industry, Vol I. AVI Publishing Co., Westport, Conn.

LARMOND, E. 1970. Methods for sensory evaluation of food. Publ. *1284*. Can. Dep. Agric.

LARMOND, E. 1973. Physical requirements for sensory testing. Food Technol. *27*, No. 11, 28,30,32.

LARMOND, E. 1976. Sensory methods—Choices and limitations. Am. Soc. Testing Materials Spec. Tech. Publ. *594*, 26–35.

MARTIN, S. L. 1973. Selection and training of sensory judges. Food Technol. *27*, No. 11, 22,24,26.

MARTINDALE'S EXTRA PHARMACOPOEIA, 25th Edition, 1967. Pharmaceutical Press, London.

MASUOKA, Y. 1976. Quality evaluation of herbs and spices in military food system. Tech. Rep. *FEL-54*. U.S. Army Natick Res. Dev. Command.

McGRATH, R. J. 1969. Organization, philosophy and psychology of quality control. Food Prod. Dev. *2*, No. 6, 34–38.

MOSKOWITZ, H. R. 1974. Combination rules for judgements of odor quality difference. J. Agric. Food Chem. *22*, 740–743.

SCHULTZ, H. G. 1971. Sources of invalidity in the sensory evaluation of foods. Food Technol. *25*, No. 3, 53,56–57.

SPENCER, H. W. 1971. Techniques in the sensory analysis of flavours. Flavour Ind. *2*, 293–302.

STAHL, W. H. and EINSTEIN, M. A. 1973. Sensory testing methods. In Encyclopedia of Industrial Chemical Analysis. John Wiley & Sons, New York.

STEWART, R. A. 1971. Sensory evaluation and quality assurance. Food Technol. *25*, No. 4, 103–106.

WELNER, G. 1972. Flavour materials and their quality control. In Quality Control in the Food Industry, Vol 3. S. M. Herschdoerfer (Editor). Academic Press, London.

Appendix

ENGLISH, FRENCH AND GERMAN TERMS USED IN SENSORY DESCRIPTIVE ANALYSIS

Terms Related to IMPACT

ETHEREAL	Ethère	Ätherische
FRESH	Frais	Frisch
LIGHT	Léger	Leicht
POWERFUL	Puissant	Stark; kraftvoll
SOFT	Tendre; delicat	Weich
STRONG	Fort	Stark
WEAK	Faible	Schwach

Terms Related to BODY of Odor or Flavor

BLAND	Fade	(Ein)Schmeichelnd
CREAMY	Crèmeux, onctueux	Sahnig
FLAT	Plat	Flach
FULL	Ayant du corps	Voll
INSIPID	Insipide	Geschmacklos
MELLOW	Moelleux, fendant	Reif
MODERATE	Equilibre	Mässig
RICH	Riche	Kraftig (fett)
ROUNDED	Arrondi	Abgerundet
SMOOTH	Lisse	Glatt, eben
THICK	Épais	Dick
THIN	Sans corps	Dunn

Terms Related to SPECIFIC ATTRIBUTES and EFFECTS

Mouth Feel

ASTRINGENT	Astringent	Zusammenziehend
BURNING	Brûlant	Brennend
CLOYING	Rassasiant	Ubersattigend
COOLING	Rafraîchissant	Kuhlend
DRYING	Dessèchant	Trockend
NUMBING	Engourdissant, paralysant	Erstarrend
TINGLING	Picotant	Pricklend
WARMING	Rechauffant	Wärmend

Associated with Burning

BURNT	Gout de brûle sent le brûle	Gebrannt
EMPYRHEUMATIC	Pyrogèneux	Brenzlich
SMOKY	Gout fume, sent le fume	Rauchig, gerächert
TAR-LIKE	Gout de goudron, sent le goudron	Wie Teer (tarry-terrig)

Physical Effects in the Nose

ACRID	Acre	Beissend, scharfätzend
BITING	Mordant	Beissend, scharf scheinend
HARSH	Rapeux, apre	Herb, rauh
INTOXICATING	Suffocant	Berauschend
IRRITANT	Irritant	Irritierend
LACHRYMATORY	Lacrymogène	Tränenerrengend
PENETRATING	Pénétrant	Durchdringend
PUNGENT	Piquant	Beissend stechend
SHARP	Prononcé, aigu	Scharf

Basic Tastes

ACIDIC	Acide	Säurehaltig
ALKALINE	Alcalin	Laugensalzig
BITTER	Amer	Bitter
METALLIC	Métallique	Metallisch
SALTY	Salé	Salzig

SOAPY	Gout ou odeur de savon	Seifig
SOUR	Sur, aigre	Sauer
SWEET	Doux	Süss

Persistence

LINGERING	Durable	Nachhaltend
PERSISTENT	Persistent	Hartnäckig

Terms Related to Other Specific Attributes or ODOR and FLAVOR EXPERIENCES BY ASSOCIATION

Like	Rappelant ou Goût de	Wie
ANISE-like	L'anis	Anis
ALLIACEOUS	Alliacé; de l'ail	Zwiebel
ALDEHYDIC	Aldéhydique	Aldehydig
AMMONIACAL	Ammoniacal	Ammoniakalisch
BUTTERY	de beurre	Butterahnlich
BREAD-like	de pain	Brotartig
BOILED CABBAGE-like	de choux bouilli	Gekochter Kohl
CARDBOARD-like	de carton	Pappartig
CAMPHORACEOUS	le camphre	Kampferhaltig
CINEOLIC	Cinéolique	Eukalyptus
CLOVE-like	le clou de girofle	Gerwurznelke
CREAMY	Crémeux	Sahnig
CITRUS-like	d'agrume	Zitrone
CARAMEL-like	de caramel	Caramel
CHEESE-like	de fromage	Käsig
DUSTY	de poussiere	Staubig
EARTHY	Terreux, de terre	Erdig
FOETID	Fétide	Übelriechend
FLORAL	Fleuri	Blumenhaft
FRUITY	Fruité	Fruchtig
FISHY	de poisson	Fischartig
GRASS-like	d'herbe	Grasig
HAY-like	de foin	Heu
HERBACEOUS	Herbacé	Kräutartig
HERBY	Herbeux	Kräuterartig
LEAFY	de verdure	Blattreich
LEMMONY	de citron	Zitronenhaft

LEGUMINOUS	de legume	Hülsenträgend
LICORICE-like	de reglisse	Sussholz
MEATY	de viande	Fleischig
MEDICINAL	Médicinal, pharmaceutique	Medizinisch
MOLDY (MOULDY)	Moisi	Schimmelig
MUSTY	sent le renfermé	Modrig
MINTY	de menthe	Minzartig
NUTTY	de noisette	Nussartig
OILY	Huileux	Ölig
ORIENTAL	Oriental, exotique	Morgenlandisch
PINE-like	de pin	Kieferartig
PLASTIC-like	de plastique	Plastik
RANCID	Rance	Ranzig
RESINOUS	Résineux	Harzig
SPICY	Épicé	Pikant
SOAPY	Savoneux, de savon	Seifig
TEA-like	de thé	Tee
TERPENEY	de terpenes, de peinture	Terpenhaltig
THYMOLIC	Thymolique	Thymolhaltig
WOODY	de bois	Hölzern
YEASTY	de levure	Hefig

Terms Related to HEDONIC EFFECTS

NAUSEATING	Nauseabond(e), acoeurant	Ekelerregend
OBJECTIONABLE	Repoussant, inacceptable	Unzulässig, antössig
PLEASING	Agreable	Gefällig
SOOTHING	Enivrant	Besänftigend
STIMULATING	Stimulant	Anregend
UNPLEASANT	Désagréable	Unangenehm
AROMATIC	Aromatique	Wutzig aromatisch
FRAGRANT	Parfume	Wohlriechend aromatisch

References:

HARPER, R. 1975. Some chemicals representing particular odour qualities. Chem. Senses Flavour, *1*, 353–357.

HARPER, R. 1975. Terminology in the sensory analysis of food. Int. Flavours Food Additives, *6*, 215–216.

Bibliographical Index

Allspice (pimento), 163
Anise, 163
Aromatic profiles, 24–25
Asafetida, 180

Baked goods, 425
Bakery technology, 425
Basil, 71
Bay, sweet, 71
Bay, West Indian, 163
Beverages, soft drinks, 506

Capsicum, 163
Caraway, 163
Cardamom, 163
Cassia, 163–164
Celery, 163
Cinnamon, 163–164
Citrus oils, 194
Cocoa, 220
Coffee, 221
Coriander, 164
Cumin, 164

Dill, 164

Essential oils, 294

Fish, 411
Flavorings, components, 23
  general, 212
Flavors, in baked goods, 425
  classification, 22–23
  creation, 378–380
  development, 378–380
  general, 25, 366–367
  in meat processing, 411
  legislation, 380–382
Fruit juice products, 260–261

Garlic, 180
Ginger, 164–165
Grapefruit, 195

Herbs, 70, 250–252
Horseradish, 165

Ice cream, 490–491

Laurel (sweet bay), 71
Lemon, 195–196
Lime, 196

Mace, 165
Mandarin, 196
Marjoram, 71
Meat, flavors in, 412–413
  processing, 411
Mint, 71
Mushrooms, 180
Mustard, 165

Nutmeg, 165

Odors, classification, 22–23
Onion, 180–181
Orange, 196–197
Origanum, 72

Paprika, 166
Parsley, 165
Pepper, 166
Pickles and sauces, 467–468
Pimento (allspice), 166
Profiles, aromatic, 24–25

Quality assurance, 520–522

Rosemary, 72

Saffron, 166
Sensory analysis, 520–522
Snack foods, 433
Soups, 478
Spices, 162, 250–252
Star anise, 163
Sugar confectionery, 450–451

Tangerine, 196
Tea, 222–223
Thyme, 72
Turmeric, 166

Vanilla, 206–207

Zedoary oil, 279

Subject Index

Acetals in flavors, 329
Acetates, 316–317
Acids in flavors, 302, 304–309
Acidulants in soft drinks, 494
Adulteration of spices, 228
African chillies, 76
African ginger, 92–93
Agar jellies, 443
Alcohols in flavors, 302, 304–308, 340
Aldehydes in flavors, 302–303, 315, 325–328
Alkane thiols, 340
Alleppy cardamoms, 111–115
Alliaceous vegetables, 167–178
  evaluation, 519
Allspice (pimento), 74, 151–154
  essential oil, 263
    components, 153–154
    profile, 153
Almond flavor, 352, 417
Ambrette oil, 263
American saffron, 158–159
Amines in flavors, 303, 338
2-Amino benzoates, 323
Amyris oil, 277
Anesthetic effects in flavors, 519
Angelica oil, 263
Angostura oil, 263
Anise, 74, 117–118
  essential oil, 263
Anise, China star, 74, 118–122
  essential oil, 278
    components, 122
    profile, 121–122
Anthranilates, 323
Anticaking agents, 363
Antioxidants, 233
Apple flavor, 348
Appraisal of flavorings, 520
Apricot flavor, 348
Aromatic barks, 74, 137–145
Aromatic fruits, 74, 101–117
Aromatic profiles, 13–18, 27 *et seq.*
  impact ratings, 27
Aromatic spices, in pickles, 465–466
Aromatic vegetables, 167–181
Artemisia oil, 263
Asafetida, 178–179
  essential oil, 263
  profile, 178–179

Baked goods, 414–426
  baked-in flavors, 416
  coatings, 423–424
  covertures, 423–424
  dusted-on flavors, 420, 431
  fermentation effects, 414
  fillings, 421–423
  flavors, 421
  heat-resistant flavors, 417
  savory flavors, 418
  sprayed-on flavors, 420
Bakery products, 414
Baking processes, 429
Balm oil, 263
Balsam Peru oil, 264

Balsam Tolu oil, 279
Banana flavor, 348
Basil, sweet, 27, 43–47, 74
  essential oil, 264
    components, 45, 47
    profiles, 43, 45–46
Bastard saffron, 158
Baumé degrees, 497
Bay laurel (sweet bay), 28–29
  essential oil, 264
    components, 29
    profile, 28–29
Bay, West Indian, 28, 74, 154–155
  essential oil, 264
    components, 155
    profile, 154–155
Benzoates, 322
Bergamot oil, 264
Beverages, materials, 212–223
  soft drinks, 493–506
Birch bark oil, 265
Bird chillies, 78
Biscuits, 414
Bitter almond oil, 265
Bitter fennel, 133
Bitterness, evaluation, 519
Bitter orange, 189–190
  essential oil, 274
    profile, 190
  flavor, 351
Blackberry flavor, 348
Black currant flavor, 348
Black mustard, 98
Black pepper, 73, 78–86
  commercial varieties, 82
  essential oil, 276
    components, 86–87
    GLC tracings, 84–85
    profile, 83
  in pickles, 464
  in seasonings, 406–407
Bois de rose oil, 265
Boletus mushroom, 179
Botanical classification of essential oils, 280
Botanical sources of essential oils, 280
Bottling syrups, 501
Bramble flavor, 348
Brandy flavor, 353
Brazilian *Mentha arvensis* oil, 66–67
British Essence Manufacturers' Association (BEMA), 377
Brix, degrees, 497
  value, 258
Buchu oil, 265
Butter cream, 422
Butter flavor, 352
  in ice cream, 487
Butterscotch flavor, 353
Butyrates, 318–319

Cacao beans, 212–213
  essential oil, 265
  flavor, 213
    components, 214
  nomenclature, 212
Cake mixes, 366, 424
Camomile
  English oil, 265
  German oil, 265
  Roman oil, 265
Camphor tree oil, 265
Cananga oil, 266
Canned products
  chicken breasts, 400
  corned beef, 400
  herring, 402
  meat, 399
  pork luncheon meat, 400
  stewed steak in gravy, 399
  whole chicken, 400
Caproates, 320–321
Caprylates, 321
Capsaicin, 76, 464
Capsicum, 73, 75–78
  classification, 76
  oleoresin, 464
  pungency, 76
Caramels, 440
Caraway, 74, 122–124
  essential oil, 266
    components, 124
    profile, 124
Carbohydrates as anticaking agents, 363
Carbonation of soft drinks, 502
  base formula, 511
Carboxylic acids in flavors, 302, 304–309
Cardamom, 74, 113–115, 417
  essential oil, 266
    components, 115
    profile, 114

Carriers for flavor, 510
Carrot oil, 266
Cascarilla oil, 266
Cassia, Chinese, 74, 141–145
  essential oil, 286
    profile, 144
    redistilled, 144
Cayenne pepper, 75, 78, 476
Cedar, white oil, 266
Celery, 74, 125
  essential oil, 266
    components, 125
    profile, 125
  herb, 125–126
Cheese flavor, 353
  in biscuits, 422
Chemicals in flavorings, classification, 299
  pungent, 343
  structural relationships, 302–303, 310–314
  structure and odor, 300
Cherry bark oil, 266
Cherry flavor, 349
Chervil oil, 266
Chewing gum, 445–446
  flavors, 446
Chili powder, 465
Chillies, African, 76, 78
China cassia, 144–145
China star anise, 118, 266
Chinese *Mentha arvensis* oil, 65–67
Chives, 167
Chocolate, 214, 448–450
  chips, 425
  confectionery, 434–450
    flavors, 449–450
  coverture, 423
  flavor, 215, 354, 449–450
  in ice cream, 485
  production, 448
Chromatographic analysis, 19, 21, 84–85, 96–97, 108–109, 372
Chromatographic deterpenation of essential oils, 291
Cinnamates, 323
Cinnamon bark, 74, 137–145, 417
  classification, 138–139
  essential oil, 267
    components, 141
    profile, 140, 144
  in pickles, 465
  in seasonings, 406–407
Cinnamon leaf, 74, 145–151
  essential oil 267
    components, 151
    profile, 150
Citronella oil, 267
Citrus fruits, 182–197
Citrus juices, 254
Citrus oils, 182–197
  concentrated, 291
  flavoring strength, 291
  profiles, 182–194
  terpeneless, 291–292
Clove, 74, 145–150
  bud, 145–149
    essential oil, 267
      components, 149
      profile, 148
  in pickles, 466
  in seasonings, 406–407
  leaf, 148
    essential oil, 267
    rectified essential oil, 149
  stem, 148
    essential oil, 267
  tannins, 229
Clover oil, 267
Coatings for biscuits, 423
Cochin ginger, 97
Cocoa, 212–216
  chocolate, 214
  components, 214
  extract, 215
  flavor, 213, 215
  powder, 214
Coconut flavor, 352
Coffee, 216–218
  caffeine, 218
  decaffeinated, 218
  flavor, 216, 340, 353
  instant, 217–218
Cognac flavor, 353
  essential oil, 267
Cola beverages, 503
Color units in oleoresins, 464, 467
Colored spices, 74, 155–162
  in pickles, 466
Colors, in soft drinks, 496
  in seasonings, 394
Common field mushroom, 179
Complex molecules, 302
Concentrated citrus oils, 291

Condiments, 73
Cookies, 414
Coriander, 74, 126–128
  essential oil, 267
    components, 128
    profile, 128
  in seasonings, 406–407
Corn mint (*Mentha arvensis*), 27, 65–67
  essential oil, 273
    components, 67
    profile, 65–66
Coverture for biscuits, 423
Crackers, 414
  cheese, 429
Cream, butter, 422
  dairy, 421
  flavor, 353
  soda, 504
  synthetic, 421
Creme de menthe flavor, 353
Crystal beverages, formulation, 505
Cubeb oil, 267
Culinary herbs, classification, 26–27
  definition, 26
Culinary mint, 70, 462
Cumin, 74, 128–130
  essential oil, 268
    components, 130
    profile, 129–130
  in curry powder, 476
Curcumin oil, 279
Curry powder, formulation, 476
  in soups, 475
Custard filling for biscuits, 421, 423

Dalmatian sage, 27, 52–54
  essential oil, 277
    components, 54
    profile, 52
Davana oil, 268
Decanoates, 321
Decylates, 321
Deep frying, 429
Descriptive analysis, 13–19, 27
  vocabulary, 523–526
Deterpenated essential oils, manufacture, 289–292
  terpeneless citrus oils, 291
Devils dung (asafetida), 178
Dextrose, 394
Dill pickles, 459
Dillseed, 74, 130–133
  essential oil, 268
    components, 133
    profile, 130–132
  Indian, 74, 133
Dillweed, 130–133
  essential oil, 268
    profile, 132
  in pickles, 463
Diluents in quality control, 510
Dispersed flavorings, 360
Distillation, 282–292
  steam, 282
  water, 283
  water and steam, 282
Disulfides, 303, 340
Diterpenes, 284
Dodecanoates, 321
Dry cake mixes, 424
Dry flavors, stabilized, 365
Dry processed spices, 241–243

East Indian sandalwood oil, 277
Edible fungi, 167
Edible oils, 358
Elderflower oil, 268
Elemi oil, 268
Emulsifiers, 495
Emulsions, flavoring, 358
  spice oil, 240
Encapsulated flavorings, 362
  spices, 243–245
English peppermint oil, 63
English sage, 27, 55–56
  profile of essential oil, 55–56
Environment for quality assurance, 507
Enzymes, garlic flavor, 170
  in spices, 228
  mustard flavor, 98, 342
  onion flavor, 169
Erigeron oil, 268
Essential oils, 261–295
  blends, 292–293
  botanical sources, 280
  classification, 280
  constitution, 284
  description, 262

deterpenation, 289
distillation methods, 282–285
function in plants, 281
isolates, 293–294
labeling, 292
principal (tabulated), 263–279
processing, 285
quality control, 293, 513
rectified, 286–289
redistilled, 286
terpeneless, 289
terpenoid hydrocarbons, 284–285
uses, 232–234, 263–279–293, 438
Esters in flavors, 302, 308–314, 316–323, 368
Estragon (tarragon), 27, 49–51, 268, 279
essential oil, 268
components, 51
profile, 51
in pickles, 463
Ethers in flavors, 302–303
Ethyl alcohol (ethanol), 356
Ethyl vanillin, 206
Ethylene oxide, 340
in spice sterilization, 229
Eucalyptus oil, 269
Eugenol-containing spices, 145–155
Evaluation, flavor, 514
liquid materials, 513
odor, 512–513
profiles, 13–21
seasonings, 403–404
sprays, 513
Exotic basil, 43
Extrusion, 430

Fat-based spice products, 245
Fennel, 74, 133–135
bitter, 133
sweet, 133
essential oil, 269
components, 135
profile, 135
Fenugreek, 74, 115–117
oleoresin, profile, 116
Filth in spices, 227
Fir oil, 269
Fish products, 401–403
processing, 392
Flash points, solvents, 355
Flavor and Extract Manufacturers' Association (FEMA) GRAS lists, 377–378
Flavors/flavorings, advantages and disadvantages in use, 11–12
artificial, 300
changes in natural, 5
characterization, 370–371
classification, 4, 6, 299
components, 7, 18
creation, 368–369
criteria for use, 387–389
definition, 7, 374
development, 368, 372
dispersed, 360
dry stabilized, 365
dusted-on, 420, 431
emulsions, 358, 500
enhancers, 394, 475
evaluation, 514–517
fermentation effects, 414
flavor pods, 425
formulation, 346–347
GRAS components, 377
heat resistant, 417
imitation, 9, 295, 348–355, 394, 428
impact, 4
in baked goods, 416
in creams and fillings, 421
in foods, 3 387 *et seq.*
in frozen yogurt, 491
in ice cream, 481
in ice lollies, 488
in meat products, 410–411
in snack products, 431
in soft drinks, 495–498, 499
in soups, 477
in sugar boilings, 438
in topping syrups, 488
in vinegars, 453
legislative constraints, 373
liquid, 344
manufacture, 345
microencapsulated, 362–364
natural, 4, 299
natural vs synthetic, 9
nature identical, 299
organic chemicals for use in, 300–344
powdered, 359
processing constraints, 378–379
profiles, 13

regulation, 299
research, 371
soluble essences, 499
spray-cooled, 366
spray-dried, 363
sprayed-on, 420
storage, 501
uses, problems, 11
Foam-mat drying, 260
Foam products, 443
Fondant, 421, 510
creams, 442
filling for baked goods, 423
Food and Drug Administration (FDA) GRAS lists, 377
Food colors, 394
Formates, 316
Freeze drying, 260
Frozen custard, 479
Frozen goods, 479
Fruit, citrus, 182–197
comminutes, 259
flavors in ice cream, 486
flavors in yoghurt, 491
imitation flavors, 348–351
juices, 252–261
blended, 258
concentrated, 256
dehydrated, 259–260
depectinized, 259
extraction, 252–255
fortified, 258
in soft drinks, 498
manufacture, 256
preservation, 255–256
with other natural flavors (WONF), 258
Fudge, 440
Functional relationships of alcohols, acids and esters, 310–314
Fungi, 179–180
flavor, 179

Galangal oil, 269
Galbanum oil, 269
Garden mint, 27, 70, 462
Garlic, 167, 175–178
essential oil, 175, 269
components, 178
profile, 177
flavor development, 175
fresh, profile, 176–177
powder, profile,; 76
Geranium oil, 270
Ginger, 73, 86–94, 96–98, 417
ale, 504
beer, 504
commercial grades, 93
essential oil, 270
components, 94
GLC tracing, 96–97
profile, 93–94
profile, ground, 92
Glazes, 421, 423
Glycerol (glycerin), 358
Glyceryl triacetate, 358
Grape flavor, 349
Grapefruit, 193–194
essential oil, 270
components, 194
profile, 194
flavor, 351
GRAS substances, 377–378
Greek sage, 27, 54–55
profile, oil, 54
Gums, 443

Hazelnut flavor, 352
Heat-resistant flavors, 365
Hemlock oil, 270
Heptanoates, 321
Heptylates, 321
Herbs, 223–250
aromatic profiles of essential oils, 27, *et seq.*
classification, 26–27
containing cineole, 27–32
menthol, 56-70
sweet alcohols, 43–51
thujone, 51–56
thymol/carvacrol, 32–43
definition, 26
flavoring effects, 408
in snack foods, 427
in soups, 473
Hexanoates, 320–321
High boiled confectionery, 435
Honey flavor, 354, 417
Honeycomb mushroom, 179
Hop oil, 270

Horsemint oil, 271
Horseradish, 73, 100
  components, 100–101
Hydrolyzed vegetable protein, 394
Hygiene, 227, 482
Hyssop oil, 271

Ice cream, 479–489
  flavors in, 481–488
  process technology, 480
  ripple, 487
  sherbet, 480
  types, 480
  water, 480
Imitation flavorings, 9, 295–367
  advantages and disadvantages in use, 11–13
  beverages, 353
  citrus fruits, 351–352
  classification of flavor chemicals, 299–300
  components, 348–355
  dairy products, 352
  fruit, 348–351
  nature of, 295–296, 298
  nut, 352
  specific flavors, 353–355
Immortelle oil, 271
Indian dill, 74, 133
Infrared drying, 429
Inorganic anticaking agents, 363
International Organization of Flavor Industries (IOFI) 373–374
  membership list, 375–376
International Standards Organization (ISO), 22
Ionones in flavors, 329
Isobutyrates, 319
Isolates from essential oils, 293–294
Isopropanol (isopropyl alcohol), 357
Isothiocyanates, 342
Isovalerates, 319–320
Italian peppermint oil, 62–63

Jamaican ginger, 92
Jamaican pepper, 151
Japanese *Mentha arvensis* oil, 67
Jasmine oil, 271
Jellies, 443
Juniperberry oil, 271

Ketones in flavor, 302–303, 329–337
Kola flavor, 503

Labdanum oil, 271
Lactones in flavor, 302–303, 315, 324
Lampong black pepper, 83
Laurates, 321
Laurel (sweet bay), 27–29
  essential oil, 271
Lavender oil, 272
  spike, 272
Lavandin oil, 271
  abrialis, 271
Leek, 167
Legislation, food additives, 373
  GRAS substances, 377–378
  mixed system of regulation, 377
  negative lists, 374
  positive lists, 376
  restrictions on flavoring materials, 373
Lemon, 183–186
  essential oil, 272
    components, 186
    profile, 185
  flavor, 351
Lemon balm oil, 263
Lemon oil, profile, 185
Lemongrass oil, 272
Lemon verbena oil, 272
Lime, 192–193
  essential oil, 272
    components, 193
    profile, 192–193
  flavor, 351
Linaloe oil, 273
Liquid flavorings, 344–345
Lovage, 74
  essential oil, 273
Low boiled confectionery, 439

Mace, 74, 111–112
  essential oil, 273

components, 111
profile, 111
Maillard reaction, 410
Malabar black pepper, 83
Malt vinegar, 452
Mandarin (tangerine), 190–191
essential oil, 273
components, 191
profile, 191
flavor, 351
Maple flavor, 417
Marigold oil, 278
Marjoram, sweet, 27, 33, 47–49
essential oil, 273
components, 49
profile, 47, 49
Marshmallow, 443
Mayonnaise, 456
Meat, analogs, 408–409
canned products, 399
cured, 398
extenders, 408–409
flavorings, 410–411
pies, 400
pork pie, 400
process technology, 392–401
sausages, 394–398
seasonings, 392, 403, 404–408
steak and kidney pie, 401
Melon flavor, 349
Mentha arvensis oil, 273
Mercaptans in flavor, 340
Mexican lime oil, 193
Mexican saffron, 158–159
Mexican sage, 27, 41–43
essential oil, profile, 41–43
Microcrystalline cellulose, 363
Microencapsulated flavors, 364
Microwave cooking, 429
Milk, 511
Milk whey as anticaking agent, 363
in seasonings, 394
Mint, 27, 56-70, 462
apple, 57
corn, 57, 65–67, 273
culinary, 57, 70, 462
curley, 57
garden, 57, 70, 462
peppermint, 56, 57-64, 276
spearmint, 57, 67-70
Monarda oil, 271
Monosodium glutamate (MSG), 475
Mushroom, flavor in, 179
extract, 478
Mustard, 73, 95, 98-100
black, 95
enzymes in flavor formation, 98, 342
essential oil, 273
artificial, 98, 342
components, 98
powder, 100
prepared, 455
Myristates, 322
Myrosin, 98
Myrrh oil, 273
Myrtle oil, 274

Natural colors, stability, 162
Natural extracts, 427
Natural flavoring substances, 9, 11, 299
Nature-identical flavoring substances, 299
Nectars, 259
Neutral soup formulation, 403, 510
New product development, 389
finished product, 389
process, 389
unit operations, 390-391
Nigerian ginger, 96
NItro compounds, 303
Nitrogen-containing compounds in flavor, 302, 338
Nonanoates, 321
Noncarbonated beverages, 504
Nut flavors, 352, 417
Nutmeg, 74, 101–110
butter, 110
essential oil, 274
components, 107
GLC tracing, 108–109
profile, 105–106
in curry powder, 474
Nuts, 428

Ocimum species oil, 274
Octanoates, 321
Odor evaluation, 512–514

Oil canals, 262
Oil cells, 262
Oleoresins, 234–238
  black pepper, 464
  capsicum, 464, 519
  dilution for quality control, 520
  fenugreek, 116–117
  production, 234
  red pepper, 464
  residual solvent limits, 235
  turmeric, 467
  uses, 237–238
Onion, 167, 168–175
  essential oil, 274
    components, 174–175
    profile, 174
  extract, 477
  flavor development, 168
  lachrymatory property, 172
  powder, 169–171
    profile, 173
  pyruvic acid as quality index, 169
Orange, 186–190
Orange, bitter, 189–190
  essential oil, 274
    components, 190
    profile, 190
  flavor, 351
Orange, sweet, 187–189
  essential oil, 275
    components, 189
    profile, 187
  flavor, 352
Oregano (Mexican sage), 27, 41–43
  profile of essential oil, 41–42
Organic chemicals in flavorings, 300–344
  acetals, 329
  acids, 302, 304, 309
  alcohols, 302, 304–308
  aldehydes, 303, 315, 323, 325–328
  amines, 303, 328
  carboxylic acids, 302, 304, 309
  classification, 302
  disulfides, 303, 341
  esters, 302, 308–315, 316–323
  ethers, 303
  functional relationships of alcohols, acids and esters, 310–311
  ionones, 329
  isothiocyanates, 342
  ketones, 303, 329, 330–337
  lactones, 303, 315, 324
  mercaptans, 341
  nitrocompounds, 303
  phenols, 302
  pungent compounds, 342–343
  structural relationships, 302–303
  sulfides, 303, 338, 340–341
  terpenoid hydrocarbons, 303, 338 339–340
  thioalcohols, 303
  thiocyanates, 341
Origanum, 27, 33, 35–36
  essential oil, 275
    components, 36
    profile, 36
Orris root oil, 275
Over-run, 481

Padi straw mushroom, 179
Palmarosa oil, 275
Paprika, 74, 78, 156, 464
Parsley, 74, 135–137
  essential oils, 275
    comparative profiles, 137
    herb, 137
    seed, 135–137
Pastilles, 443
Patchouly oil, 275
Peach aldehyde, 323
  flavor, 349
Peanut flavor, 352
Pear flavor, 349
Pectin jellies, 443, 511
Pelargonates, 321
Pennyroyal oil, 275
Pepper, 73, 78–86
  black, 82
  commercial varieties, 82
  essential oil, 276
    components, 86
    GLC tracings, 84–85
    profile, 83, 86
  in seasonings, 406–407
  oil, 278
  white, 83
Peppermint, 27, 57–64
  cultivation, 58–61
  essential oil, 276
    components, 58, 63–64
    profiles, 61–62, 63

rectified oils, 287
flavor, 354
Peru Balsam oil, 264
Peruvian pepper oil, 278
Petitgrain oil, 276
Phenols in flavors, 301–302, 304
in spices, 145–155
Phenylacetates, 322
Pickles, 452–468
classification, 454
dill, 459
flavored vinegar, 452
malt vinegar, 452
mustard, 456
piccalilli, 456
processing technology, dill pickles, 459
fruit sauces, 454
mayonnaise, 456
prepared mustard, 455
salad dressing, 456
tomato ketchup, 457
Worcestershire sauce, 458
seasonings, 460–467
Pimento (allspice), 74, 151–154
essential oil, 276
components, 153–154
profile, 153
in curry powder, 476
in seasonings, 406–407
in soups, 474
Pine oil, 269
dwarf oil, 276
Scotch oil, 276
white oil, 276
Pineapple flavor, 356
Piperine, in oleoresin pepper, 464
Plum flavor, 350
Polysulfides in flavor, 340
Popcorn, 428
Popsicles, 480
Potato, mashed in quality control, 404, 510
Potato chips (crisps), 366, 427
Poultry processing, 392
Powder flavorings, 359
Preparation of samples for quality assurance, 510
Preservatives, 394, 502
Pretzels, 428
Profiles, aromatic, 13
descriptive, 15
flavor, 16
vocabulary used in, 523–527
Proof spirit, definitions, 356
Propionates, 317–318
Propylene glycol, 357
Protein, structured products, 408–409
vegetable, 394
Pungency, 75, 517
chemical compounds, 342–344
evaluation, 517
in oleoresin black pepper, 464–465
in oleoresin capsicum, 464
in oleoresin red pepper, 464
in pickles, 463
in spices, 75–101
Scoville units, 464, 518

Quality assurance, 507–520
carriers and diluents, 510
environment for tests, 507
methodology, 508
sample preparation, 510

Raspberry flavor, 350
Rectified essential oils, 286
Red currant flavor, 350
Redistilled essential oils, 286
Red pepper, oleoresin, 464
Red thyme, 27, 33–35
essential oil, 279
components, 35
profile, 33
white oil, 35
Regulation of flavoring substances, 299
RESALOK flavors, 417
5′-Ribonucleotides, 475–476
Ripple ice cream, 487
Roller drying, 259
Root beer, 503
Rose oil, 277
flavor, 354
Rosemary, 27, 29–31
essential oil, 276
components, 31

GLC tracing, 21
profile, 20, 31
in soups, 474
Rosewood oil, 265
Rum flavor, 353

Saccharin, 338
Safflower, 74, 158–159
Saffron, 74, 156
components, 156
flavor, 158
Sage, 27, 51–56
Dalmatian, 52–54, 56
English, 55–56
essential oil, 277
comparative profiles, 56
Greek, 54–55, 56
Mexican, 41–43
Spanish, 27, 31–32, 56
Salad cream, 456
Salad dressing, 457
Salt, 394, 408, 432, 472
Samples, preparation for quality assurance, 512
Sandalwood oil, 277
Sassafras oil, 277
Satsuma, 190
Sauces, 452–468
classification, 454
flavoring, 455, 457, 458, 460
fruit, 454
thick white, 404
Worcestershire, 458
Sausages, 394–398
breakfast, 396
burgers, 398
cooked, 397
fermented, 397
frankfurters, 397
fresh, 396
seasonings, 406–407
weiners, 397
Savory, sweet, 27, 39–41
essential oil, 277–278
components, 41
profile, 39
summer, 39
winter, 39
Schinus molle oil, 278
Schizolysigenous oil glands, 262
Scoville, test for pungency, 518
units, 464
Seasonings, 392–394, 404–408
blending, 246
building formulation, 247
evaluation, 403
formulation, 247, 406–407, 474, 476
manufacture, 404–405
storage, 467
use of, 403
Secretory oil ducts, 262
Selective solvent extraction, 290
Sensory analysis, test methods, 508
Sesquiterpenes, 284
Sherbet, 480
Shiitake mushroom, 179
Sicilian lemon oil, 183
tangerine oil, 191
Silica (silicic acid), anticaking agent, 363
in deterpenation of citrus oils, 291
Sinalbin, 98
Sinigrin, 98
Smoked fish, 401
Snack foods, 427–433
flavoring application, 431
flavorings, 427–428
process technology, 428
seasonings, 433
Sodium benzoate, 502
caseinate, 394
Soft drinks, 493–506
acidulants, 494
classification, 493
cola (kola), 503
colors, 496
cream soda, 504
emulsifiers, 495
flavorings, 495
ginger ale, 504
ginger beer, 504
root beer, 503
specialty products, 503
sweeteners, 495
syrups, 496
Solvents, flash points, 355
in extraction, 236
in flavorings, 347, 355
residual limits, 235

Soups, 469–478
  canned, 469
  dry mixes, 471
  flavors, 472
  process technology, 469
  quick frozen, 472
  seasoning formulations, 474
Spanish sage, 27, 31–32
  essential oil, 277
    components, 32
    profile, 32
Spanish thyme, 33
Spearmint, 27, 67–70
  essential oil, 278
    components, 70
    profile, 69–70
    rectified oil, 288
  flavor, 354
Spices, 223–250
  adulteration, 228
  advantages and disadvantages in use, 244
  classification, 73–74
  colored, 74, 155–162, 466–467
  definition, 73
  dry processed (dispersed), 241–243
  emulsions, 240
  encapsulated, 243–245, 394
  enzymes in, 228
  essential oils, 232–234, 394
  extraction, 234–238
  fat-based, 245,
  filth in, 227
  ground, 223, 225, 394
  hygienic quality, 227
  in curry powder, 476
  in meat products, 403–408
  in snack foods, 427
  in soups, 474
  liquid flavorings, 238–241
  microencapsulated, 243
  oleoresins, 234–238, 394
  products, advantages and disadvantages in use, 244
    photomicrographs, 239
  pungent, 73–74, 517–518
  seasonings, 246–250
    formulations, 247, 406–407, 474, 476
  soluspices, 240
  solvents in extraction, 236–238
  stability, 228
  sterilization of, 229–232
  tannins, 229
  variability, 225–226
Spike lavender oil, 278
Spray-cooled flavors, 366
Spray drying, 260
Spruce oil, 270
Stabilized dry flavors, 365
Star anise, China, 74, 76, 118–122
  essential oil, 278
    components, 122
    profile, 121–122
Starches as anticaking agents, 363
Starch-deposited confectionery, 441–444
Stem ginger, 87
Stock seasonings, 473–474
Storax (styrax) oil, 278
Strawberry aldehyde, 323
  flavor, 351
Structural relationships of flavoring organics, 302–303
Structured proteins, 408–409
Sucrose, 394
Sugar confectionery, 434–451
  classification, 434
  depositing, 438
  flavorings, 446–448
  forming and finishing, 437–438
  graining and doctoring, 436–437
  high boiled, 435, 511
  low boiled, 439
  production, 435–438
  starch-deposited, 441
  tempering and kneading, 437
Sugars in bottling syrups, 472, 510
Sulfides in flavor, 303, 340
Sulfur-containing compounds, 302, 338, 340–341
Sulfur dioxide, 502
Sweet basil, 27, 43–47
  essential oil, 264
    components, 45–46
    profile, 43, 45–46
Sweet bay (laurel), 27–29
  essential oil, 264, 271
    components, 29,
    profile, 28–29
Sweet birch oil, 265
Sweet fennel, 133
Sweet marjoram, 27, 33, 47–49
  essential oil, 273
    components, 49
    profile, 47, 49

Sweet orange, 187–189
  essential oil, 275
    components, 189
    profile, 188
  flavor, 352
Sweet savory, 27, 39–41
  essential oil, 277–278
    components, 41
    profile, 39
Sweeteners in soft drinks, 495
Synthetic chemicals in flavors, 348–355
  legislation, 374
Syrups, bottling, 501
  composition table, 497
  flavors, 488
  preparation, 497

Tabasco, 78
Tagetes oil, 278
Tangerine (mandarin), 190–191
  essential oil, 273
    components, 191
    profile, 191
  flavor, 351
Tannins in spices, 229
Tarragon (estragon), 27, 49–51, 74
  essential oil, 268, 279
    components, 51
    profile, 51
  in pickles, 463
  in soups, 474
Taste, definition, 7
  modifiers, 472
  panels, 517
Tea, 218–220
  black, 219
  flavor, 219
  green, 219
  instant, 220
  varieties, 218–219
Terpeneless essential oils, 291
Terpenoid hydrocarbons in flavor, 284, 302–303, 338–340
Thioalcohols in flavor, 303
Thiocyanates in flavor, 341–342
Thiosulfinates, 169
Thiosulfonates, 169
Thuja oil, 266
Thyme, red Spanish, 27, 33–35
  essential oil, 279
    components, 35
    profile, 33–35
Toffees, 440
Tolu balsam oil, 279
Tomato ketchup, 457
Topping syrups, 488
Triacetin, 358
Trichomes, 362
Truffle, 179
Tuberose oil, 279
Turmeric, 74, 159–161
  essential oil, 279
    components, 161
    profile, 161
  in curry powder, 476
  in pickles, 466–467
  in soups, 474
  oleoresin, 467
Twaddell degrees of sugar syrups, 497

Umbelliferous fruits, 117–137
  flavor relationships, 119
Undecanoates, 321
Undecylates, 321
Unit operations affecting flavor 390–391

Vacuum drying, 259
Valerates, 319–320
Valerianates, 319–320
Vanilla, 198–207, 417
  components, 201, 204, 296–298
  curing process, 199
  ethyl vanillin (vanbeenol), 206
  extract, 204, 484
  flavor, 203, 355, 484–485
  in ice cream, 483
  paste, 484
  profiles, 200–201, 206
  standards, 205
  sugar, 484
  vanillin, 205
Variability of spices, 225–226
Vinegar, flavored, 453
  in soups, 472
  malt, 452

Violet flavor, 355
Vittae, 117, 262
Vocabulary for sensory analysis, 523–527

Water, 355
  ices (Popsicles), 480
  in soft drinks, 494
West Indian bay, 74
West Indian lime oil, 192
West Indian mace, 111
West Indian nutmeg, 105
West Indian sandalwood oil, 277
Whiskey flavor, 353
White pepper, 73, 82
White sauce, thick, 404
White thyme oil, 33, 35
Wild cherry bark oil, 266
Wild marjoram, 27, 33, 37–39
  essential oil, 275
    components, 39
    profile, 37–38
Wild thyme, 33

Yeast, extract, 392, 394
  in fruit juice spoilage, 255
Ylang Ylang oil, 279
Yogurt, 490

Other AVI Books

BAKERY TECHNOLOGY AND ENGINEERING
2nd Edition *Matz*
BEVERAGES: CARBONATED AND NONCARBONATED
Woodroof and Phillips
BYPRODUCTS FROM MILK
2nd Edition *Webb and Whittier*
CEREAL TECHNOLOGY
Matz
COMMERCIAL VEGETABLE PROCESSING
Luh and Woodroof
CONVENIENCE AND FAST FOOD HANDBOOK
Thorner
DAIRY TECHNOLOGY AND ENGINEERING
Harper and Hall
ECONOMICS OF NEW FOOD PRODUCT DEVELOPMENT
Desrosier and Desrosier
EGG SCIENCE AND TECHNOLOGY
Stadelman and Cotterill
ELEMENTARY FOOD SCIENCE
Nickerson and Ronsivalli
ENCYCLOPEDIA OF FOOD TECHNOLOGY
Johnson and Peterson
FOOD COLLOIDS
Graham
FOOD FLAVORINGS: COMPOSITION, MANUFACTURE AND USE
2nd Edition *Merory*
FOOD MICROBIOLOGY: PUBLIC HEALTH AND SPOILAGE ASPECTS
deFigueiredo and Splittstoesser
FOOD OILS AND THEIR USES
Weiss
FRUIT AND VEGETABLE JUICE PROCESSING TECHNOLOGY
2nd Edition *Tressler and Joslyn*
FUNDAMENTALS OF DAIRY CHEMISTRY
2nd Edition *Webb, Johnson and Alford*
ICE CREAM
2nd Edition *Arbuckle*
POTATO PROCESSING
3rd Edition *Talburt and Smith*
POULTRY PRODUCTS TECHNOLOGY
2nd Edition *Mountney*
PRACTICAL BAKING
3rd Edition *Sultan*

PRACTICAL FOOD MICROBIOLOGY AND TECHNOLOGY
2nd Edition *Weiser, Mountney and Gould*
PRINCIPLES OF FOOD CHEMISTRY
deMan
SCHOOL FOODSERVICE
Van Egmond
SEED PROTEINS
Inglett
SUGAR CHEMISTRY
Shallenberger and Birch
TECHNOLOGY OF WINE MAKING
3rd Edition *Amerine, Berg and Cruess*
THE MEAT HANDBOOK
3rd Edition *Levie*
THE TECHNOLOGY OF FOOD PRESERVATION
4th Edition *Desrosier and Desrosier*
TWENTY YEARS OF CONFECTIONERY AND CHOCOLATE PROGRESS
Pratt